The Spectral Analysis of Time Series

Probability and Mathematical Statistics

A Series of Monographs and Textbooks

---

1. Thomas Ferguson. Mathematical Statistics: A Decision Theoretic Approach. 1967
2. Howard Tucker. A Graduate Course in Probability. 1967
3. K. R. Parthasarathy. Probability Measures on Metric Spaces. 1967
4. P. Révész. The Laws of Large Numbers. 1968
5. H. P. McKean, Jr. Stochastic Integrals. 1969
6. B. V. Gnedenko, Yu. K. Belyayev, and A. D. Solovyev. Mathematical Methods of Reliability Theory. 1969
7. Demetrios A. Kappos. Probability Algebras and Stochastic Spaces. 1969
8. Ivan N. Pesin. Classical and Modern Integration Theories. 1970
9. S. Vajda. Probabilistic Programming. 1972
10. Sheldon M. Ross. Introduction to Probability Models. 1972
11. Robert B. Ash. Real Analysis and Probability. 1972
12. V. V. Fedorov. Theory of Optimal Experiments. 1972
13. K. V. Mardia. Statistics of Directional Data. 1972
14. H. Dym and H. P. McKean. Fourier Series and Integrals. 1972
15. Tatsuo Kawata. Fourier Analysis in Probability Theory. 1972
16. Fritz Oberhettinger. Fourier Transforms of Distributions and Their Inverses: A Collection of Tables. 1973
17. Paul Erdös and Joel Spencer. Probabilistic Methods in Combinatorics. 1973
18. K. Sarkadi and I. Vincze. Mathematical Methods of Statistical Quality Control. 1973
19. Michael R. Anderberg. Cluster Analysis for Applications. 1973
20. W. Hengartner and R. Theodorescu. Concentration Functions. 1973
21. Kai Lai Chung. A Course in Probability Theory, Second Edition. 1974
22. L. H. Koopmans. The Spectral Analysis of Time Series. 1974

In Preparation

L. E. Maistrov. Probability Theory: A Historical Sketch

William F. Stout. Almost Sure Convergence

Z. Govindarajulu. Sequential Statistical Procedures

E. J. McShane. Stochastic Calculus and Stochastic Models

The Spectral Analysis of Time Series

L. H. KOOPMANS

Department of Mathematics and Statistics
University of New Mexico
Albuquerque, New Mexico

ACADEMIC PRESS New York and London 1974
A Subsidiary of Harcourt Brace Jovanovich, Publishers

ACADEMIC PRESS, INC.
111 Fifth Avenue, New York, New York 10003

United Kingdom Edition published by
ACADEMIC PRESS, INC. (LONDON) LTD.
24/28 Oval Road, London NW1

Library of Congress Cataloging in Publication Data

Koopmans, Lambert Herman, Date
The spectral analysis of time series.

(Probability and mathematical statistics series)
Bibliography: p.
1. Time–series analysis. 2. Spectral theory (Mathematics) I. Title.
QA280.K64 1974 519.2′32 73-7441
ISBN 0–12–419250–5

AMS (MOS) 1970 Subject Classification: 62M15

PRINTED IN THE UNITED STATES OF AMERICA

TO MY PARENTS

Contents

Preface

This book is intended to provide an introduction to the techniques and theory of the frequency domain (spectral) analysis of time series. It has been written for use both as a textbook and for individual reading by a rather diverse and varied audience of time series analysis "users." For this purpose, the style has been kept discursive and the mathematical requirements have been set at the minimum level required for a sound understanding of the theory upon which the techniques and applications rest. It is essential even for the reader interested only in the applications of time series analysis to have an understanding of the basic theory in order to be able to tailor time series models to the physical problem at hand and to follow the workings of the various techniques for processing and analyzing data. Acquiring this understanding can be a stimulating and rewarding endeavor in its own right, because the theory is rich and elegant with a strong geometric flavor. The geometric structure makes possible useful intuitive interpretations of important time series parameters as well as a unified framework for an otherwise scattered collection of seemingly isolated results. Both features are exploited extensively in the text.

The book is suitable for use as a one-semester or two-quarter course for students whose mathematical background includes calculus, linear algebra and matrices, complex variables through power series, and probability and statistics at the postcalculus level. For students with more advanced mathematical preparation, additional details and proofs of several of the results stated in the text are given in the appendices.

The basic geometry of vector spaces used throughout the book is summarized in Chapter 1 and the various (nonprobabilistic) models possessing spectral decompositions required in later chapters are presented as applications of the geometric theory. The univariate, continuous-time models used in spectral analysis are introduced in Chapter 2 and the discrete-time models are given in Chapter 3 along with a discussion of the sampling of time series. Chapter 4 contains a general discussion of linear filters while Chapter 6 is concerned with a variety of special purpose filters in discrete time (digital filters). Multivariate time series models are introduced in Chapter 5 and a number of examples illustrating the use and interpretation of the multivariate spectral parameters are given. The standard finite parameter time series models are presented in Chapter 7 along with a discussion of linear prediction and filtering.

The statistical theory of spectral analysis is covered in Chapters 8 and 9. The distributions of spectral estimators are derived in Chapter 8 and are applied to the calculation of confidence intervals and hypothesis tests for the more important spectral parameters. The properties of spectral estimators as point estimates are considered in Chapter 9. This chapter also contains a discussion of the experimental design of spectral analyses and of the various computational methods for estimating spectra. The necessary tables for the hypothesis tests, confidence intervals, and experimental design methods covered in the text are provided in the appendix to Chapter 9.

This book contains no (formal) sets of exercises. It is my philosophy that a course in time series analysis should be tailored to the students' needs and this is best reflected in the kinds of activities required of them. In this regard, the exercises should be determined by the interests and preparation of the audience. For graduate students in mathematics and statistics, mathematical exercises will be appropriate, and several will be suggested to the instructor in the form of enlargements on the theory in the text and the appendices. Students with more applied interests should devote most of their effort to familiarizing themselves with the methods and computer programs for performing the analyses described in the text and to applying these techniques to simulated time series and to actual data from their fields of study. There is absolutely no substitute for practical experience in learning this subject. In fact, even the more theoretically oriented students of time series analysis should undertake some activities of this nature.

Acknowledgments

I was originally motivated to write this book by my own experience with trying to learn enough about the subject to design and analyze time series experiments for geophysical data during my employment at Sandia Corporation from 1958 to 1964. The selection of topics and the method of presentation are very much influenced by this experience. The book first took shape as a series of lecture notes written for a course I presented at Sandia in the summer of 1968 at the invitation of Melvin Merritt and William Perritt. Their encouragement and Sandia Corporation's support are much appreciated.

The major part of the present version of the book was written while I was on sabbatical leave at the University of California, Santa Cruz. I am indebted to the Mathematics Board of the University for providing the perfect environment for this endeavor.

Several friends, colleagues, and students have made contributions to the book by providing illustrations, critical comments on the manuscript, and encouragement. For this help I would like to thank Roger Anderson, Baldwin van der Bijl, Dirk Dahlgren, Jack Reed, John Rhodes, and Pamela Wilson.

Special thanks are due to David Brillinger and Robert Wijsman who read the original manuscript and made many valuable suggestions. My colleague, H. T. Davis, read portions of the manuscript, constructed the computer programs for plotting some of the graphs, and generally, through our association over the past few years, has provided me with encouragement and stimulation for writing this book.

I am indebted to the Literary Executor of the late Sir Ronald A. Fisher, F.R.S., to Dr. Fred Yates, F.R.S., and to Oliver and Boyd, Edinburgh, for permission to reprint an abridgment of Table III from their book *Tables for Biological, Agricultural and Medical Research.*

Finally, to simply say that my wife Sharon typed the manuscript would only begin to indicate her contribution to this book. Her hours of work and devotion are tantamount to coauthorship. Without her encouragement the pleasure I got from writing it would have been far less.

CHAPTER

1 Preliminaries

1.1 INTRODUCTION

The first goal of this chapter will be to provide the reader with a preliminary idea of the scope of applicability of time series analysis. The physical processes that models of time series are designed for will be illustrated and some of the basic features of the models will be introduced. The central feature of all models is the existence of a spectrum by which the time series is decomposed into a linear combination of sines and cosines. Actually, several kinds of spectral or Fourier decompositions are used in time series analysis and it is somewhat of a problem to remember them clearly. Fortunately, they all have properties in common which are essentially geometric in character. Moreover, the same geometry, which is basically the geometry of vector spaces, plays a central role in the construction and interpretation of the important stochastic time series model to be introduced in Chapter 2. We will use this geometry wherever possible to unify and simplify the theory. A summary of the relevant geometry is given in Section 1.3.

A summary of probability topics used in the book but not readily available in standard texts is provided in Section 1.4.

1.2 TIME SERIES AND SPECTRA

Time series analysis is primarily concerned with the study of the time variations of physical processes. If the "state" of the process can be represented by a vector of real numbers (measurements) with one or more components at each relevant time point, then the variations of the process over time can be represented by a vector of real-valued functions

$$\mathbf{x}(t) = \begin{pmatrix} x_1(t) \\ x_2(t) \\ \vdots \\ x_p(t) \end{pmatrix}. \tag{1.1}$$

The value of $x_j(t)$ at any time t is called the *amplitude* of that component at that time and the units of measurement are amplitude units. It is overly optimistic to think that a complex process can be completely described in this way. However, we will assume that the observed functions or *time series* characterize some interesting facet of the process and that an analysis of the time series will provide useful information about this aspect. The validity of this assumption clearly depends on the skill of the experimenter in selecting the right kinds of measurements.

The analysis of time series will depend upon the construction of one or more mathematical models which "generate" time functions of the type under observation. The models will be constructed in such a way that the parameters of the models can be identified with or readily related to the important characteristics of the physical process. Thus, procedures designed to obtain information from the observed time functions about the model parameters will also provide information about the underlying process.

To indicate the properties the models will have, we will look at some examples of time series to which they have been applied. Some time series of the type we will be interested in are graphed in Figs. 1.1–1.6. Typically, time

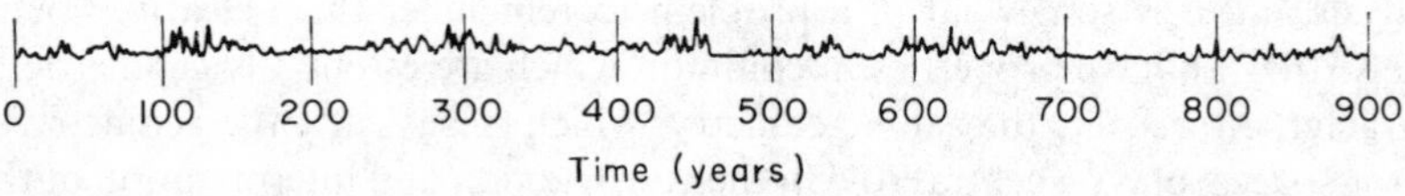

Fig. 1.1 *Geological (varve) series. Nine hundred years of data recorded at yearly intervals (interpolated to yield continuous curve). Amplitude units—millimeters.* Source: Anderson and Koopmans (1963); copyright by the American Geophysical Union.

series are recorded either continuously in time by an analog device, as in Figs. 1.2 and 1.5, or at equally spaced time points as in Figs. 1.1, 1.3, and 1.4. Thus we will require both *continuous-time* and *discrete-time* models. Moreover, we will distinguish between *univariate time series*, for which the vector function (1.1) has only one component ($p = 1$) and *multivariate time series* for which $p \geq 2$. For the time being we will restrict our discussion to a univariate time series $x(t)$.

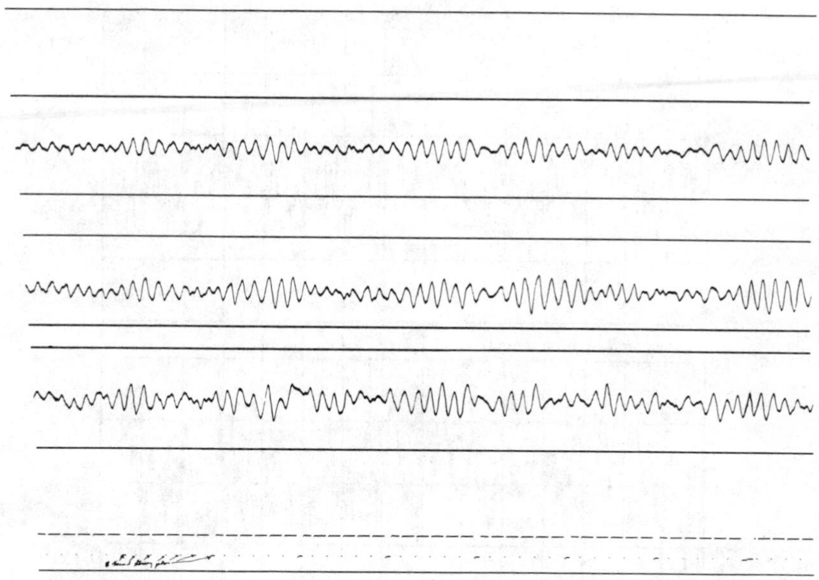

Fig. 1.2 *Three-dimensional seismic noise record. Continuously recorded data representing the vertical component of earth motion at three recording sites. Amplitude units—millimicrons.* Source: Sandia Corporation, 1959.

The time series in Figs. 1.1–1.5 have a common property of *persistence* or "ongoingness" and even a seeming "unchangingness" in character. One can imagine them as having originated at some time in the (distant) past and as continuing into the future with roughly the same general characteristics throughout their entire history. This description is less palatable for the encephalogram records which clearly begin and end with the birth and death of the individual and also show some form of evolution during the person's lifetime. However, by restricting attention to relatively short time periods in which the person is in a given state of mental activity or mental condition such as illustrated in Fig. 1.5, a model which embodies the properties of persistence and "ongoingness" will produce time functions with the appropriate characteristics over the given time periods. The model can then be used to characterize and study the different states of mental activity.

The record of aluminum production graphed in Fig. 1.6 is quite different from the others. It displays a distinctive trend which, because of the logarithmic amplitude scale, would seem to indicate a nearly exponential growth in production. However, another important feature of this graph is the yearly fluctuations around this growth trend. There appear to be cycles in growth which could have important economic implications and would certainly be

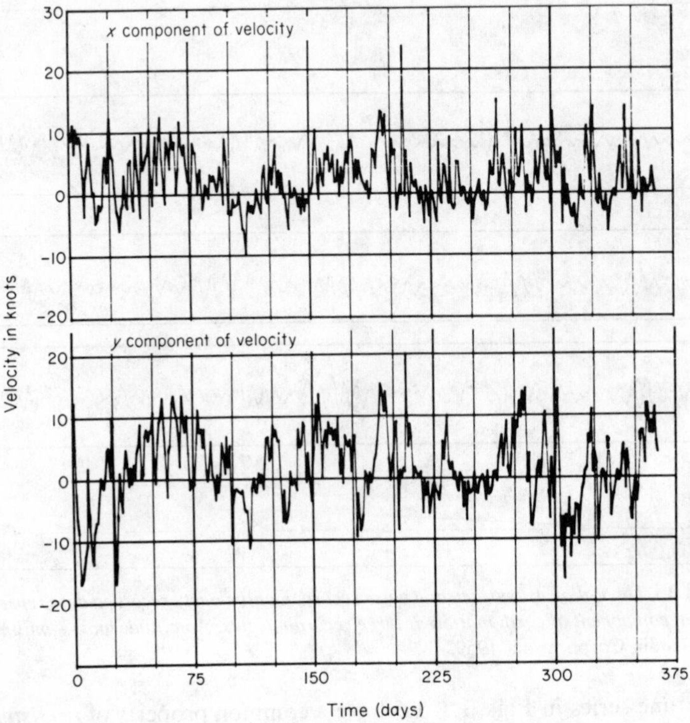

Fig. 1.3 *Two components of wind velocity data recorded at 1000-feet altitude at half-day intervals (interpolated to yield continuous curve).* Source: N.O.A.A., Las Vegas, Nevada.

useful to take into account for purposes of predicting future production. A reasonable procedure would be to fit, say, an exponential function to the data and study the residual from trend as a time series in its own right. The logarithmic scale compresses the fluctuations for the most recent time period making it difficult to see whether a model with the properties of persistence and "unchangingness" would adequately describe this residual series. If not, by a time-varying scale transformation, these properties could be attained to a good degree of approximation. Then, the characteristics of the yearly fluctuations can be studied by the techniques we will develop. Thus, our methods can be used to obtain valuable information about a wide variety of time series by preprocessing the data to separate out the persistent residual for analysis. The trend term is usually treated by standard least squares techniques [see, e.g., Hannan (1970)]. We will concentrate on models which are designed primarily to provide information about the long term, "steady state" behavior of the residual term.

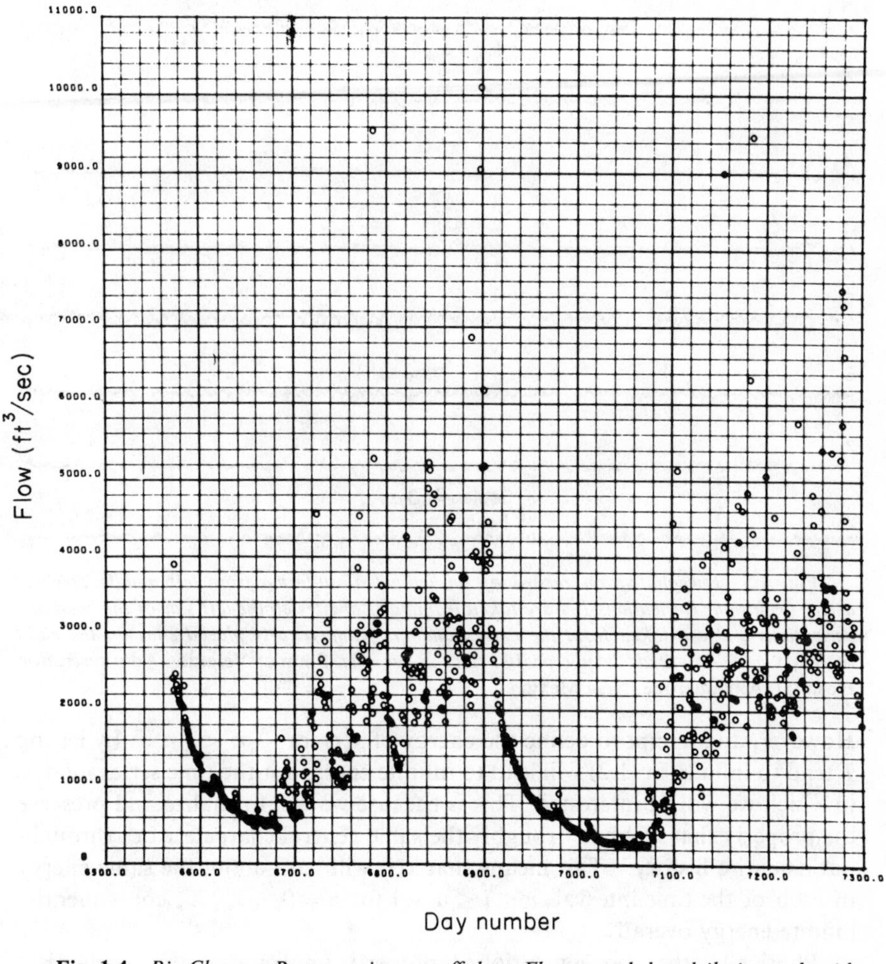

Fig. 1.4 *Rio Chagres, Panama, river runoff data. Flow recorded at daily intervals with day number 1 being January 1, 1907.* Source: J. W. Reed (1971).

The Basic Parameter of the Model—Power

In our model, we will idealize time to extend from the infinite past to the infinite future. That is, each time series $x(t)$ will be defined for $-\infty < t < \infty$. A useful measure of the "activity" of the time series for the interval of time, $-T \le t \le T$, is the mathematical version of *energy*

$$\text{energy of } x(t) \text{ for } -T \le t \le T = \int_{-T}^{T} x^2(t)\, dt. \tag{1.2}$$

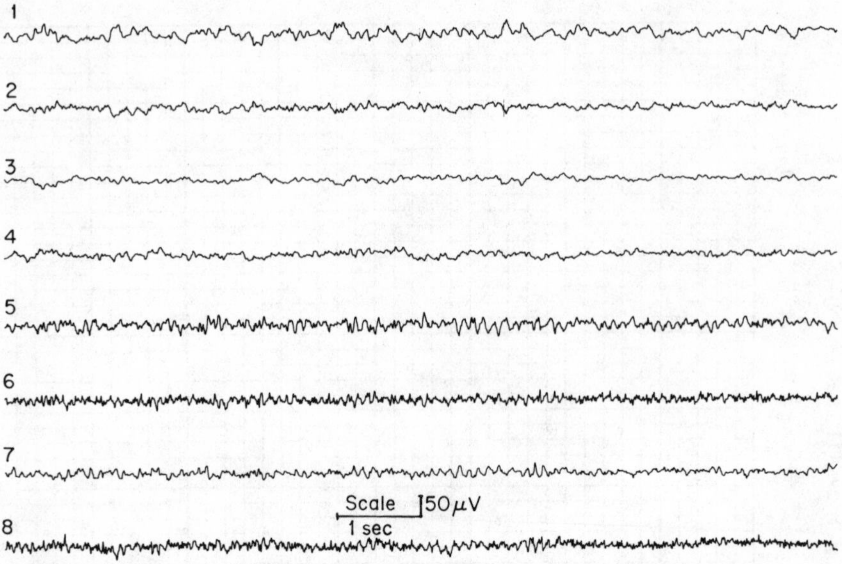

Fig. 1.5 *Encephalogram record of an individual suffering from left middle cerebral artery occlusion. Derivations 1–4 are taken from pairs of sites on the left side of the head and derivations 5–8 are taken from the corresponding, symmetrically placed sites on the right (normal) side. Amplitude in microvolts, time in seconds.* Source: Veterans Administration Hospital, Albuquerque, New Mexico.

However, an attempt to define the energy of $x(t)$ for $-\infty < t < \infty$ by letting T tend to infinity in (1.2) will lead to infinite energy for the time series models in which we will be interested. This is because we will formalize and preserve the property that $x(t)$ has "roughly the same general characteristics throughout its entire history." This means that $x(t)$ will have about the same energy in each of the time intervals $n \le t \le n+1$ for $n = 0, \pm 1, \ldots$; consequently, infinite energy overall.

Physically, the above mentioned property implies that the mechanism generating the time series does not change significantly in time. This time invariance will be embodied in the stochastic time series model to be defined in Chapter 2 as the property of *stationarity*.

A reasonable substitute for energy as a measure of "activity" is the average energy per unit time

$$\frac{(\text{energy of } x(t) \text{ for } -T \le t \le T)}{2T} = \frac{1}{2T} \int_{-T}^{T} x^2(t)\, dt. \tag{1.3}$$

This quantity has the dimension of the physical parameter, power, described in mechanics (where it is actually the time rate of energy expenditure) and the terminology is carried over to designate the basic parameter of time series

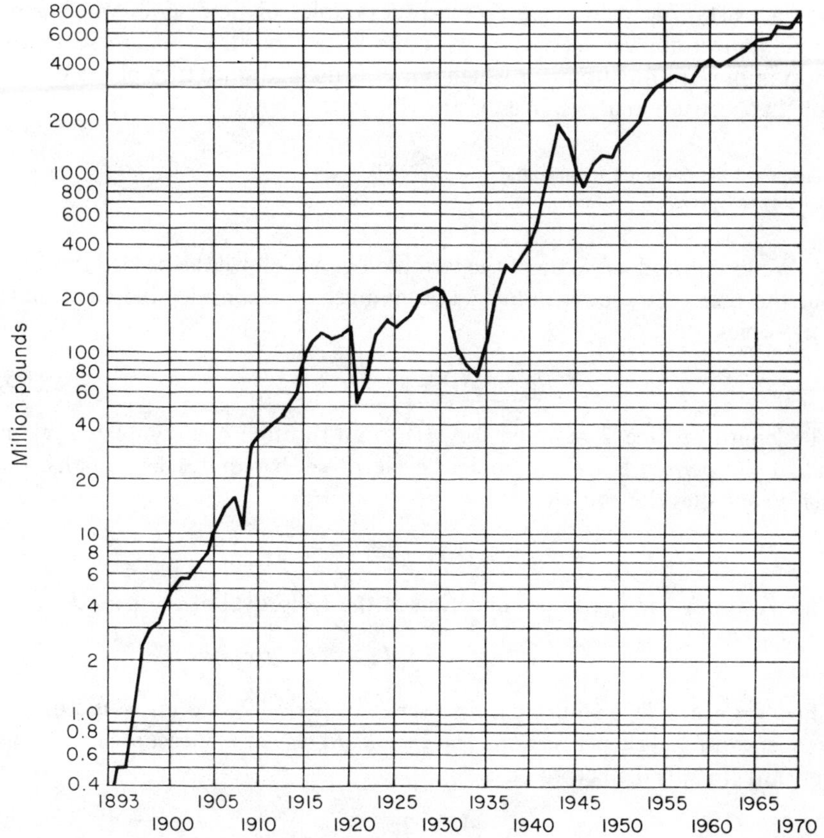

Fig. 1.6 *United States primary production of aluminium from 1893 to 1970 in millions of pounds. Yearly recorded data interpolated to yield continuous curve.* Source: The Aluminum Association.

models. Passing to the limit in (1.3), the *power in the time series* $x(t)$ is defined to be

$$\text{power of } x(t) = \lim_{T \to \infty} \frac{1}{2T} \int_{-T}^{T} x^2(t)\, dt. \tag{1.4}$$

The implication of this definition is that the limit exists, in some sense, and is finite. Note that power has the units of squared amplitude [of $x(t)$].

Power is clearly not the only possible measure of "activity," but is distinguished by the fact that a rich and physically meaningful mathematical theory can be associated with it. Nonpersistent (i.e., transient) time series for which the total energy [the limit of (1.2) as $T \to \infty$] is finite, all have zero power, thus constitute a rather uninteresting class of time series for which power exists. As we will see presently, examples of time series which have finite, nonzero power are $x(t) = \sin \lambda t$ and $y(t) = \cos \lambda t$. In fact, these are the most important functions for which power is defined since they play a central role in spectrum analysis. We discuss this next.

Sines, Cosines, Complex Exponentials, and Power Spectra

We first introduce some of the terminology which will be needed throughout the text. Let λ be a nonnegative number and consider the elementary time series

$$x(t) = \sin \lambda t. \tag{1.5}$$

The length of time T required for $x(t)$ to go through one complete cycle is called the *period.* Since a complete cycle of $\sin \vartheta$ requires 2π radians, the period satisfies the equation

$$\lambda T = 2\pi.$$

The *frequency* f in cycles per unit time is the reciprocal of the period

$$f = 1/T.$$

Thus, for example, a sinusoidal time series of period 2 seconds would have a frequency of $\frac{1}{2}$ cycle per second. The *angular frequency* λ, measured in radians per unit time, is defined to be

$$\lambda = 2\pi f.$$

This is easily seen to be the parameter which appears in (1.5). Hereafter, without fear of confusion, we will drop the term "angular" and call λ, simply, *frequency.*

By introducing two additional parameters A and φ, called the *amplitude* and *phase*, respectively, a large variety of elementary time series with the same frequency as (1.5) can be generated

$$x(t) = A \sin(\lambda t + \varphi), \qquad -\infty < t < \infty. \tag{1.6}$$

The amplitude A, which would be more properly called the "maximum" amplitude by virtue of our previous use of this term, is a nonnegative number measured in the amplitude units appropriate to the study at hand. The phase is a dimensionless parameter which measures the displacement of the sinusoid relative to the given time origin. Because of the periodic repetition of the

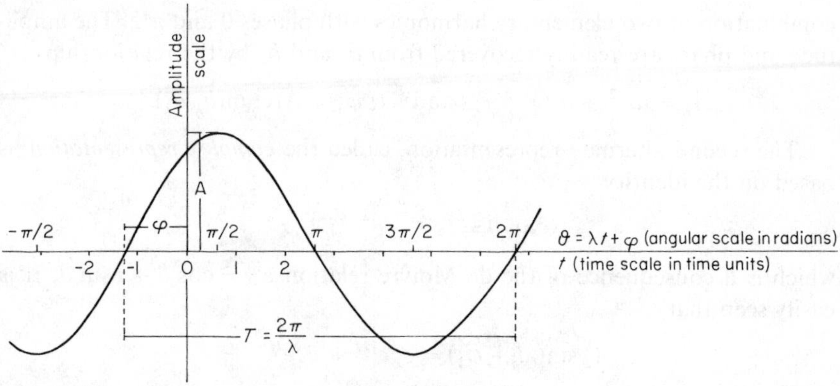

Fig. 1.7 *Graph of A* sin(λt + φ) *with parameters indicated.*

sinusoid, the phase can be restricted to the range $-\pi < \varphi \leq \pi$. These parameters are indicated in Fig. 1.7.

In the same sense that light of various colors is composed of a blend of monochromatic components and musical tones are formed by a superposition of pure harmonics, time series can be constructed by composing a number of "monochromatic" or *harmonic* functions $A_\lambda \sin(\lambda t + \varphi_\lambda)$ with varying frequencies, amplitudes, and phases, i.e.,

$$x(t) = \sum_\lambda A_\lambda \sin(\lambda t + \varphi_\lambda), \qquad -\infty < t < \infty. \tag{1.7}$$

The symbol $\sum_\lambda$ represents any of a number of possible summing operations, including integrals, which will be considered more concretely later when we discuss specific models. Expression (1.7) is called the *spectral representation of the time series* and, as we will see, every time series model we will study has an explicit spectral representation of this type.

Two alternate forms for the spectral representation are important. The first, based on the trigonometric identity $\sin(\alpha + \beta) = \sin\alpha\cos\beta + \cos\alpha\sin\beta$ is

$$x(t) = \sum_\lambda (a_\lambda \cos\lambda t + b_\lambda \sin\lambda t), \qquad -\infty < t < \infty, \tag{1.8}$$

where

$$a_\lambda = A_\lambda \cos\varphi_\lambda, \qquad b_\lambda = A_\lambda \sin\varphi_\lambda.$$

This is called the *real* or *cartesian* representation. Since $\cos\lambda t = \sin(\lambda t + (\pi/2))$, each harmonic $A_\lambda \sin(\lambda t + \varphi_\lambda)$ is represented as a linear combination of two elementary harmonics with phases 0 and $\pi/2$. The amplitude and phase are readily recovered from a_λ and b_λ by the relationships

$$A_\lambda = (a_\lambda{}^2 + b_\lambda{}^2)^{1/2} \qquad \text{and} \qquad \varphi_\lambda = \operatorname{Arctan}(b_\lambda/a_\lambda).\dagger$$

The second alternate representation, called the *complex representation*, is based on the identity

$$\sin \vartheta = (e^{i\vartheta} - e^{-i\vartheta})/2i,$$

which is a consequence of the de Moivre relation $e^{i\vartheta} = \cos\vartheta + i\sin\vartheta$. It is easily seen that

$$A_\lambda \sin(\lambda t + \varphi_\lambda) = c_\lambda e^{i\lambda t} + \bar{c}_\lambda e^{i(-\lambda)t},$$

where

$$c_\lambda = A_\lambda e^{i\varphi_\lambda}/2i.$$

Thus, if we adopt the notation,

$$c_{-\lambda} = \bar{c}_\lambda,$$

(1.8) can be written in the form

$$x(t) = \sum_{\pm\lambda} c_\lambda e^{i\lambda t}, \qquad -\infty < t < \infty. \tag{1.9}$$

The symbol $\pm\lambda$ indicates that the sum is taken over both positive and negative frequencies. Wherever (1.7) has a harmonic component with frequency λ, the complex representation (1.9) will have two harmonic components, one with frequency λ and the other with frequency $-\lambda$. Since $|c_\lambda| = \frac{1}{2}A_\lambda$, in effect the original amplitude is divided equally between the two complex harmonics $e^{i\lambda t}$ and $e^{-i\lambda t}$. The condition $c_{-\lambda} = \bar{c}_\lambda$ guarantees that $x(t)$ will be real-valued.

The complex representation (1.9) is notationally the most compact of the three and, as we will see in Chapter 4, it has significant advantages from the viewpoint of describing linear filters. Consequently, even at the expense of introducing the somewhat artificial notion of negative frequencies, we will restrict ourselves to the complex spectral representation hereafter.

† Here and throughout the book the following extension of the usual principal component of the arctangent function (with range $-\pi/2 < \arctan < \pi/2$), is used:

$$\operatorname{Arctan}\left(\frac{b}{a}\right) = \begin{cases} \arctan(b/a), & a > 0, \\ \arctan(b/a) + \pi, & a < 0, \quad b \geq 0, \\ \arctan(b/a) - \pi, & a < 0, \quad b < 0, \\ \pi/2, & a = 0, \quad b \geq 0, \\ -\pi/2, & a = 0, \quad b < 0. \end{cases}$$

This definition extends the arctangent to the range $(-\pi, \pi]$ and has the effect of making $\operatorname{Arctan}(b/a) = \arg(c)$, where $c = a + ib$.

The Power Spectrum

In order to be somewhat more precise, we will, for the moment, consider a time series with only finitely many frequencies

$$x(t) = \sum_{j=-n}^{n} c_j e^{i\lambda_j t},$$

where $\lambda_0 = 0$, $\lambda_{-j} = -\lambda_j$, and $c_{-j} = \bar{c}_j$. We will calculate the power of $x(t)$ using this spectral representation. We first compute the average energy for the time interval $-T \leq t \leq T$:

$$\frac{1}{2T}\int_{-T}^{T} x^2(t)\,dt = \frac{1}{2T}\int_{-T}^{T}\left(\sum_{j=-n}^{n} c_j e^{i\lambda_j t}\right)\overline{\left(\sum_{k=-n}^{n} c_k e^{i\lambda_k t}\right)}\,dt$$

$$= \sum_{j=-n}^{n}\sum_{k=-n}^{n} c_j \bar{c}_k\left(\frac{1}{2T}\int_{-T}^{T} e^{i(\lambda_j - \lambda_k)t}\,dt\right). \tag{1.10}$$

However,

$$\frac{1}{2T}\int_{-T}^{T} e^{i\lambda t}\,dt = \begin{cases} 1, & \lambda = 0, \\ (e^{i\lambda T} - e^{-i\lambda T})/2i\lambda T = \sin \lambda T/\lambda T, & \lambda \neq 0. \end{cases}$$

It follows that

$$\lim_{T\to\infty} \frac{1}{2T}\int_{-T}^{T} e^{i\lambda t}\,dt = \begin{cases} 1, & \lambda = 0, \\ 0, & \lambda \neq 0. \end{cases} \tag{1.11}$$

Using this result in (1.10), we obtain

$$\lim_{T\to\infty} \frac{1}{2T}\int_{-T}^{T} x^2(t)\,dt = \sum_{j=-n}^{n} |c_j|^2. \tag{1.12}$$

It is easily seen by this calculation that $|c_j|^2$ is the power of the harmonic term $c_j e^{i\lambda_j t}$, i.e., the power of $x(t)$ at frequency λ_j. Thus, (1.12) can be stated more intuitively as

$$\text{power of } x(t) = \sum_{j=-n}^{n} [\text{power of } x(t) \text{ at frequency } \lambda_j].$$

This important equation means that the various frequency components of the time series contribute their power *additively* to the total power of $x(t)$. That is, there is no interaction between different harmonic components in the sense that the amount of power contributed by one harmonic is independent of the amplitudes, phases, and frequencies of the other harmonics making up the time series. This is a consequence of expression (1.11) which we will return to in Section 1.3.

It is obvious from (1.12) that the power in the time series is known once the power in each harmonic component is known. The power at each frequency, as a function of frequency, is called the *power spectrum of the time series*. The power spectrum provides a great deal more information about the time series than simply the total power. It exhibits all of the nuances and variations of power with frequency. These variations often have important interpretations for physical time series. For example, because of the additivity property, the power of $x(t)$ in any set of frequencies S can be obtained by "summing" the power at each frequency in S,

$$[\text{power of } x(t) \text{ for frequency set } S] = \sum_{\lambda_j \in S} [\text{power of } x(t) \text{ at } \lambda_j].$$

Thus, as if often the case in "real" time series, if different phenomena contribute power to different frequency ranges, the power attributable to each phenomenon can be "sorted out" and evaluated by means of the power spectrum. In this book we will view the power spectrum as the fundamental "parameter" and object of study for time series. Examples and applications of the power spectrum will be given as our discussion develops beginning in Chapter 2.

Now, recall that we started this section with elementary harmonic functions and built up more complicated time series by forming linear combinations of these harmonics. This led to the spectral representation (1.9) and then to the definition of the power spectrum. In applications, the "building up" process is reversed: We start with a physical time series such as one of those pictured in Figs. 1.1–1.6. We will assume that this time series can be characterized by one of the mathematical models to be developed in Chapters 2–7. Each model will have a well-defined power spectrum with properties similar to those of the elementary time series given above. The principal goal of spectrum analysis is then to decompose the power of the given series into its harmonic components or, more precisely, to estimate the power spectrum from the available data. The estimated spectrum can then be used to gain information about the mechanism which generated the data. The techniques for estimating spectra will be covered in Chapters 8 and 9.

Concerning the Rest of the Chapter

The rest of this chapter covers the background material required for the more mathematical discussions in the book. The remainder of the book, starting with Chapter 2, has been written in a sufficiently self-contained fashion that by taking certain statements on faith and ignoring most references to Chapter 1 and to the Appendixes, a working knowledge of the theory and its applications can be obtained with relatively modest mathematical preparation. Section 1.3 contains material that is conceptually somewhat more difficult than that used in later discussions and the reader who is not particularly interested in the geometric setting of time series analysis can proceed to Chapter 2 at this point. A summary of probability topics used in the book but not readily available in standard texts is provided in Section 1.4.

1.3 SUMMARY OF VECTOR SPACE GEOMETRY

A *real* (*complex*) *vector space* is composed of a collection of *vectors* $\mathbf{x}, \mathbf{y}, \ldots,$ the collection of real (complex) numbers called *scalars*, and two operations: vector addition, denoted $\mathbf{x} + \mathbf{y}$, and scalar multiplication, denoted $\alpha\mathbf{x}$, which satisfy the system of axioms given, for example, by Halmos (1948, p. 1). The identity (zero) vector is denoted by $\mathbf{0}$ and the inverse of $\mathbf{x}$ by $-\mathbf{x}$. Lengths, distances and angular measure are defined for vectors by means of an *inner product* $\langle\mathbf{x}, \mathbf{y}\rangle$ which assigns scalar values to pairs of vectors according to the axioms: (i) $\langle\mathbf{x}, \mathbf{y}\rangle = \langle\mathbf{y}, \mathbf{x}\rangle$, (ii) $\langle\alpha\mathbf{x} + \beta\mathbf{y}, \mathbf{z}\rangle = \alpha\langle\mathbf{x}, \mathbf{z}\rangle + \beta\langle\mathbf{y}, \mathbf{z}\rangle$, (iii) $\langle\mathbf{x}, \mathbf{x}\rangle \geq 0$ with equality if and only if $\mathbf{x} = \mathbf{0}$. With the "only if" portion of this property deleted, the inner product is said to be *indefinite*. A vector space for which an inner product is defined is called an *inner product space*. A measure of length is provided by the *norm*; $\|\mathbf{x}\| = (\langle\mathbf{x}, \mathbf{x}\rangle)^{1/2}$. The natural *distance* between vectors $\mathbf{x}$ and $\mathbf{y}$ is the length of the difference vector, $\|\mathbf{x} - \mathbf{y}\|$.

The cosine of the angle $\vartheta_{x,y}$ between $\mathbf{x}$ and $\mathbf{y}$, generalized from the usual two-dimensional situation as discussed by Halmos (1948, pp. 86–88), is defined to be $\cos \vartheta_{x,y} = \langle\mathbf{x}, \mathbf{y}\rangle / \|\mathbf{x}\| \|\mathbf{y}\|$. Thus, it is geometrically sensible to call two vectors *orthogonal*, written $\mathbf{x} \perp \mathbf{y}$, if $\langle\mathbf{x}, \mathbf{y}\rangle = 0$. A collection of nonzero vectors $\{\mathbf{x}_1, \mathbf{x}_2, \ldots\}$ will be called *orthogonal* if $\mathbf{x}_i \perp \mathbf{x}_j$ for all $i \neq j$ and *orthonormal* if, in addition, $\|\mathbf{x}_i\| = 1$ for all i. A useful property of an orthogonal set of vectors is the *Pythagorean theorem*

$$\left\| \sum_i \mathbf{x}_i \right\|^2 = \sum_i \|\mathbf{x}_i\|^2.$$

Two important inequalities valid for any inner product space are the *Minkowski inequality*

$$\|\mathbf{x} + \mathbf{y}\| \leq \|\mathbf{x}\| + \|\mathbf{y}\|,$$

and the *Schwarz inequality*

$$|\langle\mathbf{x}, \mathbf{y}\rangle| \leq \|\mathbf{x}\| \|\mathbf{y}\|.$$

Equality holds in both cases if and only if $\mathbf{x} = \alpha\mathbf{y}$ for some scalar α or if $\mathbf{y} = \mathbf{0}$. With the "only if" part of this statement deleted, the two inequalities are also valid when the inner product is indefinite. Note that the Schwarz inequality implies the well-known property of cosines; $|\cos \vartheta_{x,y}| \leq 1$.

A finite set of vectors $\{\mathbf{x}_1, \mathbf{x}_2, \ldots, \mathbf{x}_n\}$ is called *linearly independent* if the only solution to the equation $\alpha_1 \mathbf{x}_1 + \alpha_2 \mathbf{x}_2 + \cdots + \alpha_n \mathbf{x}_n = \mathbf{0}$ is $\alpha_1 = \alpha_2 = \cdots = \alpha_n = 0$. An orthogonal set is necessarily linearly independent.

If a vector $\mathbf{z}$ can be represented as a linear combination, $\mathbf{z} = \beta_1 \mathbf{x}_1 + \beta_2 \mathbf{x}_2 + \cdots + \beta_n \mathbf{x}_n$, of the elements of a linearly independent set, then the coefficients $\beta_1, \beta_2, \ldots, \beta_n$ are unique. When the set is orthogonal, these coefficients have the explicit representation

$$\beta_i = \langle \mathbf{z}, \mathbf{x}_i \rangle / \|\mathbf{x}_i\|^2.$$

These numbers are called the *Fourier coefficients* of $\mathbf{z}$ relative to the orthogonal set.

If every element of a vector space can be represented as a linear combination of the elements of a finite set of vectors $\{\mathbf{x}_1, \mathbf{x}_2, \ldots, \mathbf{x}_n\}$, this set is said to *generate* the space. If the set is also linearly independent it is called a *basis* for the vector space and if it is an orthogonal (orthonormal) set it is said to be an *orthogonal* (*orthonormal*) *basis*. The Fourier coefficients relative to an orthonormal basis satisfy an important equation called the *Parseval relation*:

$$\|\mathbf{z}\|^2 = \sum_{i=1}^{n} |\langle \mathbf{z}, \mathbf{x}_i \rangle|^2.$$

Infinite-Dimensional Inner Product Spaces—Hilbert Spaces

Every basis for a finite-dimensional vector space has the same number of elements. This number is called the dimension of the space. For inner product spaces of dimension n, it is easily established that any set of n nonzero orthogonal vectors is a basis. This will not be true of all inner product spaces, however.

Most of the vector spaces we will be interested in are not of finite dimension. [See Halmos (1951, p. 29) for a discussion of dimension for Hilbert spaces.] This creates the problem of developing a theory for general inner product spaces which preserves the more important features of finite-dimensional theory. To extend the above development, for example, the theory should be extended to admit infinite linear combinations of orthogonal vectors and infinite sums. Infinite sums are defined as limits of finite sums in analysis and to carry this idea over to inner product spaces, a concept of limit is required.

The above defined concept of distance leads to a natural definition of convergence. Let $\{\mathbf{x}_n : n = 1, 2, \ldots\}$ be a sequence of vectors in the space. The sequence is said to *converge* to an element $\mathbf{x}$ (of the space), denoted

$\mathbf{x}_n \to \mathbf{x}$, if $\lim_{n\to\infty} \|\mathbf{x}_n - \mathbf{x}\| = 0$. From the Schwarz inequality it follows that if $\mathbf{x}_n \to \mathbf{x}$ then $\langle \mathbf{x}_n, \mathbf{z} \rangle \to \langle \mathbf{x}, \mathbf{z} \rangle$ for every $\mathbf{z}$. (The arrow symbol will be used rather indiscriminately to indicate convergence. The type of convergence will be evident from the context.) Similarly, $\mathbf{x}_n \to \mathbf{x}$ implies $\|\mathbf{x}_n\| \to \|\mathbf{x}\|$. These properties are termed *continuity* of the inner product and norm, respectively.

A sequence for which $\lim_{m,n\to\infty} \|\mathbf{x}_m - \mathbf{x}_n\| = 0$ is called a *Cauchy sequence*. By using the Minkowski inequality it is easily seen that every convergent sequence is a Cauchy sequence. If, conversely, every Cauchy sequence converges to an element of the space, the *space* is said to be *complete*. A complete inner product space is called a Hilbert space after the celebrated mathematician David Hilbert (1862–1943). All finite-dimensional inner product spaces are complete, thus are Hilbert spaces. Other examples will be given shortly.

Hilbert spaces preserve many of the important properties of finite-dimensional spaces given above. For example, to define the infinite linear combination of an orthonormal set $\{\mathbf{x}_1, \mathbf{x}_2, \ldots\}$ relative to scalar coefficients $\beta_1, \beta_2, \ldots$ satisfying the condition $\sum_{k=1}^{\infty} |\beta_k|^2 < \infty$, define the finite partial sums $\mathbf{z}_n = \sum_{k=1}^{n} \beta_k \mathbf{x}_k$. For $m > n$ show by the Pythagorean theorem that $\|\mathbf{z}_m - \mathbf{z}_n\|^2 = \sum_{k=n+1}^{m} |\beta_k|^2 \to 0$ as $m \to \infty$ and $n \to \infty$. Thus the partial sums form a Cauchy sequence and, by completeness, there exists a vector $\mathbf{z}$ in the space such that $\mathbf{z}_n \to \mathbf{z}$. The infinite linear combination is now defined to be equal to $\mathbf{z}$; $\mathbf{z} = \sum_{k=1}^{\infty} \beta_k \mathbf{x}_k$. It is easy to show that $\beta_k = \langle \mathbf{z}, \mathbf{x}_k \rangle$. Then, continuity of the norm implies the infinite-dimensional Parseval relation; $\|\mathbf{z}\|^2 = \sum_{k=1}^{\infty} |\langle \mathbf{z}, \mathbf{x}_k \rangle|^2$.

From the viewpoint of a number of important applications we will make of a particular Hilbert space—the space of square integrable periodic functions (see Example 1.2)—it is of interest to ask what conditions must be imposed on an orthonormal sequence in order that *every* vector in a Hilbert space of countably infinite dimension can be represented as an infinite linear combination of the elements of the sequence with square-summable coefficients. The *orthonormal sequence* is called *complete* if whenever $\langle \mathbf{z}, \mathbf{x}_j \rangle = 0$ for $j = 1, 2, \ldots$, then $\mathbf{z} = \mathbf{0}$. It can then be shown that the orthonormal sequence will have the desired property if and only if it is complete. This, in turn, is equivalent to the validity of the Parseval relation (Halmos, 1951, p. 27).

Linear Subspaces and Orthogonal Projections

A subset $\mathcal{M}$ of a Hilbert space is called a *linear manifold* if whenever $\mathbf{x} \in \mathcal{M}$ and $\mathbf{y} \in \mathcal{M}$, then $\alpha\mathbf{x} + \beta\mathbf{y} \in \mathcal{M}$ for all scalars α, β. A linear manifold which contains the limit of every Cauchy sequence formed from its elements is called a *linear subspace*. A linear subspace of a Hilbert space is itself a Hilbert space.

A set of vectors $\{\mathbf{x}_\lambda : \lambda \in \Lambda\}$ is said to *generate* a linear subspace $\mathcal{M}$ if $\mathcal{M}$ is the smallest subspace containing all of the elements of the set. It is easily argued that $\mathscr{M}$ consists of all finite linear combinations of the $\mathbf{x}_\lambda$'s along with the limits of all Cauchy sequences formed from these linear combinations.

If $\mathscr{M}$ is a linear subspace and $\mathbf{x}$ is an element of the Hilbert space not in $\mathscr{M}$, then the *distance from* $\mathbf{x}$ *to* $\mathscr{M}$ is defined to be $\inf_{\mathbf{y} \in \mathscr{M}} \|\mathbf{x} - \mathbf{y}\|$. An important property of Hilbert space is the existence of a unique element $\mathbf{z} \in \mathscr{M}$ for which this (minimum) distance is achieved. We will say that a vector $\mathbf{y}$ is *orthogonal to a subspace* $\mathscr{N}$, written $\mathbf{y} \perp \mathscr{N}$, if $\mathbf{y} \perp \mathbf{v}$ for every $\mathbf{v} \in \mathscr{N}$. Now, the element $\mathbf{z}$ which achieves the minimum distance from $\mathbf{x}$ to $\mathscr{M}$ can be characterized by the properties: (i) $\mathbf{z} \in \mathscr{M}$ and (ii) $(\mathbf{x} - \mathbf{z}) \perp \mathscr{M}$. For this reason, $\mathbf{z}$ is called *the orthogonal projection of* $\mathbf{x}$ *on* $\mathscr{M}$ and will be denoted by

$$\mathbf{z} = \mathscr{P}(\mathbf{x} | \mathscr{M}).$$

An application of the Pythagorean theorem to $\mathbf{z}$ and $\mathbf{x} - \mathbf{z}$ yields the inequality

$$\|\mathscr{P}(\mathbf{x} | \mathscr{M})\| \leq \|\mathbf{x}\|. \tag{1.13}$$

When $\mathscr{M}$ is generated by a set of elements $\{\mathbf{x}_\lambda : \lambda \in \Lambda\}$, the condition (ii) is easily seen to be equivalent to $(\mathbf{x} - \mathbf{z}) \perp \mathbf{x}_\lambda$, thus $\langle \mathbf{x} - \mathbf{z}, \mathbf{x}_\lambda \rangle = 0$, for all $\lambda \in \Lambda$. This produces the equations

$$\langle \mathbf{z}, \mathbf{x}_\lambda \rangle = \langle \mathbf{x}, \mathbf{x}_\lambda \rangle, \qquad \lambda \in \Lambda,$$

which, along with the condition $\mathbf{z} \in \mathscr{M}$, completely determines $\mathbf{z}$. We will call this the *criterion for determining orthogonal projections.*

When the Hilbert space is generated by a countable collection of orthogonal vectors, $\{\xi_n : n = 1, 2, \ldots\}$ and $\mathscr{M}$ is generated by a subset $\{\xi_n : n \in J\}$, it is easily checked that the orthogonal projection has the simple form

$$\mathscr{P}(\mathbf{x} | \mathscr{M}) = \sum_{n \in J} \frac{\langle \mathbf{x}, \xi_n \rangle}{\|\xi_n\|^2} \xi_n. \tag{1.14}$$

This expression and the following properties of projections will be needed in Chapter 7.

The idea of a linear transformation will be familiar to the reader (Halmos, 1948, p. 33). A linear transformation A on a Hilbert space into itself is called *continuous* if whenever $\mathbf{x}_n \to \mathbf{x}$, then $A(\mathbf{x}_n) \to A(\mathbf{x})$. An orthogonal projection (with $\mathscr{M}$ fixed) is a linear transformation which is continuous because of (1.13): $\|\mathscr{P}(\mathbf{x}_n | \mathscr{M}) - \mathscr{P}(\mathbf{x} | \mathscr{M})\| = \|\mathscr{P}(\mathbf{x}_n - \mathbf{x} | \mathscr{M})\| \leq \|\mathbf{x}_n - \mathbf{x}\|$.

If $\mathscr{M}$ and $\mathscr{N}$ are linear subspaces of a Hilbert space, then $\mathscr{M}$ is said to be orthogonal to $\mathscr{N}$, written $\mathscr{M} \perp \mathscr{N}$, if $\mathbf{x} \perp \mathbf{y}$ for every $\mathbf{x} \in \mathscr{M}$ and $\mathbf{y} \in \mathscr{N}$. The direct sum of $\mathscr{M}$ and $\mathscr{N}$, $\mathscr{M} \oplus \mathscr{N} = \{\mathbf{x} + \mathbf{y} : \mathbf{x} \in \mathscr{M}, \mathbf{y} \in \mathscr{N}\}$, is the smallest linear subspace containing both subspaces (Halmos, 1951, p. 25). It can now be shown that

$$\mathscr{P}(\mathbf{x}|\mathscr{M} \oplus \mathscr{N}) = \mathscr{P}(\mathbf{x}|\mathscr{M}) + \mathscr{P}(\mathbf{x}|\mathscr{N}). \tag{1.15}$$

If $\mathscr{M}$ and $\mathscr{N}$ are subspaces with $\mathscr{M} \subset \mathscr{N}$, then the set of elements $\mathscr{M}_{\mathscr{N}}{}^{\perp}$ in $\mathscr{N}$ which are orthogonal to $\mathscr{M}$, *called the orthogonal complement of* $\mathscr{M}$ *in* $\mathscr{N}$, is a linear subspace such that $\mathscr{N} = \mathscr{M} \oplus \mathscr{M}_{\mathscr{N}}{}^{\perp}$ (Halmos, 1951, p. 26). Applying the above property of projections to this expression, it is possible to derive the relation

$$\mathscr{P}(\mathscr{P}(\mathbf{x}|\mathscr{N})|\mathscr{M}) = \mathscr{P}(\mathbf{x}|\mathscr{M}). \tag{1.16}$$

Some Specific Hilbert Spaces

A number of the properties of time series models that we will need in the book are simply the properties of vector spaces summarized above applied to particular Hilbert spaces. A rather diverse collection of results is unified and given a useful geometric setting by the recognition of this fact. To stress this point, we will present these topics as examples of the general theory.

Example 1.1 *Discrete-Time Periodic Time Series and the Finite Fourier Transform*

Let $x(t)$, $t = 0, \pm 1, \ldots$ be a discrete-time, real-valued time series (sequence) and let N be a positive integer. Then the time series is said to be *periodic of period N* if

$$x(t) = x(t + N)$$

for every integer t. *A* definition of inner product for periodic time series will be motivated by the expression (1.4) for power. The discrete analog of this expression is

$$\text{power of } x(t) = \lim_{L\to\infty} \frac{1}{2L+1} \sum_{t=-L}^{L} x^2(t). \tag{1.17}$$

Now, a periodic time series is completely determined by its values over a single cycle. Thus, we can identify $x(t)$, $t = 0, \pm 1, \ldots$ with the N-dimensional vector $\mathbf{x} = (x(1), x(2), \ldots, x(N))$. Moreover, it is easily argued that for periodic time series, the expression (1.17) is simply the time average over a single cycle:

$$\text{power of } x(t) = \frac{1}{N} \sum_{t=1}^{N} x^2(t).$$

The collection of all N-tuples of complex numbers, $\mathbf{w} = (w_1, w_2, \ldots, w_N)$, with the complex scalars is a vector space under coordinatewise addition and scalar multiplication: $\mathbf{w} + \mathbf{y} = (w_1 + y_1, \ldots, w_N + y_N)$, $\alpha\mathbf{w} = (\alpha w_1, \ldots, \alpha w_N)$. The function

$$\langle \mathbf{w}, \mathbf{y} \rangle = \frac{1}{N} \sum_{j=1}^{N} w_j \bar{y}_j$$

is an inner product for this space. Consequently, the class of discrete-time, real-valued time series can be viewed as a subset of this Hilbert space and the inner product is defined so as to make the power of $x(t) = \|\mathbf{x}\|^2$.

An expression which will be needed in this example and at other points later in the text is recorded here for reference purposes. It is a straightforward consequence of the formula for the sum of a finite geometric series and the trigonometric relation $(e^{i\vartheta} - e^{-i\vartheta})/2i = \sin \vartheta$:

$$\sum_{j=a}^{b} e^{i\lambda j} = \begin{cases} e^{i\lambda(b+a)/2} \dfrac{\sin \lambda((b-a+1)/2)}{\sin(\lambda/2)}, & \lambda \neq 0, \\ b - a + 1, & \lambda = 0, \end{cases} \tag{1.18}$$

for any integers $a \leq b$.

Let $\lambda_\nu = 2\pi\nu/N$ and let $[x]$ denote the largest integer not exceeding x. By means of expression (1.18) it is easy to show that the N vectors

$$\mathbf{z}_\nu = (e^{i\lambda_\nu}, e^{i2\lambda_\nu}, \ldots, e^{iN\lambda_\nu}), \qquad -[(N-1)/2] \leq \nu \leq [N/2],$$

form an orthonormal set, thus a basis for the vector space. It follows that every vector has the representation

$$\mathbf{w} = \sum_{\nu=-[(N-1)/2]}^{[N/2]} \alpha_\nu \mathbf{z}_\nu,$$

where α_ν is the Fourier coefficient, $\alpha_\nu = \langle \mathbf{w}, \mathbf{z}_\nu \rangle$. The corresponding expressions for the coordinates of the vector $\mathbf{x}$ are

$$x(t) = \sum_{\nu=-[(N-1)/2]}^{[N/2]} \alpha_\nu e^{i\lambda_\nu t}, \qquad t = 1, 2, \ldots, N, \tag{1.19}$$

and

$$\alpha_\nu = \frac{1}{N} \sum_{t=1}^{N} x(t) e^{-i\lambda_\nu t}, \qquad -[(N-1)/2] \leq \nu \leq [N/2]. \tag{1.20}$$

Since $e^{i\lambda_\nu t}$ is periodic of period N in both ν and t, both of these expressions remain valid if the range of values of t in (1.19) and ν in (1.20) are extended to all integers. Then (1.19) is the spectral representation of the discrete periodic time series $x(t)$, $t = 0, \pm 1, \ldots$ in the form (1.9). The expression (1.20) by which $x(t)$ is transformed into the frequency domain—i.e., into a function of the frequencies λ_ν—is called the *finite Fourier transform*. In a stochastic setting, this transform will play a significant role in the distribution theory for spectral estimators in Chapter 8.

The Parseval relation yields the expression

$$\|\mathbf{x}\|^2 = \sum_{\nu=-[(N-1)/2]}^{[N/2]} |\alpha_\nu|^2. \tag{1.21}$$

But $|\alpha_\nu|^2 = \|\alpha_\nu \mathbf{z}_\nu\|^2$ is the power of the time series (vector) with components $\alpha_\nu e^{i\lambda_\nu t}$. Thus (1.24) represents the spectral decomposition of the total power of a discrete-time periodic time series into its frequency components analogous to (1.12). The power spectrum of $x(t)$ has value $|\alpha_\nu|^2$ at frequency λ_ν for $-[(N-1)/2] \le \nu \le [N/2]$ with periodic extension of period N outside of this range.

Example 1.2 *Continuous-Time Periodic Time Series, Fourier Series, and the Space* ℓ_2

A real- or complex-valued, continuous-time function $x(t)$, $-\infty < t < \infty$, is said to be periodic of period T $(T > 0)$ if

$$x(t + T) = x(t)$$

for all t. For the present, we will assume that $T = 2\pi$. As in Example 1.1, it can be argued that $x(t)$ can be restricted to the interval $(-\pi, \pi]$ and that

$$\lim_{T\to\infty} \frac{1}{2T}\int_{-T}^{T} x(t)\overline{y(t)}\,dt = \frac{1}{2\pi}\int_{-\pi}^{\pi} x(t)\overline{y(t)}\,dt,$$

when the integral is defined.

Let $\mathscr{L}_2(-\pi, \pi)$, denote the collection of all complex-valued functions $\mathbf{x} = \{x(t)\colon -\pi < t \le \pi\}$ for which $\int_{-\pi}^{\pi} |x(t)|^2\,dt < \infty$, where the integral is the Lebesgue integral. If vector addition and scalar multiplication are defined coordinatewise, $\mathscr{L}_2(-\pi, \pi)$ is a Hilbert space with inner product

$$\langle \mathbf{x}, \mathbf{y}\rangle = \frac{1}{2\pi}\int_{-\pi}^{\pi} x(t)\overline{y(t)}\,dt.$$

(In fact, this is an indefinite inner product and in order to form an inner product space it would be necessary to take as elements the collection of equivalent classes under the equivalence relation $\mathbf{x} \equiv \mathbf{y}$ if $\|\mathbf{x} - \mathbf{y}\| = 0$. This distinction will be ignored here and in future cases of indefinite inner products.)

Now it is known from classical analysis that if $\mathbf{z}_n = \{e^{int}\colon -\pi < t \le \pi\}$, then $\{\mathbf{z}_n\colon n = 0, \pm 1, \ldots\}$ is a complete orthonormal sequence in $\mathscr{L}_2(-\pi, \pi)$. It follows from the above summary that $\mathbf{x} \in \mathscr{L}_2(-\pi, \pi)$ if and only if there exists a sequence of complex numbers $\{\beta_n\colon n = 0, \pm 1, \ldots\}$ with $\sum_{n=-\infty}^{\infty} |\beta_n|^2$

$< \infty$ such that $\mathbf{x} = \sum_{n=-\infty}^{\infty} \beta_n \mathbf{z}_n$. The β_n's are the Fourier coefficients, $\beta_n = \langle \mathbf{x}, \mathbf{z}_n \rangle$. In coordinate notation,

$$x(t) = \sum_{n=-\infty}^{\infty} \beta_n e^{int}, \tag{1.22}$$

where

$$\beta_n = \frac{1}{2\pi} \int_{-\pi}^{\pi} x(t) e^{-int}\, dt. \tag{1.23}$$

Expression (1.22) is the *Fourier series* representation of $x(t)$, which is a spectral representation of the form (1.9). The Parseval relation is

$$\frac{1}{2\pi} \int_{-\pi}^{\pi} |x(t)|^2\, dt = \sum_{n=-\infty}^{\infty} |\beta_n|^2.$$

Denote by ℓ_2 the class of all complex sequences $\boldsymbol{\alpha} = \{\alpha_n : n = 0, \pm 1, \ldots\}$ for which $\sum_{n=-\infty}^{\infty} |\alpha_n|^2 < \infty$. With complex scalars and coordinatewise definitions of addition and scalar multiplication this sequence space is a Hilbert space with inner product

$$\langle \boldsymbol{\alpha}, \boldsymbol{\beta} \rangle = \sum_{n=-\infty}^{\infty} \alpha_n \bar{\beta}_n.$$

Expressions (1.22) and (1.23) characterize a one-to-one mapping $\mathscr{F}$ from $\mathscr{L}_2(-\pi, \pi)$ onto ℓ_2 which can be shown to preserve inner products. The process of applying this mapping is often described as *Fourier transformation*.

Fourier transformation has an important property from the viewpoint of later applications. Define the *convolution* of functions $x(t)$ and $y(t)$ in $\mathscr{L}_2(-\pi, \pi)$ by

$$z(t) = \frac{1}{2\pi} \int_{-\pi}^{\pi} x(u) y(t-u)\, du.$$

By the Schwarz inequality, $z(t)$ is in $\mathscr{L}_2(-\pi, \pi)$ and thus corresponds uniquely to a sequence in ℓ_2. It is not difficult to show that this sequence is $\{\alpha_n \beta_n : n = 0, \pm 1, \ldots\}$, where $\boldsymbol{\alpha}$ and $\boldsymbol{\beta}$ are the sequences corresponding to $x(t)$ and $y(t)$, respectively. If we denote convolution by $*$ and adopt the notational convention, $\boldsymbol{\alpha\beta} = \{\alpha_n \beta_n : n = 0, \pm 1, \ldots\}$, this relationship can be represented symbolically as

$$\mathscr{F}(\mathbf{x} * \mathbf{y}) = \mathscr{F}(\mathbf{x}) \mathscr{F}(\mathbf{y}).$$

If the *discrete convolution* of two sequences is defined by $\boldsymbol{\gamma} = \boldsymbol{\alpha} * \boldsymbol{\beta}$, where $\gamma_n = \sum_{m=-\infty}^{\infty} \alpha_m \beta_{n-m}$, then it is also easy to show that

$$\mathscr{F}(\mathbf{xy}) = \mathscr{F}(\mathbf{x}) * \mathscr{F}(\mathbf{y}).$$

Here, **xy** is the coordinatewise product of the functions **x** and **y**: $\mathbf{xy} = \{x(t)y(t): -\pi < t \leq \pi\}$.

Now, $\mathscr{F}$ is invertible and these two relationships can be shown to imply the same two relationships for the inverse of $\mathscr{F}$. The inverse mapping can also be legitimately described as Fourier transformation because of its representation by (1.22). Thus, these two expressions can be summarized by the following statement for the spaces ℓ_2 and $\mathscr{L}_2(-\pi, \pi)$: *Fourier transformation converts convolution into multiplication and multiplication into convolution.* A final consequence of these relationships and the corresponding properties for multiplication is that *convolution is commutative and associative*: $\mathbf{x} * \mathbf{y} = \mathbf{y} * \mathbf{x}$ and $(\mathbf{x} * \mathbf{y}) * \mathbf{z} = \mathbf{x} * (\mathbf{y} * \mathbf{z})$.

In a number of places in the text, we will need the analogs of expressions (1.22) and (1.23) for *periodic functions of period T.* If $x(t)$ is such a function, then $y(u) = x(Tu/2\pi)$ has period 2π and (1.22) and (1.23) apply. Reversing the transformation we obtain the expressions

$$x(t) = \sum_{n=-\infty}^{\infty} \alpha_n e^{i\lambda_n t}, \qquad -T/2 < t \leq T/2, \tag{1.24}$$

and

$$\alpha_n = \frac{1}{T}\int_{-T/2}^{T/2} x(t)e^{-i\lambda_n t}\,dt, \qquad n = 0, \pm 1, \ldots, \tag{1.25}$$

where $\lambda_n = 2\pi n/T$. In the terminology of Section 1.2, $x(t)$ has power $|\alpha_n|^2$ at frequency λ_n. Thus, the power is distributed at equally spaced points on the frequency axis with spacing $\Delta\lambda = 2\pi/T$.

Example 1.3 *Almost Periodic Functions in Continuous and Discrete Time*

A function $x(t)$, $-\infty < t < \infty$, is said to be *almost periodic* if for every $\varepsilon > 0$ it is possible to find a positive number l such that every interval of the t-axis of length l contains at least one number τ such that

$$|x(t + \tau) - x(t)| < \varepsilon$$

for $-\infty < t < \infty$ (Riesz and Nagy, 1955, pp. 254–256).

A periodic function repeats itself at intervals of T time units, thus, by induction, at spacings of nT for $n = \pm 1, +2, \ldots$. It follows that for all n $|x(t + nT) - x(t)| = 0$, $-\infty < t < \infty$. Thus, periodic functions are almost periodic with the τ's equal to the quantities nT. The similarity between periodic and almost periodic functions suggests that a spectral representation resembling Fourier series should exist for them. This representation will again be based on the definition of the appropriate Hilbert space.

Let $\mathbf{x} = \{x(t): -\infty < t < \infty\}$ and $\mathbf{y} = \{y(t): -\infty < t < \infty\}$ be continuous, complex-valued almost periodic functions and define vector addition and scalar multiplication coordinatewise as before. Define

$$\langle \mathbf{x}, \mathbf{y} \rangle = \lim_{T \to \infty} \frac{1}{2T} \int_{-T}^{T} x(t)\overline{y(t)}\, dt.$$

This quantity is an inner product for the class of continuous almost periodic functions. The resulting inner product space is not complete, however, and ideal elements must be added to the space to make it complete (Riesz and Nagy, 1955, p. 331). The result is the Hilbert space of almost periodic functions.

It is easily shown by means of expression (1.11) that the class of functions $\mathbf{z}_\lambda = \{e^{i\lambda t}: -\infty < t < \infty\}$ for $-\infty < \lambda < \infty$ is an orthonormal set. Since this class is uncountably infinite the dimension of the Hilbert space is uncountable and there is no possibility of expressing every element of the space as an infinite linear combination of a *fixed* countable set of vectors as in Example 1.2. However, if $\mathbf{x}$ is a continuous almost periodic function, there will be an increasing sequence of (real) numbers $\ldots, \lambda_{-1}, \lambda_0, \lambda_1, \ldots$ ($\lambda_0 = 0$), such that the Fourier coefficients $\alpha_\lambda = \langle \mathbf{x}, \mathbf{z}_\lambda \rangle$ are zero when λ is not an element of this sequence. Let $\beta_k = \alpha_{\lambda_k}$, $k = 0, \pm 1, \ldots$. Now, it is easily established that

$$\left\| \mathbf{x} - \sum_{k=-\infty}^{\infty} \beta_k \mathbf{z}_{\lambda_k} \right\|^2 = \|\mathbf{x}\|^2 - \sum_{k=-\infty}^{\infty} |\beta_k|^2. \tag{1.26}$$

The fundamental theorem of almost periodic functions asserts that the Parseval relation $\|\mathbf{x}\|^2 = \sum_{k=-\infty}^{\infty} |\beta_k|^2$ is valid for all continuous almost periodic functions. This and (1.26) imply that (in coordinate form)

$$x(t) = \sum_{k=-\infty}^{\infty} \beta_k e^{i\lambda_k t}. \tag{1.27}$$

By the definition of inner product and the above discussion,

$$\begin{aligned} \alpha_\lambda &= \lim_{T \to \infty} \frac{1}{2T} \int_{-T}^{T} x(t) e^{-i\lambda t}\, dt \\ &= \begin{cases} \beta_k, & \lambda = \lambda_k, \quad k = 0, \pm 1, \ldots, \\ 0, & \text{otherwise}. \end{cases} \end{aligned} \tag{1.28}$$

Note that the Parseval relation implies $\sum_{k=-\infty}^{\infty} |\beta_k|^2 < \infty$. In fact if $\ldots, \lambda_{-1}, \lambda_0, \lambda_1, \ldots$ ($\lambda_0 = 0$) is any increasing sequence of real numbers and $\{\beta_k: k = 0, \pm 1, \ldots\}$ is any sequence of complex numbers for which $\sum_{k=-\infty}^{\infty} |\beta_k|^2 < \infty$, then an element of the Hilbert space will be represented by expression (1.27) as a limit of partial sums. Moreover, (1.28) will be satisfied. This suggests an alternate definition of the Hilbert space of almost periodic functions as the class of all functions possessing spectral representations of the form (1.27) for sequences of frequencies λ_k and square summable coefficients β_k. The frequencies and coefficients are uniquely determined by (1.28).

The *convolution* of two almost periodic functions

$$z(\tau) = \lim_{T\to\infty} \frac{1}{2T} \int_{-T}^{T} x(\tau - t)y(t)\,dt$$

is again almost periodic. Thus, in particular, a function of importance in the general time series model of Chapter 2,

$$C(\tau) = \lim_{T\to\infty} \frac{1}{2T} \int_{-T}^{T} x(t+\tau)x(t)\,dt,$$

will be almost periodic if $x(t)$ is.

The spectral representation for *discrete almost periodic functions* $\mathbf{x} = \{x(t)\colon t = 0, \pm 1, \ldots\}$ parallels the above theory closely. The inner product is now

$$\langle \mathbf{x}, \mathbf{y} \rangle = \lim_{L\to\infty} \frac{1}{2L+1} \sum_{t=-L}^{L} x(t)\overline{y(t)},$$

and it is easily shown by means of (1.18) that the sequences $\mathbf{z}_\lambda = \{e^{i\lambda t}\colon t = 0, \pm 1, \ldots\}$ for $-\pi < \lambda \le \pi$ form an orthonormal set. Thus, for every discrete almost periodic function $x(t)$ there will exist an increasing sequence of frequencies $\ldots, \lambda_{-1}, \lambda_0, \lambda_1, \ldots$, with $\lambda_0 = 0$ and $-\pi < \lambda_k \le \pi$ for all k, and a square-summable sequence of complex numbers $\{\beta_k\colon k = 0, \pm 1, \ldots\}$ such that

$$x(t) = \sum_{k=-\infty}^{\infty} \beta_k e^{i\lambda_k t},$$

where

$$\lim_{L\to\infty} \frac{1}{2L+1} \sum_{t=-L}^{L} x(t)e^{-i\lambda t} = \begin{cases} \beta_k, & \lambda = \lambda_k, \quad k = 0, \pm 1, \ldots, \\ 0, & \text{otherwise.} \end{cases}$$

In particular, if $x(t)$ is a discrete periodic sequence with spectral representation (1.19) and (1.20), then since

$$\lim_{L\to\infty} \frac{1}{2L+1} \sum_{t=-L}^{L} x(t)e^{-i\lambda t} = \frac{1}{N} \sum_{t=1}^{N} x(t)e^{-i\lambda t}$$

when $\lambda = 2\pi v/N$, $-[N/2] \le v \le [N/2]$, and is otherwise zero, the almost periodic spectral representation (1.27) and (1.28) reduces to (1.19) and (1.20).

Example 1.4 *A Spectral Representation Based on Finite Energy. Fourier Integrals*

By extending the definition given in Section 1.2, the total energy of a complex-valued time series would be $\int_{-\infty}^{\infty} |x(t)|^2\,dt$. Taking the integral to be the Lebesgue integral, the class of all functions for which this quantity is finite is a Hilbert space under coordinatewise vector addition and scalar multiplication with inner product

$$\langle x(t), y(t) \rangle = \frac{1}{2\pi} \int_{-\infty}^{\infty} x(t)\overline{y(t)}\, dt.$$

This space is denoted by $\mathscr{L}_2(-\infty, \infty)$ or, simply, $\mathscr{L}_2$.

The functions $e^{i\lambda t}$ are periodic, thus have infinite energy and are not in this Hilbert space. However, properly interpreted, an important spectral representation of the elements of $\mathscr{L}_2$ still exists.

Consider first a complex-valued function $x(t)$ for which $\int_{-\infty}^{\infty} |x(t)|\, dt < \infty$. Define the generalized Fourier coefficients,

$$g(\lambda) = \langle x(t), e^{i\lambda t} \rangle = \frac{1}{2\pi} \int_{-\infty}^{\infty} x(t) e^{-i\lambda t}\, dt. \tag{1.29}$$

This function, called the *Fourier integral* (*Fourier transform*) of $x(t)$, is a well-defined bounded function, since

$$|g(\lambda)| \le \frac{1}{2\pi} \int_{-\infty}^{\infty} |x(t) e^{-i\lambda t}|\, dt = \frac{1}{2\pi} \int_{-\infty}^{\infty} |x(t)|\, dt.$$

It is also easily established that $g(\lambda)$ is continuous. Now, by analogy with our earlier examples, it would seem plausible that the appropriate spectral representation for $x(t)$ would be

$$x(t) = \int_{-\infty}^{\infty} g(\lambda) e^{i\lambda t}\, dt. \tag{1.30}$$

However, this integral need not exist, since although $g(\lambda)$ is bounded, it need not be integrable. Of course, if $\int_{-\infty}^{\infty} |g(\lambda)|\, d\lambda < \infty$ then the roles of $g(\lambda)$ and $x(t)$ can be reversed and (1.30) will be well defined.

A more satisfactory situation holds for functions in $\mathscr{L}_2$ [see Goldberg (1961)]. Briefly, for $x(t) \in \mathscr{L}_2$ form

$$g_N(\lambda) = \frac{1}{2\pi} \int_{-N}^{N} x(t) e^{-i\lambda t}\, dt.$$

These functions are well defined in the sense given above and it can be shown that this sequence of functions is a Cauchy sequence in $\mathscr{L}_2$. By completeness, there exists $g(\lambda) \in \mathscr{L}_2$ such that

$$\|g_N(\lambda) - g(\lambda)\|^2 = \frac{1}{2\pi} \int_{-\infty}^{\infty} |g_N(\lambda) - g(\lambda)|^2\, d\lambda \to 0$$

as $N \to \infty$. The Fourier integral is now defined to be this limit;

$$g(\lambda) = \frac{1}{2\pi} \int_{-\infty}^{\infty} x(t)e^{-i\lambda t}\, dt.$$

Since $g(\lambda) \in \mathscr{L}_2$, we can repeat this procedure and define

$$x_N(t) = \int_{-N}^{N} g(\lambda)e^{i\lambda t}\, dt.$$

This is again a Cauchy sequence in $\mathscr{L}_2$ and, moreover, it can be shown that its limit is $x(t)$. That is, in the sense of $\mathscr{L}_2$ limit,

$$x(t) = \int_{-\infty}^{\infty} g(\lambda)e^{i\lambda t}\, d\lambda.$$

This is the desired spectral representation (1.30).

The Parseval relation for $\mathscr{L}_2$ functions is

$$\frac{1}{2\pi} \int_{-\infty}^{\infty} |x(t)|^2\, dt = \int_{-\infty}^{\infty} |g(\lambda)|^2\, d\lambda.$$

If $g(\lambda)$ and $h(\lambda)$ are the Fourier integrals of $x(t)$ and $y(t)$, respectively, it can be shown that

$$\frac{1}{2\pi} \int_{-\infty}^{\infty} x(t)y(\tau - t)\, dt = \int_{-\infty}^{\infty} g(\lambda)h(\lambda)e^{i\tau\lambda}\, d\lambda.$$

The left-hand side of this expression is defined to be the *convolution* of $x(t)$ and $y(t)$ and will be denoted by $x * y(\tau)$. By the Schwarz inequality it can be seen that $\int_{-\infty}^{\infty} |g(\lambda)h(\lambda)|\, d\lambda < \infty$. It follows that $x * y(\tau)$ is a bounded, continuous function. If, in addition, $\int_{-\infty}^{\infty} |g(\lambda)h(\lambda)|^2\, d\lambda < \infty$, then $x * y(\tau)$ will also be in $\mathscr{L}_2$ and we will have

$$g(\lambda)\, h(\lambda) = \frac{1}{2\pi} \int_{-\infty}^{\infty} e^{-i\lambda\tau}\, x * y\, (\tau)\, d\tau.$$

In the notation for Fourier transformation established in Example 1.2 this can be written

$$\mathscr{F}(x * y) = \mathscr{F}(x)\mathscr{F}(y).$$

The second relationship,

$$\mathscr{F}(xy) = \mathscr{F}(x) * \mathscr{F}(y),$$

also holds when $\mathscr{F}$ is given by (1.29). Thus, in a somewhat more restricted sense, Fourier transformation again converts convolutions to products and conversely.

1.4 SOME PROBABILITY NOTATIONS AND PROPERTIES

For convenience, some probability topics needed in later chapters but not commonly available in standard texts are collected here. A good reference for the probability and statistics required in this book is the book by Tucker (1962).

Here, Ω, $\mathscr{S}$, P denotes a probability space consisting of a set (Ω) of elements ω, the collection $\mathscr{S}$ of events, and a probability P. $E(X)$ or EX will denote the expectation of a (real-valued) random variable X with respect to P. A *complex-valued random variable*, Z, can be defined by its representation in cartesian form, $Z = X + iY$, where X and Y are real-valued random variables. Then, the distribution of Z is determined by the joint distribution of X and Y. Similarly, the joint distribution of several complex random variables $Z_j = X_j + iY_j$, $j = 1, 2, \ldots, n$, is governed by the joint distribution of $X_1, \ldots, X_n, Y_1, \ldots, Y_n$.

Expectation retains its *linearity property*,

$$E\sum_{j=1}^{n} a_j Z_j = \sum_{j=1}^{n} a_j EZ_j,$$

when the coefficients and random variables are complex valued. In particular, $EZ = EX + iEY$. This linearity property extends to random vectors and matrices $\mathbf{X} = [X_{i,j}]$, where this notation means that the real- or complex-valued random variable $X_{i,j}$ is the i, jth element of the matrix. Expectation is defined componentwise; $E\mathbf{X} = [EX_{i,j}]$. Then if $\mathbf{A}$ and $\mathbf{B}$ are matrices of real or complex constants and $\mathbf{A}^*$ denotes the conjugate transpose of $\mathbf{A}$,

$$E\mathbf{A}^*\mathbf{X}\mathbf{B} = \mathbf{A}^*E\mathbf{X}\mathbf{B}.$$

This linearity property for the expectation of random matrices will be needed in Chapter 5.

Let $\mathbf{V}$ and $\mathbf{W}$ denote $n \times 1$ dimensional random vectors and define $\boldsymbol{\mu}_V = E\mathbf{V}$, $\boldsymbol{\mu}_W = E\mathbf{W}$, and $\boldsymbol{\Sigma}_{V,W} = E(\mathbf{V} - \boldsymbol{\mu}_V)(\mathbf{W} - \boldsymbol{\mu}_W)^*$. Then, for example, $\boldsymbol{\Sigma}_{V,V}$ is the covariance matrix of the components of $\mathbf{V}$. The *multivariate normal distribution* for (the components of) a real random vector $\mathbf{X}$ will be denoted by the symbol $\mathfrak{N}(\boldsymbol{\mu}_X, \boldsymbol{\Sigma}_{X,X})$. We will require the idea of a *multivariate complex normal distribution* in later chapters. If $\mathbf{Z} = \mathbf{X} + i\mathbf{Y}$ is a $n \times 1$ complex random vector in cartesian form, then $\mathbf{Z}$ will be said to have a multivariate complex normal distribution if the $2n \times 1$ random vector $\binom{\mathbf{X}}{\mathbf{Y}}$ has a (real) multivariate normal distribution. There is some possible ambiguity for the joint distribution of this random vector, however. It can be shown from the properties of expectation that $\boldsymbol{\Sigma}_{Z,Z} = \boldsymbol{\Sigma}_{X,X} + \boldsymbol{\Sigma}_{Y,Y} + i(\boldsymbol{\Sigma}_{Y,X} - \boldsymbol{\Sigma}_{X,Y})$. Thus, if $\boldsymbol{\Sigma}_{Z,Z} = \mathbf{C} + i\mathbf{Q}$ is the cartesian representation for $\boldsymbol{\Sigma}_{Z,Z}$, these matrices are required to satisfy the equations

$$\boldsymbol{\Sigma}_{X,X} + \boldsymbol{\Sigma}_{Y,Y} = \mathbf{C},$$
$$\boldsymbol{\Sigma}_{Y,X} - \boldsymbol{\Sigma}_{X,Y} = \mathbf{Q}.$$

Of the various possible solutions we will select $\boldsymbol{\Sigma}_{X,X} = \boldsymbol{\Sigma}_{Y,Y} = \frac{1}{2}\mathbf{C}$ and $\boldsymbol{\Sigma}_{X,Y} = \boldsymbol{\Sigma}'_{Y,X} = -\frac{1}{2}\mathbf{Q}$, where $'$ denotes transpose. (These are valid solutions, since it can be shown that $\mathbf{C}' = \mathbf{C}$ and $\mathbf{Q}' = -\mathbf{Q}$.) It follows that $\mathbf{Z}$ will have the multivariate complex normal distribution with mean $\boldsymbol{\mu}_Z = \boldsymbol{\mu}_X + i\boldsymbol{\mu}_Y$ and covariance matrix $\boldsymbol{\Sigma}_{Z,Z}$, denoted $\mathbf{Z} \approx \mathfrak{N}_C(\boldsymbol{\mu}_Z, \boldsymbol{\Sigma}_{Z,Z})$, if and only if

$$\begin{pmatrix}\mathbf{X}\\ \mathbf{Y}\end{pmatrix} \approx \mathfrak{N}\left(\begin{pmatrix}\boldsymbol{\mu}_X\\ \boldsymbol{\mu}_Y\end{pmatrix}, \frac{1}{2}\begin{pmatrix}\mathbf{C} & -\mathbf{Q}\\ \mathbf{Q} & \mathbf{C}\end{pmatrix}\right).$$

An important property of the complex normal distribution is the *Isserlis theorem*: Let $\mathbf{Z} = (Z_1, Z_2, Z_3, Z_4)$ have the multivariate complex normal distribution with mean $\boldsymbol{\mu}_Z = \mathbf{0}$ and arbitrary covariance matrix. Then,

$$E(Z_1 Z_2 Z_3 Z_4) = E(Z_1 Z_2)E(Z_3 Z_4) + E(Z_1 Z_3)E(Z_2 Z_4) + E(Z_1 Z_4)E(Z_2 Z_3).$$

[See Blackman and Tukey (1959, p. 100) for the version for real-valued random variables.]

Example 1.5 *The Hilbert Space $L_2(P)$*

Consider the collection of all complex-valued random variables on a probability space Ω, $\mathscr{S}$, P for which

$$E|X|^2 < \infty. \tag{1.31}$$

By the inequality $E|X| \leq 1 + E|X|^2$ it follows that EX exists. With coordinatewise addition and scalar multiplication (by complex scalars) the class of random variables satisfying (1.31) and the condition $EX = 0$ constitutes a Hilbert space with inner product

$$\langle X, Y\rangle = EX\overline{Y}.$$

This space is denoted by $L_2(P)$. We will occasionally distinguish this space from the Hilbert space of zero-mean, real-valued random variables satisfying $EX^2 < \infty$, with real scalars and inner product $\langle X, Y\rangle = EXY$, by designating the first, complex $L_2(P)$ and the second, real $L_2(P)$.

Note that every random variable satisfying (1.31) becomes an element of $L_2(P)$ by replacing X by $X - EX$. The definition of covariance for random variables entails precisely this substitution;

$$\mathrm{Cov}(X, Y) = E(X - EX)\overline{(Y - EX)} = \langle X - EX,\ Y - EY\rangle.$$

Thus many of the more useful properties of correlation theory are simply the geometric properties of $L_2(P)$. For example (assuming $EX = EY = 0$ again), the correlation coefficient in the notation of $L_2(P)$ inner product is

$$\rho = \frac{\mathrm{Cov}(X, Y)}{(\mathrm{Var}(X)\mathrm{Var}(Y))^{1/2}} = \frac{\langle X, Y\rangle}{\|X\|\,\|Y\|}.$$

As we noted in the previous section, this is simply the cosine of the "angle" between the random variables X and Y. This observation and the intuition developed from the study of two-dimensional geometry make it possible to interpret geometrically the usual properties of correlation. Thus, for example, the conditions $\rho = \pm 1$ of perfect correlation are interpreted to represent the colinearity of the "vectors" X and Y, with the same or opposite orientation, while the condition $\rho = 0$ corresponds to orthogonality. Several other important properties of correlation are a consequence of the theory of orthogonal projections summarized in Section 1.3. This will be illustrated in the time series context in Chapter 5.

The concept of *mean-square convergence* of random variables will be used at several points in the text. A sequence of random variables Y_n is said to converge to Y in mean-square if $\lim_{n\to\infty} E(Y_n - Y)^2 = 0$. Because of the relation $E(X - Y)^2 = \|X - Y\|^2$, the properties of mean-square convergence are simply the properties of convergence in the distance function of $L_2(P)$. Thus, for example, a mean-square Cauchy sequence will always have a limit in $L_2(P)$.

The $L_2(P)$ spaces play a key role in time series analysis because of their relationship to the basic time series model, weakly stationary stochastic processes. We will establish this relationship in Chapter 2. The spectral representations for these models rely heavily on Hilbert space arguments as will be outlined in the appendixes to Chapters 2, 4, and 5. Moreover, the geometry of $L_2(P)$ is the natural setting for prediction theory to be studied in Chapter 7.

CHAPTER

2

Models for Spectral Analysis—The Univariate Case

2.1 INTRODUCTION

Historically, the introduction of models for time series which admit a spectral decomposition followed two lines of development. The first, originating in the study of light in physics and motivated by the work of Sir Arthur Schuster (1898, 1906) in geophysics culminated in the treatise "Generalized Harmonic Analysis" by Wiener (1930). In this remarkable work, the spectral analysis for functions with finite power was completely detailed. Wiener's theory covered both univariate and multivariate time series, and applied to stochastic as well as nonstochastic series, although at that time the nature of the stochastic series, as stochastic processes, was not well understood.

The second line of development began with a series of papers in 1932–1934 by the Russian mathematician Khintchine who introduced both stationary and weakly stationary stochastic processes and developed the correlation theory for weakly stationary processes [see Khintchine (1934)]. This development was important not only for time series analysis but was also one of the pioneering works in the modern theory of stochastic processes. Later, Kolmogorov (1971a) developed the geometric theory of weakly stationary time series and Cramér (1942) discovered the important spectral decomposition of weakly stationary processes of which we will have many opportunities to take advantage in this book.

The more recent work in time series analysis, both in the study of problems of a purely probabilistic nature and in the development of the statistical theory, has been based on the stationary models. As we will see in this and subsequent chapters, even the most commonly used examples of nonstationary models are obtained by modifying stationary stochastic processes in elementary ways. Consequently, stationary models dominate the theory and their study will occupy us for the major part of the book. The basic theory for univariate weakly stationary processes is given in this chapter and the multivariate theory is given in Chapter 5.

From the viewpoint of applying the theory to real problems, the Wiener model provides some peace of mind to the experimenter who is concerned about the validity of his model, because it applies equally well to a large class of nonstochastic as well as stochastic time series. Moreover, the class of stochastic processes to which it applies is far larger than the class of weakly stationary processes. Some indication of the scope of applicability of the Wiener theory to stochastic processes will be given at the end of this chapter.

The way we will go about estimating spectra in Chapter 8, though ostensibly geared to the stationary model, actually provides an estimate of Wiener's power spectrum. Consequently, even if some of the hypotheses for statistically estimating spectra are violated (including the assumption that we are dealing with a stochastic process!), the existence of a spectrum and the validity of the method of estimation are guaranteed insofar as the more general Wiener theory is applicable. We will have more to say about this in Chapter 9.

2.2 THE WIENER THEORY OF SPECTRAL ANALYSIS

Let $x(t)$, $-\infty < t < \infty$, be a real-valued function (time series) with the property that

$$C(\tau) = \lim_{T\to\infty} \frac{1}{2T} \int_{-T}^{T} x(t+\tau)x(t)\,dt \tag{2.1}$$

exists and is finite for every τ. The function $C(\tau)$ is called the *autocovariance function* of the time series. Then

$$C(0) = \lim_{T\to\infty} \frac{1}{2T} \int_{-T}^{T} x^2(t)\,dt$$

is the *total power* or simply, the *power* of $x(t)$, and is finite by hypothesis. Note that a simple change of variables yields

$$C(-\tau) = C(\tau). \tag{2.2}$$

It is easily established that if $y(t)$ and $z(t)$ represent (possibly) complex-valued functions, then

$$\langle y(t), z(t) \rangle = \lim_{T \to \infty} \frac{1}{2T} \int_{-T}^{T} y(t)\overline{z(t)}\, dt$$

is an indefinite inner product (Section 1.3) for the collection of all functions for which (2.1) exists for all τ. In particular, the Schwarz inequality can be applied to obtain

$$|C(\tau)| = |\langle x(t+\tau), x(t) \rangle| \leq \|x(t+\tau)\| \; \|x(t)\|.$$

However, as is easily seen, $\|x(t+\tau)\|^2 = \|x(t)\|^2 = C(0)$. Consequently, we obtain the inequality

$$|C(\tau)| \leq C(0). \tag{2.3}$$

That is, the autocovariance function is always bounded in modulus by the power.

The Spectral Representation of the Autocovariance Function

Wiener established the existence of a bounded nondecreasing function $F(\lambda)$, called the *spectral distribution function*, such that

$$C(\tau) = \int_{-\infty}^{\infty} e^{i\lambda\tau} F(d\lambda). \tag{2.4}$$

This important expression is called the *spectral representation of the autocovariance function*. The spectral distribution function determines a measure $F(A)$ called the *spectral distribution* of the time series and (2.4) is an integral with respect to this measure. This theory can be outlined exclusively in terms of the distribution function without explicitly introducing the idea of measure. However, to do so has the disadvantage that simple and intuitive properties have rather clumsy and unpleasant notational expressions. Simply think of the spectral distribution $F(A)$ as the amount of power in the harmonic components of the time series with frequencies in the set A. The spectral distribution and spectral distribution function are related by the expression

$$F(\lambda) = F((-\infty, \lambda]).$$

Thus, for example, $F(\lambda)$ is the power for frequencies less than or equal to λ. The symbol $(a, b]$ denotes the interval $\{x : a < x \leq b\}$. This notation for intervals will be used throughout the book with the exclusion and inclusion of endpoints being indicated by round and square brackets respectively.

The integral (2.4) can be reduced to more familiar terms by the Lebesgue decomposition of $F(A)$ [see Grenander and Rosenblatt (1957, p. 35)]. For models of practical interest this measure can be expressed as the sum of two components

$$F(A) = F_d(A) + F_c(A). \tag{2.5}$$

The *discrete spectral distribution* $F_d(A)$ is completely characterized by a function $p(\lambda)$, called the *spectral function*, which has the property that $p(\lambda) \geq 0$ for all λ and $p(\lambda) > 0$ only among a "discrete" set of frequencies $\ldots, \lambda_{-1}, \lambda_0, \lambda_1, \ldots$, where $\lambda_{-j} = -\lambda_j$ (thus $\lambda_0 = 0$). Then,

$$F_d(A) = \sum_{\lambda_j \in A} p(\lambda_j). \tag{2.6}$$

The value of $p(\lambda)$ at frequency λ is the spectral mass concentrated at that frequency and, thus, is related to the spectral distribution by the expression

$$p(\lambda) = F(\{\lambda\}),$$

where $\{x\}$ denotes the set consisting of the single element x.

The continuous component $F_c(A)$ is determined by the derivative of the spectral distribution function $f(\lambda) = F'(\lambda)$, where $f(\lambda)$ is called the *spectral density function*, and the amount of continuous power or continuous spectral mass in a set of frequencies A is given by

$$F_c(A) = \int_A f(u)\, du. \tag{2.7}$$

Since the derivative of a nondecreasing function is nonnegative, the spectral density function has the property

$$f(\lambda) \geq 0 \qquad \text{for} \quad \text{all } \lambda.$$

It is sometimes useful to think of the continuous power in a set A as being given by the area under the curve $y = f(\lambda)$ over the set A.

The "spectral mass" or power in a set of frequencies A when both discrete and continuous components are present in the spectrum is

$$F(A) = \sum_{\lambda_j \in A} p(\lambda_j) + \int_A f(\lambda)\, d\lambda. \tag{2.8}$$

The spectral representation of the autocovariance (2.4) can then be represented in the form

$$C(\tau) = \sum_{k=-\infty}^{\infty} e^{i\lambda_k \tau} p(\lambda_k) + \int_{-\infty}^{\infty} e^{i\lambda\tau} f(\lambda)\, d\lambda.$$

Obtaining the Spectrum from the Autocovariance

From a practical viewpoint, spectrum analysis is based on the conversion of time-indexed data into estimates of the spectrum. One important method depends on the Fourier transformation of $C(\tau)$ to obtain $F(A)$. We now discuss the sense in which this Fourier transformation can be performed; first in the special cases of "pure" spectral types.

When the continuous component is missing, i.e., when $f(\lambda) = 0$ for all λ, the time series is said to have a *discrete spectrum*. (Another commonly used term is *point spectrum*.) Then,

$$C(\tau) = \sum_{k=-\infty}^{\infty} e^{i\lambda_k \tau} p(\lambda_k). \tag{2.9}$$

Moreover,

$$\sum_{k=-\infty}^{\infty} p(\lambda_k) = C(0) < \infty.$$

Thus, since absolutely summable series are square summable,

$$\sum_{k=-\infty}^{\infty} p^2(\lambda_k) < \infty,$$

and by the theory of Example 1.3, (2.9) is the representation of $C(\tau)$ as an almost periodic function. It follows that the spectral function can be obtained from the autocovariance by the expression

$$p(\lambda_k) = \lim_{T\to\infty} \frac{1}{2T} \int_{-T}^{T} C(\tau) e^{-i\lambda_k \tau}\, d\tau. \tag{2.10}$$

In fact, this expression yields $p(\lambda)$ for all λ, since the limit is zero if λ is not one of the λ_k's. Now, $F_d(A)$ is obtained by (2.6) if desired. It is a property of almost periodic functions that $|C(\tau)|$ approaches $C(0)$ arbitrarily closely for arbitrarily large values of $|\tau|$. Thus, in a rather strong sense, $C(\tau)$ does not go to zero as $|\tau| \to \infty$. It is also the case that $C(\tau) \nrightarrow 0$ when the spectrum is of mixed type.

The situation is not quite so simple in the case of a *continuous spectrum*. [Now $p(\lambda) = 0$ for all λ.] The representation

$$C(\tau) = \int_{-\infty}^{\infty} e^{i\lambda\tau} f(\lambda)\, d\lambda \tag{2.11}$$

is valid, and

$$\int_{-\infty}^{\infty} f(\lambda)\, d\lambda = C(0) < \infty.$$

In rather pathological situations, however, it may not be possible to invert the transform (2.11) to obtain $f(\lambda)$ from $C(\tau)$. In contrast to the discrete case it is always true that $|C(\tau)| \to 0$ as $|\tau| \to \infty$. In fact, this serves as a useful theoretical criterion for distinguishing between time series with continuous and discrete or mixed spectra. However, it is possible that $|C(\tau)|$ goes to zero so slowly that the usual inversion methods fail. This is almost never the case in practice and for most models of interest we will be able to assume that

$$\int_{-\infty}^{\infty} |C(\tau)|\, d\tau < \infty \qquad \text{or} \qquad \int_{-\infty}^{\infty} |C(\tau)|^2\, d\tau < \infty.$$

Then (see Example 1.4), $f(\lambda)$ has the representation,

$$f(\lambda) = \frac{1}{2\pi} \int_{-\infty}^{\infty} C(\tau) e^{-i\lambda\tau}\, d\tau. \tag{2.12}$$

Expressions (2.11) and (2.12) are called the *Wiener–Khintchine relations*.

Although principally of theoretical importance, it is of some interest to know that there is a means for obtaining the spectrum from the autocovariance in *any* situation, discrete, continuous, or when both spectral types are present at the same time. It is shown by Doob (1953, p. 519) that if $\Lambda_1 < \Lambda_2$ and $p(\Lambda_1) = p(\Lambda_2) = 0$, then

$$F((\Lambda_1, \Lambda_2)) = \lim_{T \to \infty} \frac{1}{2\pi} \int_{-T}^{T} C(\tau)(e^{-i\Lambda_1\tau} - e^{-i\Lambda_2\tau})/i\tau\, d\tau. \tag{2.13}$$

It can then be argued that (2.13) determines $F(A)$ for all sets A. Thus, expressions (2.4) and (2.13) assert the complete equivalence of the autocovariance function and the spectral distribution. That is, if either is known, the other can be determined exactly. However, these functions display different aspects of the correlation information about the time series. It is now commonly accepted that for practical purposes the spectrum is the more useful "parameter."

The Spectrum Is an Even Function

One other property of importance is the following. Since we have taken $x(t)$ to be real-valued, $C(\tau)$ will also be real-valued and it follows that the spectrum is symmetric about $\lambda = 0$. That is, for all λ,

$$f(-\lambda) = f(\lambda), \qquad p(-\lambda) = p(\lambda). \tag{2.14}$$

This can be checked by a change of variables argument in (2.10) and (2.11). Moreover, if A is a set of numbers and $-A = \{-\lambda : \lambda \in A\}$, the set obtained by replacing each λ in A by $-\lambda$, then

$$F(-A) = F(A). \tag{2.15}$$

Properties (2.14) indicate that $f(\lambda)$ and $p(\lambda)$ are *even functions*. It is a general property of Fourier transforms that a real function has an even transform. This works in both directions. Consequently, in order for $f(\lambda)$ and $p(\lambda)$ to be real-valued, it is necessary that $C(\tau)$ be even, but this is property (2.2) of $C(\tau)$ given earlier.

Representation of Total Power

Finally, setting $\tau = 0$ in (2.8), we obtain

$$C(0) = \sum_{k=-\infty}^{\infty} p(\lambda_k) + \int_{-\infty}^{\infty} f(\lambda)\, d\lambda.$$

That is, the total power in the time series is the sum of the power in the discrete and continuous components, and each of these, in turn, is the "sum" of the power magnitudes at each frequency. This is the spectral decomposition of the power into its harmonic components analogous to (1.12). The graph of a hypothetical spectrum is given in Fig. 2.1.

Example 2.1 *The Autocovariance and Spectrum of an Almost Periodic Function*

Let $x(t) = \sum_{j=-\infty}^{\infty} c_j e^{i\lambda_j t}$ be an almost periodic function with

$$\sum_{j=-\infty}^{\infty} |c_j|^2 < \infty \qquad \text{(Example 1.3).}$$

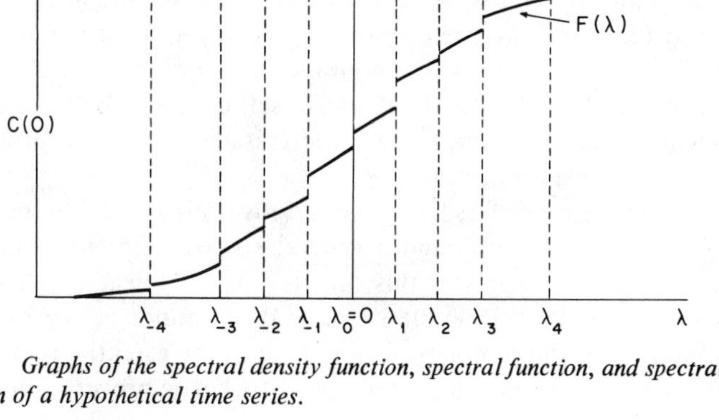

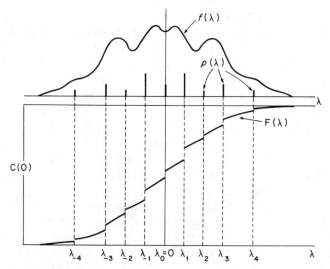

Fig. 2.1 *Graphs of the spectral density function, spectral function, and spectral distribution function of a hypothetical time series.*

Using the properties of inner product and the orthonormality of the functions $e^{i\lambda_j t}$ we can calculate the autocovariance function for this time series

$$C(\tau) = \langle x(t+\tau), x(t)\rangle = \left\langle \sum_{j=-\infty}^{\infty} c_j e^{i\lambda_j \tau} e^{i\lambda_j t}, \sum_{k=-\infty}^{\infty} c_k e^{i\lambda_k t} \right\rangle$$

$$= \sum_{j=-\infty}^{\infty} \sum_{k=-\infty}^{\infty} c_j \bar{c}_k e^{i\lambda_j \tau} \langle e^{i\lambda_j t}, e^{i\lambda_k t}\rangle$$

$$= \sum_{j=-\infty}^{\infty} |c_j|^2 e^{i\lambda_j \tau}. \tag{2.16}$$

By comparing this with (2.9) it is seen that $x(t)$ has a pure point spectrum with spectral function

$$p(\lambda) = \begin{cases} |c_j|^2, & \text{for } \lambda = \lambda_j, \ j = 0, \pm 1, \ldots, \\ 0, & \text{otherwise.} \end{cases}$$

The total power is

$$C(0) = \sum_{j=-\infty}^{\infty} |c_j|^2.$$

Note that the phase information contained in the complex coefficients $c_0, c_{\pm 1}, \ldots$ is lost when the absolute values are taken to form $C(\tau)$. Thus, knowledge of the power spectrum of $x(t)$ is not sufficient to reconstruct the time series. This is true of power spectrum analysis in general. Loss of phase information is characteristic of the indifference to time origin built into the model. The properties of time series which the model is constructed to characterize are average properties over all time and no "natural" or intrinsic time origin exists. Consequently, phase values, which measure the displacements of the harmonic components relative to a fixed time origin, are not retained.

Although we have taken the time series $x(t)$ to be a nonrandom function, nothing precludes it from being a stochastic process if the integral and limit in (2.1) and (2.2) are interpreted properly. The only restriction is that the autocovariance function must be a nonrandom function. That is, the random quantities $(1/2T) \int_{-T}^{T} x(t+\tau)x(t)\, dt$ must settle down to a fixed limit as $T \to \infty$. This actually occurs for a large class of stochastic processes. An example will be given in Section 2.9.

The Wiener theory suffers from one serious drawback from the viewpoint of model construction. Although a general spectral representation of sorts exists for $x(t)$, which expresses this function as a "linear combination" of complex exponentials (Wiener, 1930, p. 154), it cannot be used directly for the computation of the power spectrum. One of the essential reasons for the central position held by (weakly) stationary stochastic processes in time series analysis is the existence of a spectral representation for the process from which spectral parameters can be directly computed. We will have ample opportunity to appreciate the "power" of this representation in later chapters.

2.3 STATIONARY AND WEAKLY STATIONARY STOCHASTIC PROCESSES

Let $\Omega, \mathscr{S}, P$ be a probability space. An indexed family of (real- or complex-valued) random variables $\{X(t): t \in T\}$ defined on the space is called a (real or complex) *stochastic process*. If the index set T of the time variable t consists of equally spaced numbers $k\,\Delta t$, $k = 0, \pm 1, \ldots, \Delta t > 0$, the process is called a *discrete-time process*. If T is the set of all real numbers, $-\infty < t < \infty$, it is called a *continuous-time process*. For this discussion, we will restrict ourselves to real, continuous-time processes.

It is useful to think of a random experiment modeled by a stochastic process as follows: "Nature" performs a trial of the experiment governed by the probability P and obtains a value $\omega \in \Omega$. You are then allowed to observe the function $X(t, \omega)$, $-\infty < t < \infty$, where $X(t, \omega)$ is the value of the random variable $X(t)$ at ω. The functions generated by varying ω over Ω are called the *sample functions* of the process. In a practical situation for which it is reasonable to assume that a probabilistic model is appropriate for the underlying mechanism, the observed function of time is assumed to be one of the sample functions of a stochastic process. Thus, the functions graphed in Figures 1.1–1.6 would be thought of as sample functions from stochastic process models of the corresponding phenomena. By this view, a stochastic process can also be thought of as a probability distribution over the set of all possible sample functions.

The stochastic process determines the set of finite-dimensional probability distributions,

$$P(X(t_1) \in S_1, X(t_2) \in S_2, \ldots, X(t_n) \in S_n) \tag{2.17}$$

for all finite sets of time points $t_1 < t_2 < \cdots < t_n$ and real events $S_1, \ldots, S_n$. A celebrated theorem of Kolmogorov (Kolmogorov, 1933, p. 29) asserts that the reverse is also true. Namely, if we are given a collection of functions $P(t_1, \ldots, t_n; S_1, \ldots, S_n)$ indexed by finite sets of times $t_1 < t_2 < \cdots < t_n$ which are joint probability distributions in $S_1, \ldots, S_n$ and satisfy an additional condition which guarantees that these joint distributions have the same marginal distributions on common subsets of time points, then there exists a stochastic process for which these are the probabilities (2.17). In this sense, the finite-dimensional distributions completely determine the probabilistic structure of the stochastic process and any special properties desired of the process can be imposed on these distributions. For example, a stochastic process is said to be *Gaussian* if the finite-dimensional distributions are multivariate normal distributions. We will use this fact again to define stationary processes.

Stationary and Weakly Stationary Stochastic Processes

A descriptive property of the time series discussed in Section 2.2 was an "unchangingness" in time of the underlying generating mechanism. We can capture this property in the stochastic process model by making the finite-dimensional distributions time invariant. In the notation of (2.17) this property is

$$P\big(X(t_1 + \tau) \in S_1, \ldots, X(t_n + \tau) \in S_n\big) = P\big(X(t_1) \in S_1, \ldots, X(t_n) \in S_n\big) \quad (2.18)$$

for all $t_1 < \cdots < t_n$, real events $S_1, \ldots, S_n$, and τ, $-\infty < \tau < \infty$. Such a stochastic process is said to be (strictly) *stationary*. Note that the distributions depend on the relative time separations of the random variables, but not on their absolute time locations. That is, the stochastic process has the same probabilistic behavior throughout all time.

If the mean and variance of the random variables exist, it is easily established that stationarity implies the properties,

$$EX(t) = EX(0) = m, \qquad -\infty < t < \infty, \quad (2.19)$$

and

$$EX(t + \tau)X(t) = EX(\tau)X(0) = C(\tau), \qquad -\infty < t < \infty. \quad (2.20)$$

That is, the mean values are constant in time and the covariances depend upon the time displacement τ, but not on t. The function $C(\tau)$ is called the *autocovariance function* of the process.

Now, if we discard condition (2.18) and assume only that the random variables of the stochastic process have the property

$$\operatorname{Var} X(t) = C(0) < \infty \quad (2.21)$$

and satisfy conditions (2.19) and (2.20), then the process is said to be *weakly stationary*.

A weakly stationary process need not, in general, be stationary. Important exceptions are the (real or complex) Gaussian processes. Since the joint multivariate normal distributions of a weakly stationary Gaussian process depend only on the mean vector and covariance matrix of the random variables and these functions have properties (2.19) and (2.20), the joint distributions will have property (2.18). Thus, stationarity and weak stationarity are equivalent for Gaussian processes.

For reasons to be explained in the next section, we will develop the theory under the assumption that $m = 0$. Then condition (2.21) implies that $X(t) \in L_2(P)$ for all t. (See Section 1.4.) Consequently, weakly stationary processes inherit the geometric properties of this Hilbert space.

2.4 THE SPECTRAL REPRESENTATION FOR WEAKLY STATIONARY STOCHASTIC PROCESSES—A SPECIAL CASE

We will first look at the spectral representation of a stochastic almost periodic function

$$X(t) = \sum_{j=-n}^{n} Z_j e^{i\lambda_j t}, \qquad -\infty < t < \infty, \tag{2.22}$$

where $\lambda_0, \lambda_{\pm 1}, \ldots, \lambda_{\pm n}$ are fixed frequencies and the Z_j's are complex-valued random variables.

This expression actually defines a stochastic process through its spectral representation. However, without further conditions the process can be both complex-valued and nonstationary. We will now indicate the properties the Z_j's must have in order that $X(t)$ be a real-valued, weakly stationary process. This will provide some insight into the properties of the general spectral representation to be given in the next section.

In order for $X(t)$ to be real-valued, the λ_j's must be symmetrically placed about $\lambda = 0$, i.e.,$\lambda_{-j} = \lambda_j$, and the random variables Z_j must have the property

$$Z_{-j} = \bar{Z}_j, \qquad j = 0, \pm 1, \ldots, \pm n. \tag{2.23}$$

Since $X(t)$ is to be a weakly stationary stochastic process, the Z_j's must satisfy conditions sufficient to guarantee properties (2.19) and (2.20). Taking the expectation of both sides of (2.22), the first condition implies that

$$m = \sum_{j=-n}^{n} EZ_j e^{i\lambda_j t}.$$

Now, this is an ordinary (nonrandom) almost periodic function and the coefficients EZ_j are the Fourier coefficients of the function on the left side,

$$EZ_j = \lim_{T \to \infty} \frac{1}{2T} \int_{-T}^{T} m e^{-i\lambda_j t}\, dt.$$

Since m is constant, we obtain

$$EZ_0 = m \qquad \text{and} \qquad EZ_j = 0 \qquad \text{for} \quad j \neq 0.$$

To determine the implication of condition (2.20), we will compute $E\overline{X(t+\tau)X(t)}$. Since $X(t)$ is real-valued, $X(t) = \overline{X(t)}$. Thus,

$$X(t+\tau)X(t) = \left(\sum_{j=-n}^{n} Z_j e^{i\lambda_j(t+\tau)}\right)\overline{\left(\sum_{k=-n}^{n} Z_k e^{i\lambda_k t}\right)}$$

$$= \sum_{j=-n}^{n} \sum_{k=-n}^{n} Z_j \bar{Z}_k e^{i(\lambda_j - \lambda_k)t} e^{i\lambda_j \tau}$$

$$= \sum_{j=-n}^{n} |Z_j|^2 e^{i\lambda_j\tau} + \sum_{j=-n}^{n} \sum_{k\neq j} Z_j \bar{Z}_k e^{i\lambda_j\tau} e^{i(\lambda_j-\lambda_k)t}.$$

Then taking expectations,

$$EX(t+\tau)X(t) = \sum_{j=-n}^{n} E|Z_j|^2 e^{i\lambda_j\tau} + \sum_{j=-n}^{n} \sum_{k\neq j} (EZ_j \bar{Z}_k e^{i\lambda_j\tau}) e^{i(\lambda_j-\lambda_k)t}. \tag{2.24}$$

Viewed as an almost periodic function in t with τ held fixed, condition (2.20) implies that the left side is constant. The argument just given can be applied again to yield (since $e^{i\lambda_j\tau}$ is never zero),

$$EZ_j \bar{Z}_k = 0 \qquad \text{for} \quad j \neq k. \tag{2.25}$$

Thus, in order that $X(t)$ be a weakly stationary process, it is necessary that the random variables Z_j be *uncorrelated.*

The constant mean m can be viewed as a special form of trend. If $X(t)$ represented the amplitude of a fluctuating electrical current, m would be the (*average*) *direct current* (dc) amplitude. This terminology is traditionally carried over to nonengineering applications of time series analysis. It is common practice to deal with trend terms and the residuals from trend separately. For the present, we will restrict attention to the residual process $X(t) - m$. Subtracting m from both sides of (2.22) has the same effect as setting $m = 0$. *Thus, we will assume hereafter that* $m = 0$.

If we draw the analogy between the stochastic definition of autocovariance (2.20) and the definition (2.1) given in the Wiener theory, the *total power* is now simply the variance of the process.

From (2.24) and (2.25) we obtain the representation

$$C(\tau) = \sum_{j=-n}^{n} E|Z_j|^2 e^{i\lambda_j\tau}, \qquad -\infty < \tau < \infty. \tag{2.26}$$

With an interpretation of notation we will now show that this is the spectral representation of the autocovariance function.

For any set of real numbers A, define

$$Z(A) = \sum_{\lambda_j \in A} Z_j,$$

and

$$F(A) = \sum_{\lambda_j \in A} E|Z_j|^2.$$

Then

$$E|Z(A)|^2 = \sum_{\lambda_j \in A} \sum_{\lambda_k \in A} EZ_j \bar{Z}_k = \sum_{\lambda_j \in A} E|Z_j|^2.$$

This yields the expression

$$E|Z(A)|^2 = F(A).$$

For any two sets A and B, the same argument provides the more general relationship

$$EZ(A)\overline{Z(B)} = F(A \cap B).$$

Here $Z(A)$ can be thought of as a function which assigns a complex-valued random variable to each set A. In addition, it inherits from the Z_j's the properties

$$Z(-A) = \overline{Z(A)} \qquad \text{and} \qquad EZ(A) = 0.$$

Thus, the $Z(A)$'s are members of complex $L_2(P)$ (see Section 1.4). In the same sense that the sum (2.6) was represented by the integral (2.4), expressions (2.22) and (2.26) can be expressed in integral form as follows:

$$X(t) = \int_{-\infty}^{\infty} e^{i\lambda t} Z(d\lambda) \qquad \text{and} \qquad C(\tau) = \int_{-\infty}^{\infty} e^{i\lambda\tau} F(d\lambda).$$

2.5 THE GENERAL SPECTRAL REPRESENTATION FOR WEAKLY STATIONARY PROCESSES

The last five displayed expressions carry over to the general case and constitute the basic properties of the spectral representation for weakly stationary processes. The details will now be summarized.

Let $X(t)$, $-\infty < t < \infty$, be a real-valued weakly stationary process with $EX(t) = 0$ and autocovariance function $C(\tau) = EX(t + \tau)X(t)$. For technical reasons [see, e.g., Rozanov (1967, p. 9)] it is assumed that $C(\tau)$ is continuous at $\tau = 0$. It can then be shown that $C(\tau)$ is continuous for all τ. Stochastic processes for which this is not true have extremely unpleasant sample functions and are not generally useful as models for real phenomena.

On the basis of these assumptions there exists a complex-valued random set function $Z(A)$, called the *random spectral measure* of the process, and an interpretation of integral such that the process has *the spectral representation*

$$X(t) = \int_{-\infty}^{\infty} e^{i\lambda t} Z(d\lambda). \tag{2.27}$$

The derivation of this representation is sketched in the appendix to this chapter. We will restrict ourselves to the operational properties of the representation here.

The random spectral measure has the following properties:

$$Z(-A) = \overline{Z(A)} \tag{2.28}$$

[this is a consequence of assuming $X(t)$ to be real-valued];

$$EZ(A) = 0 \tag{2.29}$$

[since $EX(t) = 0$].

If $F(A)$ is defined by

$$F(A) = E|Z(A)|^2, \tag{2.30}$$

then $F(A)$ is a measure and the relationship

$$EZ(A)\overline{Z(B)} = F(A \cap B) \tag{2.31}$$

is valid. Note that if A and B are disjoint, $EZ(A)\overline{Z(B)} = 0$. That is, the random variables $Z(A)$ and $Z(B)$ are *uncorrelated* or, in geometric terms, *orthogonal*.

An Operational Convention

A convention for representing these last two properties in an operationally convenient form is the following: The symbol $d\lambda$, which has appeared previously in integrals such as (2.27) to indicate the variable of integration [or the type of integral as in (2.8)], is provided with two purely "formal" or intuitive interpretations. In the first interpretation $d\lambda$ will be thought of as a "very small" interval containing the number λ. Also, $d\mu$, $d\nu$, ... will be "small intervals" containing μ, ν, ..., respectively. Then (2.30) and (2.31) are expressed by the single relationship

$$EZ(d\lambda)\overline{Z(d\mu)} = \begin{cases} F(d\lambda), & \text{if } \mu = \lambda, \\ 0, & \text{if } \mu \neq \lambda. \end{cases} \tag{2.32}$$

That is, $d\mu$ and $d\lambda$ are thought of as being so small that if $\mu \neq \lambda$, they are disjoint. On later occasions, $d\lambda$ will also be interpreted as the length of the interval.

As an illustration of the use of (2.32) to evaluate an expectation involving the spectral decomposition (2.27), we will derive the spectral representation of the autocovariance function $C(\tau)$:

$$C(\tau) = EX(t+\tau)X(t) = E\left(\int_{-\infty}^{\infty} e^{i\lambda(t+\tau)}Z(d\lambda)\right)\overline{\left(\int_{-\infty}^{\infty} e^{i\mu t}Z(d\mu)\right)}$$

$$= E\int_{-\infty}^{\infty}\int_{-\infty}^{\infty} e^{i\lambda(t+\tau)}e^{-i\mu t}Z(d\lambda)\overline{Z(d\mu)}$$

$$= \int_{-\infty}^{\infty}\int_{-\infty}^{\infty} e^{i(\lambda-\mu)t}e^{i\lambda\tau}EZ(d\lambda)\overline{Z(d\mu)}.$$

Now, applying (2.32) this becomes

$$C(\tau) = \int_{-\infty}^{\infty} e^{i\lambda\tau}F(d\lambda). \tag{2.33}$$

Except for the way in which the autocovariance functions are defined, (2.4) and (2.33) are identical. Again $F(A)$ is called the *spectral distribution* but now of the stochastic process $X(t)$.

The kind of derivation which led to (2.33) will be useful in the construction of models for time series and the steps should be carefully noted. The use of different variables of integration is standard in the first step. Then, the complex conjugate is brought inside the second integral and the result is written as a "double" integral as in the calculus. The next step—the interchange of expectation and integral signs—is permitted by the theory governing (2.27). The exact limitations on this operation will be spelled out presently. Also, the complex exponentials are constants relative to the expectation and the third line of the derivation is obtained. Finally, by (2.32), the integral with respect to $d\mu$ vanishes except where $\mu = \lambda$ and we are left with (2.33). These are the basic steps in a type of calculation that will make it possible to derive with ease the spectral parameters of rather complicated time series through the use of the spectral representation of the process and expressions such as (2.32).

The limitation on the interchange of expectation and integral is given in the following important statement which we will refer to as the *basic property* of the spectral decomposition.

Basic Property *Let $g(\lambda)$ and $h(\lambda)$ be (nonrandom) complex-valued functions such that*

$$\int_{-\infty}^{\infty} |g(\lambda)|^2F(d\lambda) < \infty \qquad \textit{and} \qquad \int_{-\infty}^{\infty} |h(\lambda)|^2F(d\lambda) < \infty. \tag{2.34}$$

Then

$$\int_{-\infty}^{\infty} g(\lambda)Z(d\lambda) \qquad \textit{and} \qquad \int_{-\infty}^{\infty} h(\lambda)Z(d\lambda)$$

are well-defined, complex-valued, random variables with zero means and finite variances such that

$$E\left(\int_{-\infty}^{\infty} g(\lambda)Z(d\lambda)\right)\overline{\left(\int_{-\infty}^{\infty} h(\lambda)Z(d\lambda)\right)} = \int_{-\infty}^{\infty} g(\lambda)\overline{h(\lambda)}F(d\lambda). \tag{2.35}$$

In particular, the variance of $\int_{-\infty}^{\infty} g(\lambda)Z(d\lambda)$ *is,*

$$E\left|\int_{-\infty}^{\infty} g(\lambda)Z(d\lambda)\right|^2 = \int_{-\infty}^{\infty} |g(\lambda)|^2F(d\lambda). \tag{2.36}$$

This last expression is obtained by setting $h(\lambda) = g(\lambda)$ in (2.35).

This rather formidable statement is a consequence of the spectral representation of the process (2.27) (see the derivation in the Appendix). We will use it for two basic purposes. First, conditions (2.34) provide the justification for the interchange of expectations and integral in the above calculation. To see this, set $g(\lambda) = e^{i\lambda(t+\tau)}$ and $h(\lambda) = e^{i\lambda t}$ (t and τ are taken to be fixed). Then, since $|e^{i\vartheta}| = 1$, (2.34) is satisfied because $\int_{-\infty}^{\infty} F(d\lambda) = C(0) < \infty$. Now, the quantities in the first and last steps of the above calculation can be equated because of (2.35). Thus, this result not only justifies the intermediate steps but appears to make them unnecessary! However, not all of the calculations we will encounter are this straightforward and it will be most beneficial to have a step-by-step procedure for evaluating such expressions.

The second consequence of the basic property is that it gives a condition under which integrals of functions other than $e^{i\lambda t}$ can be defined with respect to $Z(A)$. To "be defined" means that the integral is a random variable in the collection we were dealing with—namely, those with zero mean and finite variance. The following is an example of how this property is to be used.

Example 2.2 *The Derivative of a Weakly Stationary Time Series*

If the time series had representation (2.22), the definition of the derivative would not be in question, since the basic linearity property of the derivative would yield

$$\begin{aligned} X(t) = \frac{dX(t)}{dt} &= \frac{d}{dt}\left(\sum_{j=-n}^{n} Z_j e^{\lambda_j t}\right) \\ &= \sum_{j=-n}^{n} Z_j \frac{de^{i\lambda_j t}}{dt} \\ &= \sum_{j=-n}^{n} Z_j\, i\lambda_j e^{i\lambda_j t}. \end{aligned}$$

That is, the interchangeability of derivative and sum would be retained in the stochastic setting. To define the derivative of a general weakly stationary stochastic process, it is reasonable to extend the linearity property by interchanging derivative and integral in the spectral representation (2.27),

$$\dot{X}(t) = \frac{d}{dt}\int_{-\infty}^{\infty} e^{i\lambda t} Z(d\lambda) = \int_{-\infty}^{\infty} \frac{de^{i\lambda t}}{dt} Z(d\lambda).$$

However, in order for this exchange to be valid, the resulting integral must "make sense." By the basic property, this will be the case if we assume that

$$\int_{-\infty}^{\infty} \left|\frac{de^{i\lambda t}}{dt}\right|^2 F(d\lambda) = \int_{-\infty}^{\infty} \lambda^2 F(d\lambda) < \infty.$$

Under this condition, for each t the derivative is then defined to be

$$\dot{X}(t) = \int_{-\infty}^{\infty} i\lambda e^{i\lambda t} Z(d\lambda).$$

Since the right-hand side is a random variable for each t, this is again a stochastic process. Moreover,

$$E\dot{X}(t) = 0$$

and the covariance function can be computed from (2.35),

$$\begin{aligned} C_{\dot{X}}(\tau) &= E\dot{X}(t+\tau)\dot{X}(t) \\ &= E\left(\int_{-\infty}^{\infty} i\lambda e^{i\lambda(t+\tau)} Z(d\lambda)\right)\overline{\left(\int_{-\infty}^{\infty} i\lambda e^{i\lambda t} Z(d\lambda)\right)} \\ &= \int_{-\infty}^{\infty} \lambda^2 e^{i\lambda\tau} F(d\lambda). \end{aligned}$$

This function depends only on τ, so $\dot{X}(t)$ is a weakly stationary process. The random spectral measure and spectral distribution of $X(t)$ are obtained directly from these calculations and are related very simply to the corresponding entities of the $X(t)$ process,

$$Z_{\dot{X}}(d\lambda) = i\lambda Z(d\lambda) \tag{2.37}$$

and

$$F_{\dot{X}}(d\lambda) = \lambda^2 F(d\lambda). \tag{2.38}$$

Not all weakly stationary processes can be differentiated in the sense we have defined since $\int_{-\infty}^{\infty} \lambda^2 F(d\lambda)$ can be infinite while $\int_{-\infty}^{\infty} F(d\lambda)$ is finite. We will interpret this physically in the next chapter when we discuss linear filters of which the derivative is an important special case.

Decomposition of the Spectrum into Discrete and Continuous Components

A decomposition of the spectral distribution into discrete and continuous components exists just as in the Wiener theory of the last section. Again a *spectral function* $p(\lambda)$ and *spectral density function* $f(\lambda)$ determine the discrete and continuous parts of the spectrum, respectively, and the spectral mass of the stationary time series in a set A is given by (2.8).

In terms of the convention for the use of the symbolic intervals $d\lambda$ given above, this can be restated in the form

$$F(d\lambda) = p(\lambda) + f(\lambda)\, d\lambda.$$

Intuitively, the power in $d\lambda$ consists of the discrete power at λ plus the continuous power represented by the area of a rectangle of height $f(\lambda)$ and base length $d\lambda$. Here, we encounter the interpretation of $d\lambda$ as the length of the interval of the same name.

Thus, aside from the different definitions of the autocovariance functions, the two spectral theories are quite parallel. However, as we will see in the next section, an additional decomposition of the random spectral measure exists for weakly stationary processes.

2.6 THE DISCRETE AND CONTINUOUS COMPONENTS OF THE PROCESS

There is an additional bonus to be derived from the spectral representation of the stochastic process. A decomposition of the random spectral measure $Z(A)$ into discrete and continuous parts also exists,

$$Z(A) = Z_d(A) + Z_c(A), \tag{2.39}$$

with

$$EZ_d(A)\overline{Z_c(B)} = 0 \qquad \text{for} \quad \text{all } A, B. \tag{2.40}$$

That is, *the discrete and continuous components are uncorrelated.* Then, if

$$F_d(A) = E|Z_d(A)|^2 \tag{2.41}$$

and

$$F_c(A) = E|Z_c(A)|^2, \tag{2.42}$$

it follows from (2.39) and (2.40) that

$$F(A) = F_d(A) + F_c(A).$$

This is precisely the Lebesgue decomposition (2.5). By using the convention introduced in the last section, these results can be summarized in a form convenient for use in calculations as

$$EZ_d(d\lambda)\overline{Z_c(d\mu)} = 0 \qquad \text{for all } \lambda, \mu, \tag{2.43}$$

$$EZ_d(d\lambda)\overline{Z_d(d\mu)} = \begin{cases} F_d(d\lambda) = p(\lambda), & \mu = \lambda, \\ 0, & \mu \neq \lambda, \end{cases} \tag{2.44}$$

$$EZ_c(d\lambda)\overline{Z_c(d\mu)} = \begin{cases} F_c(d\lambda) = f(\lambda)\, d\lambda, & \mu = \lambda, \\ 0, & \mu \neq \lambda. \end{cases} \tag{2.45}$$

Now substituting (2.39) into the general spectral representation (2.27) we obtain

$$X(t) = X_d(t) + X_c(t), \tag{2.46}$$

where

$$X_d(t) = \int_{-\infty}^{\infty} e^{i\lambda t} Z_d(d\lambda) \tag{2.47}$$

and

$$X_c(t) = \int_{-\infty}^{\infty} e^{i\lambda t} Z_c(d\lambda). \tag{2.48}$$

Based on (2.43), the covariance of $X_d(t)$ and $X_c(s)$ can be computed for any pair of times t and s,

$$\begin{aligned} EX_d(t)X_c(s) &= E\left(\int_{-\infty}^{\infty} e^{i\lambda t} Z_d(d\lambda)\right)\overline{\left(\int_{-\infty}^{\infty} e^{i\mu s} Z_c(d\mu)\right)} \\ &= \int_{-\infty}^{\infty}\int_{-\infty}^{\infty} e^{i(\lambda t - \mu s)} EZ_d(d\lambda)\overline{Z_c(d\mu)} = 0. \end{aligned}$$

We will say that two stochastic processes are uncorrelated if every random variable of one is uncorrelated with every random variable of the other. It follows than that processes $X_d(t)$ and $X_c(t)$ are uncorrelated.

Analysis of the Discrete Component

A more careful analysis of the discrete component is possible. If $\lambda_j, j = 0, \pm 1, \ldots$, are the points for which the spectral function $p(\lambda)$ is positive, from (2.41) we obtain

$$E|Z_d(A)|^2 = \sum_{\lambda_j \in A} p(\lambda_j).$$

Thus, if A contains none of the λ_j's, it follows that $E|Z_d(A)|^2 = 0$, thus, $Z_d(A)$ is the zero random variable. Consequently, $Z_d(A)$ is nonzero only when

A contains one or more frequencies λ_j and, in fact, if we let $Z_j = Z(\{\lambda_j\})$, it follows that $E|Z_j|^2 = p(\lambda_j)$ and

$$Z_d(A) = \sum_{\lambda_j \in A} Z_j.$$

By the symmetry property (2.28) we have

$$Z_{-j} = \bar{Z}_j.$$

Moreover,

$$\sum_{j=-\infty}^{\infty} E|Z_j|^2 = EX_d^2(t) < \infty. \tag{2.49}$$

Thus (2.47) can be written as

$$X_d(t) = \sum_{j=-\infty}^{\infty} e^{i\lambda_j t} Z_j.$$

By comparing this with (2.22) (see also Example 1.3), it is seen that $X_d(t)$ is a stochastic almost periodic function. Condition (2.49) is the Parseval relation. Thus, not only is it the case that every stochastic almost periodic function has a discrete spectrum, but also every weakly stationary process with a discrete spectrum is of this form. This makes the spectral representation somewhat more concrete.

Unfortunately, a comparable representation of the continuous component does not exist, in general. That is, we *cannot* define a "random spectral density" $Z(\lambda)$, say, such that

$$Z_c(d\lambda) = Z(\lambda)\, d\lambda \qquad \text{and} \qquad E|Z(\lambda)|^2 = f(\lambda).$$

This minor flaw in the nature of the representation is unfortunate from a notational viewpoint but inconsequential in the final analysis, since results calculated as though such a representation were valid turn out to be quite correct! A different kind of representation of the continuous component will be given in Chapter 7.

Extension of the Basic Property

The following extension of the *basic property* is easily verified: *If* $g(\lambda)$ *is a complex-valued function for which*

$$\int_{-\infty}^{\infty} |g(\lambda)|^2 F(d\lambda) < \infty,$$

then

$$\int_{-\infty}^{\infty} g(\lambda) Z(d\lambda) = \int_{-\infty}^{\infty} g(\lambda) Z_d(d\lambda) + \int_{-\infty}^{\infty} g(\lambda) Z_c(d\lambda),$$

where the random variables on the right-hand side are uncorrelated. Moreover, the first term can be written in the form

$$\int_{-\infty}^{\infty} g(\lambda)Z_d(d\lambda) = \sum_{j=-\infty}^{\infty} g(\lambda_j)Z_j,$$

where $Z_j = Z(\{\lambda_j\})$ *and the* λ_j*'s are the frequencies for which* $p(\lambda)$ *is positive. Let* $h(\lambda)$ *be a second function satisfying the condition*

$$\int_{-\infty}^{\infty} |h(\lambda)|^2 F(d\lambda) < \infty.$$

Then,

$$E\left(\int g(\lambda)Z(d\lambda)\right)\overline{\left(\int h(\mu)Z(d\mu)\right)} = \int g(\lambda)\overline{h(\lambda)}F_d(d\lambda) + \int g(\lambda)\overline{h(\lambda)}F_c(d\lambda)$$

$$= \sum_{j=-\infty}^{\infty} g(\lambda_j)\overline{h(\lambda_j)}p(\lambda_j) + \int_{-\infty}^{\infty} g(\lambda)\overline{h(\lambda)}f(\lambda)\, d\lambda.$$

In particular,

$$\mathrm{Var}\left(\int g(\lambda)Z(d\lambda)\right) = \sum_{j=-\infty}^{\infty} |g(\lambda_j)|^2 p(\lambda_j) + \int_{-\infty}^{\infty} |g(\lambda)|^2 f(\lambda)\, d\lambda. \quad (2.50)$$

2.7 PHYSICAL REALIZATIONS OF THE DIFFERENT KINDS OF SPECTRA

Physical realizations of the three types of spectra we have studied—discrete, continuous, and mixed—exist in a variety of contexts. Discrete spectra are usually generated by mechanisms that operate with extreme precision and regularity. Thus, light spectra produced by the motion of electrons between prescribed energy levels in atoms are discrete. A laser produces light at (very nearly) a single frequency. The frequency decomposition of a tone produced by the human voice or a musical instrument is discrete because only finitely many "modes" of vibration are possible in human vocal cords and the vibrating mechanism of the instrument.

Continuous spectra and continuous spectra with a superimposed discrete component (the mixed case) are by far the most commonly occurring spectral types. Continuous spectra are generated by complex mechanisms which have so many modes of vibration that the frequencies of the waves produced "run together" into a continuum. An ordinary light bulb produces a continuous light spectrum. The relative intensity of the light in various frequency ranges (as measured by the spectral density function) determines the color of the light. Red light is characterized by a preponderance of low-frequency power and blue light by a preponderance of high-frequency power. A uniform (flat) light spectrum is characteristic of white light.

Almost all forms of noise, for example, electronic static, seismic noise, computer roundoff error, etc., have continuous spectra. An interesting carry-over of terminology from the field of optics to time series is the use of the term "white noise" to describe weakly stationary stochastic processes with continuous spectra and constant spectral density functions.

The time series of interest in virtually all fields—engineering, geophysics, economics, medicine, etc.—have continuous spectra. This is not to say, however, that the spectra actually observed are always purely continuous. Due to a variety of causes, the continuous part of the spectrum often has superimposed on it one or more spectral lines (which show up in estimated spectra as narrow peaks for reasons to be explained in Section A8.2). Thus, in effect, the observed spectrum is of mixed type. In many cases, the discrete part of the spectrum is inadvertently contributed by the electronic gear used in recording, amplifying or transcribing the time series, thus has nothing at all to do with the phenomenon of interest. In fields such as economics and geophysics, regular daily, monthly, or yearly cycles introduce peaks of intense power into the spectrum which mask and "contaminate" the more subtle variations of the continuous spectrum. An excellent example of this phenomenon is provided by the river runoff data to be discussed in Chapter 6. The spectrum of this data is given in Fig. 6.12 and a graph of a section of the data itself is given in Fig. 1.4. In such cases the discrete spectrum, which can be thought of as a form of trend, must be removed by some means in order to study the continuous component of the spectrum. We will consider the problem of trend removal later in this and in succeeding chapters.

The methods for estimating spectra to be presented in Chapter 8 have the flexibility necessary to allow the time series analyst to study spectra of all three varieties. A spectrum analysis is, consequently, often the first type of analysis performed on data. Extraneous features of the data, such as unexpected peaks due to improperly operating recording equipment, will show up and their frequencies and power can be precisely pinpointed for future removal. Other forms of data processing, especially linear filtering, will be suggested by the spectrum analysis and the actual design of the filter can be based on the estimated spectrum. Spectrum analysis as a descriptive and analytic tool has few peers and a sampling of the wide variety of applications that have been and can be made will be given as the theory is developed.

2.8 THE REAL SPECTRAL REPRESENTATION

The spectral representation of the previous sections is occasionally encountered in the time series literature in real or cartesian form, and the computational procedures leading to estimates of multidimensional spectral parameters depend upon this form of the representation. The relationship between the real and complex decompositions is quite analogous to that given in Section 1.2. However, a few special features make it worthwhile to consider the stochastic process case separately.

The real representation is based on the introduction of two real-valued random spectral measures which are (nearly) the real and imaginary parts of $Z(A)$: Let

$$U(A) = Z(A) + \overline{Z(A)}, \tag{2.51}$$

$$V(A) = i\big(Z(A) - \overline{Z(A)}\big). \tag{2.52}$$

It follows that

$$Z(A) = \tfrac{1}{2}\big(U(A) - iV(A)\big).$$

Since $Z(-A) = \overline{Z(A)}$, we have

$$U(-A) = U(A), \tag{2.53}$$

$$V(-A) = -V(A). \tag{2.54}$$

In particular,

$$U(\{0\}) = 2Z(\{0\}) \qquad \text{and} \qquad V(\{0\}) = 0.$$

The random variables $U(A)$ and $V(A)$ have zero means for all A and covariances can be calculated from (2.51) and (2.52) as follows:

$$\begin{aligned} EU(A)V(B) &= i[EZ(A)\,Z(B) - EZ(A)\overline{Z(B)} + E\overline{Z(A)}Z(B) - E\overline{Z(A)}\overline{Z)B)}] \\ &= i[EZ(A)\overline{Z(-B)} - EZ(A)\overline{Z(B)} + \overline{EZ(A)\overline{Z(B)}} - \overline{EZ(A)\overline{Z(-B)}}]. \end{aligned}$$

We now use the facts that $EZ(A)\overline{Z(B)} = F(A \cap B)$ and $F(A)$ is real-valued to conclude that

$$EU(A)V(B) = 0 \qquad \text{for} \quad \text{all } A, B. \tag{2.55}$$

That is, *the real and imaginary components are uncorrelated.* The same kind of computation leads to the expressions

$$EU(A)U(B) = 2[F(A \cap B) + F(A \cap -B)], \tag{2.56}$$

$$EV(A)V(B) = 2[F(A \cap B) - F(A \cap -B)]. \tag{2.57}$$

Now, by the de Moivre formula for $e^{i\vartheta}$,

$$X(t) = \int_{-\infty}^{\infty} e^{i\lambda t} Z(d\lambda) = \tfrac{1}{2}\int_{-\infty}^{\infty} (\cos\lambda t + i\sin\lambda t)\big(U(d\lambda) - iV(d\lambda)\big).$$

This integral is broken up into three parts with the ranges $(-\infty, 0)$, $\{0\}$, and $(0, \infty)$. The imaginary component of the integral, $\sin \lambda t U(d\lambda) - \cos \lambda t V(d\lambda)$ is an odd function and is zero at $\lambda = 0$ and its integrals over the symmetric intervals $(-\infty, 0)$ and $(0, \infty)$ cancel one another. The real component is an even function and has the value $U(\{0\})/2 = Z(\{0\})$ at $\lambda = 0$. Thus, the representation can be expressed as an integral over the nonnegative frequency range as

$$X(t) = \tfrac{1}{2}U(\{0\}) + \int_{0^+}^{\infty} \cos \lambda t U(d\lambda) + \sin \lambda t V(d\lambda), \tag{2.58}$$

where 0^+ indicates that 0 is not included in the range of integration. *This is the real spectral representation of the process.*

The corresponding spectral representation of the autocovariance function can be obtained in a similar fashion by setting $A = d\lambda$ and $B = d\mu$ in (2.55)–(2.57), then by employing the type of calculation used to arrive at (2.33). Since λ and μ are restricted to the interval $(0, \infty)$ these expressions become

$$EU(d\lambda)V(d\mu) = 0 \qquad \text{for all } \lambda, \mu, \tag{2.59}$$

$$EU(d\lambda)U(d\mu) = EV(d\lambda)V(d\mu) = \begin{cases} 2F(d\lambda), & \lambda = \mu, \\ 0, & \lambda \neq \mu. \end{cases} \tag{2.60}$$

Then, by means of the trigonometric identity $\cos(\alpha - \beta) = \cos \alpha \cos \beta + \sin \alpha \sin \beta$ it is easy to show that

$$C(\tau) = p(0) + 2\int_{0^+}^{\infty} \cos \lambda\tau \, F(d\lambda),$$

where $p(\lambda)$ is the spectral function and $F(A)$ is the spectral distribution of the process.

Finally, a *real spectral distribution* $G(A)$ can be defined, which concentrates the spectral mass on the nonnegative frequency axis. If $A \subset [0, \infty)$, then $G(A)$ combines the original spectral mass from the sets A and $-A$ with the exception that the mass at $\lambda = 0$ is only counted once. In essence, the negative axis is folded over onto the positive axis with the fold at $\lambda = 0$. This is most easily expressed in terms of the relationships between the spectral functions and spectral density functions of $G(A)$ and $F(A)$,

$$p_G(\lambda) = \begin{cases} p(0), & \lambda = 0, \\ 2p(\lambda), & \lambda > 0, \end{cases} \tag{2.61}$$

$$f_G(\lambda) = 2f(\lambda), \qquad \lambda > 0. \tag{2.62}$$

These two functions completely determine $G(A)$ and the *spectral representation of the autocovariance* can be written in the final form

$$C(\tau) = \int_{0}^{\infty} \cos \lambda\tau \, G(d\lambda). \tag{2.63}$$

2.9 ERGODICITY AND THE CONNECTION BETWEEN THE WIENER AND STATIONARY PROCESS THEORIES

The Wiener and weakly stationary process definitions of the spectrum differ in only one essential—namely in the definition of the autocovariance function. In the Wiener theory, the autocovariance is defined by a *time average*

$$C_W(\tau) = \lim_{T\to\infty} \frac{1}{2T} \int_{-T}^{T} x(t+\tau)x(t)\, dt. \tag{2.64}$$

In the weakly stationary process theory, it is defined as an average with respect to the probability distribution of the process or, as it is often called, an *ensemble average*,

$$C(\tau) = EX(t+\tau)X(t).$$

When we try to apply the Wiener definition of autocovariance to a stochastic process, three questions arise: In what sense is the integral in (2.64) defined? In what sense and under what conditions does the limit exist? If a limit exists, how is it related to the autocovariance for the process $C(\tau)$? In particular, under what conditions is it equal to $C(\tau)$?

Since the definition given by Wiener pertains to ordinary functions of time it is reasonable to attempt to deal with these questions in terms of the individual sample functions of the process. This is especially true since, in particular, only one sample function (or more accurately a part of a sample function) will be available for analysis. The integral would then be an ordinary integral and the limit a numerical limit in the usual sense. If the integral and limit are defined except for an ω-set of probability zero, they are said to be defined *almost surely*. When the stochastic process is strictly stationary, the integral and limit exist in this sense under surprisingly weak and reasonable conditions from a practical viewpoint. The essential assumption is that $EX^2(t)$ be finite. This result is part of the individual ergodic theorem which is considered one of the most significant achievements in probability theory. Excellent treatments of this theorem are given by Doob (1953) and Rozanov (1967).

The third question proves to be the most difficult to answer in simple terms. If a constant limit exists, that is, a limit which does not depend upon the particular sample path, then it is easy to show that the limit must be $C(\tau)$. Stationary processes with this property are said to be *ergodic*. Processes which are *not* ergodic are easily found. As an example, suppose that A is a random variable such that

$$EA = 0 \qquad \text{and} \qquad EA^2 = p^2 \neq 0.$$

(Note that A is nondegenerate, since Var $A \neq 0$.) Define a random process as follows: For each $\omega \in \Omega$, let

$$X(t, \omega) = A(\omega), \qquad -\infty < t < \infty.$$

That is, the sample functions are all horizontal lines. This process is strictly stationary and its mean and autocovariance functions are

$$EX(t) = 0 \qquad \text{and} \qquad C(\tau) = p^2, \qquad -\infty < \tau < \infty.$$

Now for each ω,

$$\frac{1}{2T}\int_{-T}^{T} X(t + \tau, \omega)X(t, \omega)\, dt = A^2(\omega).$$

Consequently, the limit as T tends to infinity exists for each ω and, in fact, $C_W(\tau)$ is simply the random variable A^2. Thus, $C_W(\tau)$ is not constant in this case and its value depends on the particular sample function observed. Note that $EC_W(\tau) = C(\tau)$. This is true in general but is of little comfort if there is only one sample function with which to deal.

Conditions under which a general stationary process is ergodic are rather involved and difficult to apply in practice. Intuitively, in order for a process to be ergodic, the stochastic dependence between parts of the process which are separated by an interval of time must approach zero "sufficiently rapidly" as the length of the time interval increases to infinity.

One would hope that conditions of this sort could be made to hold in practice by modifying the process to remove the dc component and any other components of the process which might contribute to the kind of behavior exhibited in the example. This is indeed possible, in principle, for the following reason. For practical purposes, the *main impediment to ergodicity is the discrete component of the spectrum*. In the case of Gaussian processes and certain other special kinds of processes, it can be shown that this is the only impediment. That is, in these cases a continuous spectrum implies ergodicity. One of the important reasons for the decomposition of a stationary process into discrete and continuous components is that it is then reasonable to make the added assumption that the continuous component is ergodic. This is a common assumption for signal-plus-noise processes encountered in engineering in which the noise component is invariably taken to have a continuous spectrum.

A Class of Nonstationary Processes Possessing a Spectrum

The signal-plus-noise processes also provide one of the most important examples of a class of nonstationary stochastic processes for which a spectrum exists in the Wiener sense. A special case is detailed in the following example:

Example 2.3 *A Nonstationary Process with a Wiener Spectrum*

Let $N(t)$ be a stationary, ergodic, stochastic process with $EN(t) = 0$ and autocovariance function $C_N(\tau)$, and let $S(t)$ be a real-valued, nonrandom function for which

$$C_S(\tau) = \lim_{T\to\infty} \frac{1}{2T} \int_{-T}^{T} S(t+\tau)S(t)\,dt$$

exists and is finite for every τ. Further, assume that

$$\lim_{T\to\infty} \frac{1}{2T} \int_{-T}^{T} S(t+\tau)N(t)\,dt = 0,$$

where the limit is defined almost surely. Now, let

$$X(t) = S(t) + N(t), \qquad -\infty < t < \infty.$$

Then, the observed process $X(t)$ can be viewed as a nonstochastic signal masked by stationary random noise. This is a useful model in many applications. Except in the rather uninteresting case in which $S(t)$ is constant for all t, the process $X(t)$ is nonstationary. However, by the above assumptions and the ergodic theorem,

$$\begin{aligned} C_X(\tau) &= \lim_{T\to\infty} \frac{1}{2T} \int_{-T}^{T} X(t+\tau)X(t)\,dt \\ &= C_S(\tau) + C_N(\tau), \qquad \text{almost surely.} \end{aligned}$$

Consequently, $X(t)$ has the Wiener spectral distribution

$$F_X(A) = F_S(A) + F_N(A),$$

where $F_S(A)$ and $F_N(A)$ are defined in Sections 2.2 and 2.3, respectively. If, in addition, it is assumed that $N(t)$ is Gaussian, then the spectrum of $N(t)$ is necessarily continuous and the discrete and continuous components of $X(t)$ are determined by the spectral function and spectral density

$$p_X(\lambda) = p_S(\lambda), \qquad f_X(\lambda) = f_S(\lambda) + f_N(\lambda).$$

2.10 STATISTICAL ESTIMATION OF THE AUTOCOVARIANCE AND THE MEAN ERGODIC THEOREM

In the statistical theory of time series, the autocovariance $C(\tau)$ is viewed as a parameter of the time series which must be estimated on the basis of a finite length of data. If the process $X(t)$ is observed over the time interval

$(-T, T)$, then [setting $X(t+\tau) = 0$ whenever $|t+\tau| > T$], the random variable

$$\frac{1}{2T}\int_{-T}^{T} X(t+\tau)X(t)\,dt$$

is one of the possible estimators for $C(\tau)$. An important measure of the effectiveness of an estimator is the *mean-square error*;

$$D_T = E\left[\frac{1}{2T}\int_{-T}^{T} X(t+\tau)X(t)\,dt - C(\tau)\right]^2. \tag{2.65}$$

If $D_T \to 0$ as $T \to \infty$, it will follow that the estimator is consistent. It is easily seen that it is unbiased. [Recall from statistics that an estimator (sequence of estimators) $\hat{\vartheta}_n$ is *unbiased* for a parameter ϑ if $E\hat{\vartheta}_n = \vartheta$ and is *consistent* if $\lim_{n\to\infty} P(|\hat{\vartheta}_n - \vartheta| > \varepsilon) = 0$ for every $\varepsilon > 0$. Consistency is a consequence of the condition $\lim_{n\to\infty} E(\hat{\vartheta}_n - \vartheta)^2 = 0$ because of the Chebyshev inequality. For details see Tucker (1962, p. 98).]

Theorems which deal with the convergence of D_T as T tends to infinity are called *mean ergodic theorems*. The mathematical setting in which mean ergodic theorems are studied is the geometry of square integrable random variables. [See the discussion of $L_2(P)$ in Section 1.4.] The stochastic process is taken to be weakly stationary which makes available to us all of the theory developed earlier in this chapter. The appropriate definition of the integral and the conditions under which D_T tends to zero will be given in the special case of a Gaussian process. However, we first look at a somewhat simpler situation which is of interest in its own right.

The Convergence of $(1/2T)\int_{-T}^{T} X(t)\,dt$

Take $X(t)$ to be weakly stationary with $EX(t) = 0$. The time average of $X(t)$

$$\frac{1}{2T}\int_{-T}^{T} X(t)\,dt$$

is the time series equivalent of the sample mean, which is the usual statistical estimate of the population mean. In the statistical context, $X_1, X_2, \ldots, X_n$ are taken to be independent random variables with common mean $EX_i = m$ and common variance σ^2. Then by a well-known result from probability [Tucker (1962, p. 98)], the sample mean

$$\overline{X}_n = \frac{1}{n}\sum_{i=1}^{n} X_i,$$

has the property

$$E(\overline{X}_n - m)^2 \to 0 \quad \text{as} \quad n \to \infty. \tag{2.66}$$

In the time series context the random variables are not independent, and, for the present, time is taken to be continuous rather than discrete. However, it is still possible to evaluate the time series equivalent of (2.66) which, since $EX(t) = 0$, is

$$E\left(\frac{1}{2T}\int_{-T}^{T} X(t)\,dt\right)^2.$$

The integral of $X(t)$ is defined by means of the spectral representation of the process as was the derivative,

$$\begin{aligned}\frac{1}{2T}\int_{-T}^{T} X(t)\,dt &= \frac{1}{2T}\int_{-T}^{T}\left[\int_{-\infty}^{\infty} e^{i\lambda t}Z(d\lambda)\right] dt\\ &= \int_{-\infty}^{\infty}\left(\frac{1}{2T}\int_{-T}^{T} e^{i\lambda t}\,dt\right)Z(d\lambda),\end{aligned}$$

where, by the basic property, the last expression is a well-defined random variable provided

$$\int_{-\infty}^{\infty}\left|\frac{1}{2T}\int_{-T}^{T} e^{i\lambda t}\,dt\right|^2 F(d\lambda) < \infty.$$

However,

$$\frac{1}{2T}\int_{-T}^{T} e^{i\lambda t}\,dt = \begin{cases}\sin\lambda T/\lambda T, & \lambda \neq 0,\\ 1, & \lambda = 0.\end{cases}$$

Thus, extending the definition of $\sin\lambda T/\lambda T$ to equal 1 at $\lambda = 0$, we have $|\sin\lambda T/\lambda T| \le 1$ for all $T > 0$ and λ. It follows that the definition of the integral

$$\frac{1}{2T}\int_{-T}^{T} X(t)\,dt = \int_{-\infty}^{\infty}(\sin\lambda T/\lambda T)Z(d\lambda), \qquad 0 < T < \infty,$$

is valid for *every* weakly stationary process.

Now, by (2.36), we obtain

$$\begin{aligned}E\left(\frac{1}{2T}\int_{-T}^{T} X(t)\,dt\right)^2 &= E\left|\int_{-\infty}^{\infty}(\sin\lambda T/\lambda T)Z(d\lambda)\right|^2\\ &= \int_{-\infty}^{\infty}|\sin\lambda T/\lambda T|^2F(d\lambda)\\ &= \sum_{j=-\infty}^{\infty}|\sin\lambda_j T/\lambda_j T|^2p(\lambda_j)\\ &\quad + \int_{-\infty}^{\infty}|\sin\lambda T/\lambda T|^2f(\lambda)\,d\lambda. \qquad (2.67)\end{aligned}$$

The function $\sin \lambda T$ is bounded between -1 and 1. Thus,

$$\sin \lambda T/\lambda T \to 0 \qquad \text{as} \quad T \to \infty, \qquad \lambda \neq 0.$$

It follows that the integrand of the integral in the last line of (2.67) vanishes as $T \to \infty$ except at $\lambda = 0$ and that the integral, viewed intuitively as the area under the curve, also vanishes. An important theorem, the dominated convergence theorem from the theory of measure, justifies this conclusion. Similarly, all of the terms of the sum in (2.67), except the one corresponding to $\lambda = 0$, tend to zero and we are left with the result

$$\lim_{T \to \infty} E\left(\frac{1}{2T}\int_{-T}^{T} X(t)\, dt\right)^2 = p(0), \tag{2.68}$$

where $p(0)$ is the discrete power at zero frequency. Unless this term is zero, convergence of the estimator to the mean does not occur. In fact, repeating this argument on

$$E\left(\frac{1}{2T}\int_{-T}^{T} X(t)\, dt - Z(\{0\})\right)^2,$$

we would find the limit as $T \to \infty$ would now be zero. That is,

$$\frac{1}{2T}\int_{-T}^{T} X(t)\, dt \to Z(\{0\}) \qquad \text{as} \quad T \to \infty.$$

Thus, with a somewhat enlarged definition of consistency, $(1/2T)\int_{-T}^{T} X(t)\, dt$ is a consistent estimator of the dc component of the time series. In fact, if $EX(t) = m \neq 0$, we would have

$$\lim_{T \to \infty} \frac{1}{2T}\int_{-T}^{T} X(t)\, dt = m + Z(\{0\}).$$

Thus the sum of the random and nonrandom dc components is consistently estimated.

In practice, then, a reasonable procedure for removing the dc component, to the extent possible on the basis of data observed for the time interval $(-T, T)$, would be to replace the original time series by the *residual* time series

$$X(t) - \frac{1}{2T}\int_{-T}^{T} X(u)\, du.$$

From the viewpoint of spectral estimation, the extraction of the dc component is the most important preprocessing operation performed on a time series before estimates of the spectrum are computed. This is the most popular procedure for doing so. The rationale for removing the dc term will be discussed in Chapter 9.

Complex Demodulation

As we will see shortly, it is also desirable to have estimators for the other coefficients $Z_j = Z(\{\lambda_j\})$ in the discrete component of the spectrum so that these terms can be removed from the time series to the extent possible with a finite length of data. One estimation procedure, known as *complex demodulation*, is as follows: Let

$$X_\mu(t) = e^{-i\mu t}X(t), \qquad -\infty < t < \infty.$$

That is, "mix" the original time series with a periodic complex exponential. Then, by the spectral representation of $X(t)$,

$$X_\mu(t) = \int_{-\infty}^{\infty} e^{i(\lambda-\mu)t}Z(d\lambda),$$

and the steps leading to (2.68) can be repeated to obtain

$$\lim_{T\to\infty} \frac{1}{2T}\int_{-T}^{T} X_\mu(t)\,dt = Z(\{\mu\}).$$

Moreover, if the observed time series has a nonstochastic, almost periodic trend component, this estimator will also account for the trend coefficients. That is, if

$$Y(t) = \sum_{j=-\infty}^{\infty} c_j e^{i\mu_j t} + X(t)$$

and if we let

$$c(\mu) = \begin{cases} c_j, & \text{if } \mu = \mu_j, \quad j = 0, \pm 1, \ldots, \\ 0, & \text{otherwise,} \end{cases}$$

then for all μ,

$$\lim_{T\to\infty} \frac{1}{2T}\int_{-T}^{T} e^{-i\mu t}Y(t)\,dt = c(\mu) + Z(\{\mu\}). \tag{2.69}$$

Thus, the estimator in (2.69) will "pick up" the spectral mass at each discrete frequency μ whether this mass is of random, nonrandom, or mixed origin. By subtracting from $Y(t)$ the estimates of both the random and nonrandom discrete components based on the available data, the resulting residual process

$$Y(t) - \sum_j \left(\frac{1}{2T}\int_{-T}^{T} e^{-i\mu_j u}Y(u)\,du\right)e^{i\mu_j t} - \sum_j \left(\frac{1}{2T}\int_{-T}^{T} e^{-i\lambda_j u}\,Y(u)\,du\right)e^{i\lambda_j t}$$

will be a good approximation, in the sense of mean-square error, to the continuous component of the process $X(t)$. This is the basis for one of the more popular techniques for removing periodicities from time series. Other methods based on linear filtering will be considered in Chapter 6.

Convergence of D_T

Now, let us return to the discussion of the limiting behavior of D_T as $T \to \infty$ when $X(t)$ is a stationary Gaussian process. (A weakly stationary Gaussian process is automatically stationary.) The usual procedure is to fix a value of τ and form the process

$$Y_\tau(t) = X(t+\tau)X(t) - C(\tau).$$

Even without the Gaussian assumption, if the $X(t)$ process is assumed to have fourth-order moments which behave like the fourth moments of a stationary process, then $Y_\tau(t)$, $-\infty < t < \infty$, will be weakly stationary and the convergence of D_T can be dealt with in the same manner as was the convergence of $(1/2T)\int_{-T}^{T} X(t)\,dt$.

However, it is difficult to relate the spectrum of $Y_\tau(t)$ to that of $X(t)$ and the conditions under which $D_T \to 0$ are not easy to apply in practice. [See Hannan (1970, p. 210) for a good discussion of this.] The evaluation of $\lim D_T$ as $T \to \infty$ in the Gaussian case embodies all of the important features of the general result and, in addition, provides a simple criterion for guaranteeing that $\lim D_T = 0$.

It is shown in the appendix to this chapter that if $X(t)$ is a Gaussian process, then

$$\lim_{T\to\infty} D_T = \sum_{j=-\infty}^{\infty} (1 + e^{2i\lambda_j\tau})p^2(\lambda_j)$$
$$= \sum_{j=-\infty}^{\infty} p^2(\lambda_j) + \sum_{j=-\infty}^{\infty} p^2(\lambda_j)e^{2i\lambda_j\tau}.$$

Consequently, in order that $D_T \to 0$ for every τ, it is necessary and sufficient that

$$\sum_{j=-\infty}^{\infty} d_j e^{2i\lambda_j\tau} = 0 \qquad \text{for all } \tau,$$

where, assuming $\lambda_0 = 0$ as before, $d_0 = p^2(0) + \sum_{j=-\infty}^{\infty} p^2(\lambda_j)$ and $d_j = p^2(\lambda_j)$ for $j \neq 0$. However, since such expansions are unique, it follows that $d_j = 0$, thus

$$p(\lambda_j) = 0 \qquad \text{for all } j.$$

This implies

$$\lim_{T\to\infty} \frac{1}{2T}\int_{-T}^{T} X(t+\tau)X(t)\,dt = C(\tau). \tag{2.70}$$

That is, (2.70) holds if and only if the discrete component of the spectrum is absent.

The degree to which the mean-square error D_T can be made to approach zero by increasing the length of data is limited by the power in the discrete spectral component. This is one of the more important reasons for estimating and attempting to remove these terms from the time series.

APPENDIX TO CHAPTER 2

A2.1 The Spectral Representations of the Autocovariance and the Process

A complex-valued function of a real variable $C(\tau)$ is said to be *nonnegative definite* if for every n and every set of complex numbers $c_1, \ldots, c_n$ and real numbers $t_1, ,\ldots, t_n$,

$$\sum_{j=1}^{n} \sum_{k=1}^{n} c_j \bar{c}_k C(t_j - t_k) \geq 0.$$

It is easy to check that both autocovariance functions defined earlier are nonnegative definite. A celebrated theorem due to Bochner states that *to every continuous nonnegative definite function $C(\tau)$ for which $C(0)$ is finite there corresponds a nondecreasing bounded function $F(\lambda)$ such that*

$$C(\tau) = \int_{-\infty}^{\infty} e^{i\lambda\tau} F(d\lambda), \qquad -\infty < \tau < \infty. \tag{A2.1}$$

A proof of Bochner's theorem is given by Goldberg (1961). In the time series context, this is the spectral representation for the autocovariance function. Bochner's theorem also provides the basis for the spectral representation (2.27) of a weakly stationary process. The following sketch of this result is based on the derivation given by Cramèr (1951b).

A correspondence is established between elements of complex $L_2(P)$ (see Section 1.4) and the elements of $L_2(F)$, the set of complex-valued functions of a real variable for which

$$\int |g(\lambda)|^2 F(d\lambda) < \infty.$$

Here $F(A)$ is the spectral distribution of the process obtained from Bochner's theorem. The correspondence is first defined between elements of the process and the complex exponentials by

$$X(t) \leftrightarrow e^{i\lambda t}, \qquad -\infty < t < \infty.$$

This correspondence is extended by linearity to finite linear combinations

$$\sum_j a_j X(t_j) \leftrightarrow \sum_j a_j e^{i\lambda t_j},$$

where the a_j's are complex numbers, and then by continuity to limits of Cauchy sequences of such linear combinations. An important feature of $L_2(F)$ is that every element of this space can be obtained in this manner. That is, the complex exponentials generate $L_2(F)$ in the sense defined in Section 1.3. Consequently, this extension establishes a unique correspondence between the elements of $L_2(F)$ and the elements of $\mathscr{M}_X$, the subspace of $L_2(P)$ generated by the random variables of the stochastic process. Note also that by (A2.1),

$$\begin{aligned}\langle X(t), X(s)\rangle_P = EX(t)\overline{X(s)} &= C(t-s)\\ &= \int e^{i\lambda(t-s)}F(d\lambda)\\ &= \int e^{i\lambda t}\overline{e^{i\lambda s}}F(d\lambda) = \langle e^{i\lambda t}, e^{i\lambda s}\rangle_F,\end{aligned}$$

where $\langle\ \rangle_P$ and $\langle\ \rangle_F$ are the inner products in the Hilbert spaces $L_2(P)$ and $L_2(F)$. That is, the correspondence preserves inner products for the generating elements. This property is preserved by the formation of linear combinations and passage to the limit. Consequently, if $g(\lambda)$ and $h(\lambda)$ are any elements of $L_2(F)$ and G and H are the random variables in $\mathscr{M}_X$ such that

$$G \leftrightarrow g(\lambda) \qquad \text{and} \qquad H \leftrightarrow h(\lambda),$$

then

$$EG\overline{H} = \int g(\lambda)\overline{h(\lambda)}F(d\lambda). \tag{A2.2}$$

Now, if A is any Borel set and I_A is the set characteristic function of A defined by

$$I_A(\lambda) = \begin{cases} 1, & \lambda \in A,\\ 0, & \lambda \notin A,\end{cases}$$

then, clearly, $I_A \in L_2(F)$. Thus, there is a random variable $Z(A)$ in $\mathscr{M}_X$ such that

$$Z(A) \leftrightarrow I_A.$$

As a function of A this random set function is precisely the spectral measure of the process. It is defined for every Borel set A and actually possesses the properties of a complex-valued measure in that the expressions $Z(\varnothing) = 0$ a.s. and $Z(\bigcup_1^\infty A_i) = \sum_1^\infty Z(A_i)$ a.s. for disjoint sets A_i are valid, where the series converges in the mean-square sense. Moreover, by (A2.2),

$$EZ(A)\overline{Z(B)} = \int I_A(\lambda)\overline{I_B(\lambda)}F(d\lambda) = F(A \cap B),$$

since $I_A I_B = I_{A\cap B}$.

The spectral representation (2.27) is a result of the following observation. Every function $g \in L_2(F)$ can be represented as a limit of finite linear combinations of set characteristic functions

$$\sum_j a_j I_{A_j}(\lambda),$$

where the sets A_j are disjoint. In fact, for each j the constant a_j can be thought of as being a "typical" value of $g(\lambda)$ for $\lambda \in A_j$. Under the above correspondence,

$$\sum_j a_j I_{A_j}(\lambda) \leftrightarrow \sum_j a_j Z(A_j). \tag{A2.3}$$

The sum $\sum_j a_j Z(A_j)$ is reminiscent of the approximating sum of an integral. Consequently, upon passage to the limit it is natural to denote the limit of the right-hand side of (A2.3) by

$$\int g(\lambda) Z(d\lambda).$$

This limit enjoys the basic properties of an integral almost surely. For example, the limit is independent of the particular sequence of simple functions used to approximate $g(\lambda)$.

It follows that $\int g(\lambda)Z(d\lambda)$ is the element of $\mathcal{M}_X$ such that

$$g \leftrightarrow \int g(\lambda)Z(d\lambda).$$

Moreover, by (A2.2),

$$E\left(\int g(\lambda)Z(d\lambda)\right)\left(\overline{\int h(\mu)Z(d\mu)}\right) = \int g(\lambda)\overline{h(\lambda)}F(d\lambda).$$

This is the major statement of the basic property of the spectral representation.

Finally, since $e^{i\lambda t}$ corresponds (uniquely) both to $X(t)$ and $\int e^{i\lambda t}Z(d\lambda)$ we have

$$X(t) = \int e^{i\lambda t} Z(d\lambda).$$

An alternate derivation of this result, due basically to Kolmogorov, is given by Rozanov (1967, p. 14). See also Doob (1953, p. 527) for additional details.

A2.2 Evaluation of $\lim_{T\to\infty} D_T$ in the Gaussian Case

Recall that $X(t)$ is a stationary Gaussian process and

$$Y_\tau(t) = X(t+\tau)X(t) - C(\tau).$$

Consequently,

$$D_T = E\left(\frac{1}{2T}\int_{-T}^{T} Y_\tau(t)\,dt\right)^2.$$

First, note that the spectral representation of $X(t)$ yields

$$C(\tau) + Y_\tau(t) = \int_{-\infty}^{\infty}\int_{-\infty}^{\infty} e^{i(\lambda+\mu)t} e^{i\lambda\tau} Z(d\lambda)Z(d\mu).$$

Then

$$C(\tau) + \frac{1}{2T}\int_{-T}^{T} Y_\tau(t)\,dt = \int_{-\infty}^{\infty}\int_{-\infty}^{\infty}\left(\frac{1}{2T}\int_{-T}^{T} e^{i(\lambda+\mu)t}\,dt\right) e^{i\lambda\tau} Z(d\lambda)Z(d\mu).$$

$$= \int_{-\infty}^{\infty}\int_{-\infty}^{\infty} \frac{\sin(\lambda+\mu)T}{(\lambda+\mu)T} e^{i\lambda\tau} Z(d\lambda)Z(d\mu).$$

Squaring both sides and taking expectations, since $EY_\tau(t) = 0$, we obtain (dropping limits of integration)

$$C^2(\tau) + E\left(\frac{1}{2T}\int_{-T}^{T} Y_\tau(t)\,dt\right)^2$$

$$= E\iiiint \frac{\sin(\lambda+\mu)T}{(\lambda+\mu)T}\,\frac{\sin(\alpha+\beta)T}{(\alpha+\beta)T}\, e^{i(\lambda+\alpha)\tau} Z(d\lambda)Z(d\mu)Z(d\alpha)Z(d\beta).$$

However, the second term on the left side is D_T. Thus, moving the expectation inside on the right side,

$$C^2(\tau) + D_T = \iiiint \frac{\sin(\lambda+\mu)T}{(\lambda+\mu)T}\,\frac{\sin(\alpha+\beta)T}{(\alpha+\beta)}$$

$$\times\, e^{i(\lambda+\alpha)\tau} EZ(d\lambda)Z(d\mu)Z(d\alpha)Z(d\beta). \qquad \text{(A2.4)}$$

Now, by the Isserlis theorem for complex Gaussian random variables (Section 1.4), we obtain

$$\begin{aligned} EZ(d\lambda)Z(d\mu)Z(d\alpha)Z(d\beta) &= EZ(d\lambda)Z(d\mu)EZ(d\alpha)Z(d\beta) \\ &\quad + EZ(d\lambda)Z(d\alpha)EZ(d\mu)Z(d\beta) \\ &\quad + EZ(d\lambda)Z(d\beta)EZ(d\mu)Z(d\alpha). \end{aligned} \qquad \text{(A2.5)}$$

Substituting this expression into (A2.4), the right-hand side can be written as the sum of three integrals, say, I_1, I_2, I_3, which are labeled in the same order as the terms in (A2.5).

Recall the $Z(-A) = \overline{Z(A)}$ and

$$EZ(d\gamma)\overline{Z(d\delta)} = \begin{cases} F(d\gamma), & \text{if } \gamma = \delta, \\ 0, & \text{otherwise.} \end{cases}$$

Thus, the first term on the right side of (A2.5) is

$$EZ(d\lambda)Z(d\mu)EZ(d\alpha)Z(d\beta) = EZ(d\lambda)\overline{Z(-d\mu)}EZ(d\alpha)\overline{Z(-d\beta)}$$
$$= \begin{cases} F(d\lambda)F(d\alpha), & \text{if } \mu = -\lambda \text{ and } \beta = -\alpha, \\ 0, & \text{otherwise.} \end{cases}$$

It follows from (2.33) that

$$I_1 = \iint \left(\frac{\sin(0)T}{(0)T}\right)^2 e^{i(\lambda+\alpha)\tau} F(d\lambda)F(d\alpha)$$
$$= \int e^{i\lambda\tau} F(d\lambda) \int e^{i\alpha\tau} F(d\alpha) = C(\tau)^2.$$

Consequently, from (A2.4)

$$D_T = I_2 + I_3. \tag{A2.6}$$

The second term on the right-hand side of (A2.5) is

$$EZ(d\lambda)Z(d\alpha)EZ(d\mu)Z(d\beta) = \begin{cases} F(d\lambda)F(d\mu), & \text{if } \alpha = -\lambda \text{ and } \beta = -\mu, \\ 0, & \text{otherwise.} \end{cases}$$

Thus

$$I_2 = \iint \left(\frac{\sin(\lambda+\mu)T}{(\lambda+\mu)T}\right)^2 F(d\lambda)F(d\mu).$$

Similarly,

$$I_3 = \iint \left(\frac{\sin(\lambda+\mu)T}{(\lambda+\mu)T}\right)^2 e^{i(\lambda-\mu)\tau} F(d\lambda)F(d\mu).$$

Combining these integrals and using the decomposition of $F(A)$ into discrete and continuous components, (A2.6) becomes

$$D_T = \sum_{j=-\infty}^{\infty} \sum_{k=-\infty}^{\infty} \left(\frac{\sin(\lambda_j+\lambda_k)T}{(\lambda_j+\lambda_k)T}\right)^2 (1 + e^{i(\lambda_j-\lambda_k)\tau}) p(\lambda_j)p(\lambda_k)$$
$$+ \iint \left(\frac{\sin(\lambda+\mu)T}{(\lambda+\mu)T}\right)^2 (1 + e^{i(\lambda-\mu)\tau}) f(\lambda)f(\mu)\, d\lambda\, d\mu$$
$$+ \sum_{j=-\infty}^{\infty} \left[\int_{-\infty}^{\infty} \left(\frac{\sin(\mu+\lambda_j)T}{(\mu+\lambda_j)T}\right)^2 (2 + e^{i(\lambda_j-\mu)\tau} + e^{i(\mu-\lambda_j)\tau}) f(\mu)\, d\mu\right] p(\lambda_j). \tag{A2.7}$$

Now as $T \to \infty$, the integrals in (A2.7) tend to zero by the dominated convergence theorem. The remaining sum converges to zero except at those values of j and k for which $\lambda_j + \lambda_k = 0$. It follows that

$$\lim_{T\to\infty} D_T = \sum_{j=-\infty}^{\infty} (1 + e^{2i\lambda_j\tau}) p^2(\lambda_j).$$

CHAPTER

3

Sampling, Aliasing, and Discrete-Time Models

3.1 INTRODUCTION

In this chapter we will consider the basic operation of sampling a time series which converts a time series in continuous time to one in discrete time. This is a necessary step in the preparation of the series for manipulation by a digital computer. When the original series is modeled by a weakly stationary stochastic process, the sampled series will be a weakly stationary process in discrete time, complete with its own spectral representation and power spectrum.

The relationship between the spectrum of the continuous-time series and that of the sampled series is quite important, since estimates of the spectrum for the sampled series must be used to estimate the spectrum of the original series. Proper selection of the sampling rate will guarantee good agreement between the two spectra, thus the possibility of forming good estimates of the continuous-time spectrum. Improper selection of the sampling rate introduces the problem of aliasing whereby the agreement between the two spectra is, to varying degrees, destroyed. We will study the aliasing problem and the means by which it can be avoided. A sampling theorem is given which relates the required rate of sampling to recover all of the information in the time series to bounds on the "width" of the spectrum.

In some fields, time series occur naturally in discrete form. Hourly temperature readings on a hospitalized patient, daily closing stock averages, monthly sunspot averages, and tree ring thicknesses are examples of discrete time series. However, even these series are, more often than not, sampled versions of an underlying continuous process. Thus, though the discrete time version of the spectrum is the one of interest, it is important to know how this spectrum can be affected by features of the underlying process through aliasing.

The general features of discrete-time, weakly stationary models will be summarized in this chapter in order to unify our treatment of linear filters in Chapter 4. Specific discrete-time models used in practice, the important finite parameter models, will be discussed in Chapter 7.

3.2 SAMPLING AND THE ALIASING PROBLEM

Suppose that $X(t)$ is a time series in continuous time. We will restrict our attention to weakly stationary stochastic processes, but the results we obtain for the relationship between the sampled and unsampled spectrum will also be valid for the broader class of time series considered in Section 2.2. We also restrict our attention to *equispaced sampling* which, in any event, is the most important type of sampling in practice.

To sample $X(t)$, a positive quantity Δt, called the *sampling interval*, is required. This quantity determines the length of time between sampled values. The number of values sampled per unit time is called the *sampling rate* and is equal to $1/\Delta t$. Thus, for example, a time series which is sampled at intervals of $\Delta t = 0.02$ sec has a sampling rate of 50 observations/sec. The time series $X(t)$ is replaced, through sampling, by the discrete-time series

$$X_{\Delta t}(k) = X(k\,\Delta t), \qquad k = 0, \pm 1, \ldots. \tag{3.1}$$

That is, the value of the discrete series at time k is the value of the continuous series at the time instant $k\,\Delta t$.

Now, through sampling, certain harmonics in the spectral decomposition of $X(t)$ become indistinguishable from one another. This is illustrated in Fig. 3.1. Note that for the given sampling interval it is impossible to tell which of the two harmonics is being observed. Thus, the power attributed to the more slowly varying harmonic will be, in some sense, the combined power of

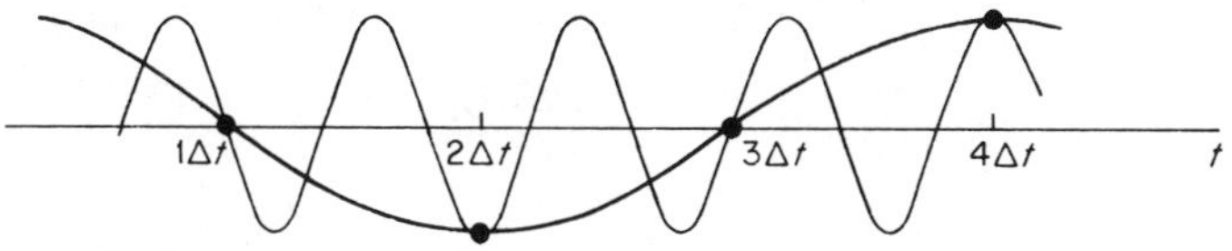

Fig. 3.1 *Illustration of two sinusoids made indistinguishable by sampling at intervals of length Δt.*

all harmonics which are made indistinguishable from it (i.e., are *aliased* with it) by sampling. In order to provide a more quantitative account of this phenomenon, we now give an intuitive derivation of the spectral representation of the sampled processs.

Recall that the continuous-time series has the spectral representation

$$X(t) = \int_{-\infty}^{\infty} e^{i\lambda t} Z(d\lambda).$$

Thus,

$$X_{\Delta t}(k) = \int_{-\infty}^{\infty} e^{i\lambda k\,\Delta t} Z(d\lambda). \tag{3.2}$$

This can be thought of as a "linear combination" in λ of the periodic functions of k

$$\varphi_\lambda(k) = e^{i\lambda k\,\Delta t},$$

with coefficients $Z(d\lambda)$. (See Example 1.1.) Fix λ and consider the corresponding periodic function at frequency $\lambda + (2\pi/\Delta t)$:

$$\begin{aligned}\varphi_{\lambda+(2\pi/\Delta t)}(k) = \exp\bigl(i(\lambda + 2\pi/\Delta t)k\,\Delta t\bigr) &= \exp\bigl(i(\lambda k\,\Delta t + 2\pi k)\bigr)\\ &= e^{i\lambda k\,\Delta t}(e^{2\pi i})^k\\ &= e^{i\lambda k\,\Delta t} = \varphi_\lambda(k).\end{aligned}$$

Thus, because we are, in effect, sampling the periodic functions $e^{i\lambda t}$ at time points Δt units apart, the function $\varphi_\lambda(k)$ is indistinguishable from the function at the higher frequency $\lambda + (2\pi/\Delta t)$. What this means is that the complex amplitude at frequency $\lambda + (2\pi/\Delta t)$, which we will denote by $Z\bigl(d\lambda + (2\pi/\Delta t)\bigr)$, will appear as a contribution to the amplitude of $\varphi_\lambda(k)$.

Now, if l is any integer, positive or negative, an identical computation yields

$$\varphi_{\lambda+(2\pi l/\Delta t)}(k) = \varphi_\lambda(k).$$

Thus, all of the amplitudes $Z\bigl(d\lambda + (2\pi l/\Delta t)\bigr)$, $l = 0, \pm 1, \ldots$, contribute to the coefficient of $\varphi_\lambda(k)$. For λ in the frequency range $-\pi/\Delta t < \lambda \le \pi/\Delta t$, let $Z_{\Delta t}(d\lambda)$ represent this coefficient. Then by collecting all of coefficients of $\varphi_\lambda(k)$ in the "sum" (3.2) we can write

$$X_{\Delta t}(k) = \int_{-\pi/\Delta t}^{\pi/\Delta t} e^{i\lambda k\,\Delta t} Z_{\Delta t}(d\lambda), \tag{3.3}$$

where

$$Z_{\Delta t}(d\lambda) = \sum_{l=-\infty}^{\infty} Z\bigl(d\lambda + (2\pi l/\Delta t)\bigr). \tag{3.4}$$

This is the spectral representation of the sampled process. This purely intuitive argument can be made mathematically acceptable. Details are left to the reader.

If μ is any number and A any set of numbers, the set $A + \mu$ is defined to be

$$A + \mu = \{\lambda + \mu : \lambda \in A\}.$$

Then (3.4) can be viewed as a special case of the expression

$$Z_{\Delta t}(A) = \sum_{l=-\infty}^{\infty} Z(A + (2\pi l/\Delta t)), \tag{3.5}$$

where $Z_{\Delta t}(A)$ is the random spectral measure for the discrete process $X_{\Delta t}(k)$ and (3.5) is the basic relationship between the spectral measures of the original process and its sampled version. We now investigate the relationship between the spectra implied by this expression.

If A is a subset of the interval $(-\pi/\Delta t, \pi/\Delta t]$, then the sets $A + (2\pi l/\Delta t)$, $l = 0, \pm 1, \ldots$, are all disjoint and property (2.31) of spectral measures yields

$$\begin{aligned} E|Z_{\Delta t}(A)|^2 &= \sum_{l=-\infty}^{\infty} \sum_{m=-\infty}^{\infty} EZ(A + (2\pi l/\Delta t))\overline{Z(A + (2\pi m/\Delta t))} \\ &= \sum_{l=-\infty}^{\infty} F(A + (2\pi l/\Delta t)). \end{aligned}$$

Thus, defining the *spectral distribution of the sampled process* by

$$F_{\Delta t}(A) = E|Z_{\Delta t}(A)|^2,$$

we obtain

$$F_{\Delta t}(A) = \sum_{l=-\infty}^{\infty} F(A + (2\pi l/\Delta t)). \tag{3.6}$$

That is, the power in the sampled time series concentrated in the set of frequencies $A \subset (-\pi/\Delta t, \pi/\Delta t]$ is the accumulation of power of the original process from all of the sets of frequencies $A + (2\pi l/\Delta t)$. For any λ in $(-\pi/\Delta t, \pi/\Delta t]$, the frequencies $\lambda + (2\pi l/\Delta t)$, $l = \pm 1, \pm 2, \ldots$, are called the *aliases* of λ.

The spectral distribution $F_{\Delta t}(A)$ is determined by a *spectral function* $p_{\Delta t}(\lambda)$ and a *spectral density function* $f_{\Delta t}(\lambda)$ which are related to the corresponding functions of the original process through (3.6) as follows:

$$p_{\Delta t}(\lambda) = \sum_{l=-\infty}^{\infty} p(\lambda + (2\pi l/\Delta t)), \tag{3.7}$$

$$f_{\Delta t}(\lambda) = \sum_{l=-\infty}^{\infty} f(\lambda + (2\pi l/\Delta t)), \qquad -\pi/\Delta t < \lambda \le \pi/\Delta t. \tag{3.8}$$

Expression (3.8) is obtained by differentiating the series for the spectral distribution function term by term, provided the series of derivatives converges. This will be the case for all spectral densities of practical interest.

Thus, all of the spectral lines (points of discrete power) in the original spectrum will appear as lines in the spectrum of the sampled process in the range $(-\pi/\Delta t, \pi/\Delta t]$, and the continuous power will be "folded" into this range to make up the continuous component of the sampled process spectrum. This is illustrated in Fig. 3.2. The upper limit of the range of the sampled

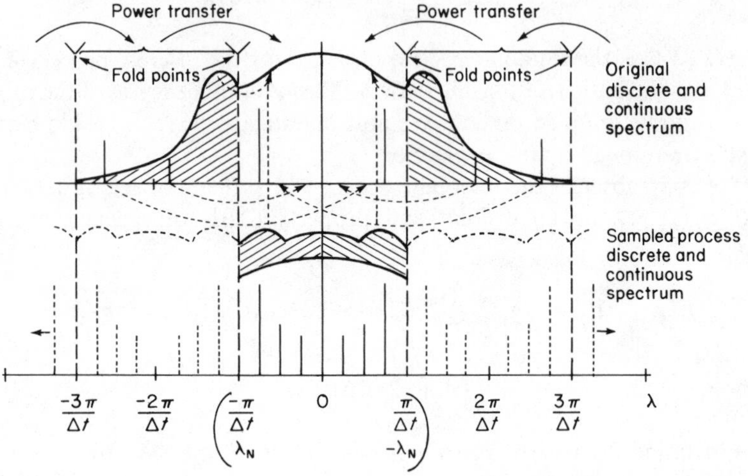

Fig. 3.2 *Illustration of the folding of the power of the continuous-time series spectrum into the range* $(-\lambda_N, \lambda_N]$ *to form the sampled time series spectrum. Aliased power is cross-hatched.*

process spectrum is called the *Nyquist folding frequency*. It can be evaluated by the expression

$$\lambda_N = \pi/\Delta t \qquad \text{(radians per unit time)}$$

or (3.9)

$$f_N = 1/2\,\Delta t \qquad \text{(cycles per unit time).}$$

Note that in terms of λ_N, the aliases of a frequency λ in the range $-\lambda_N < \lambda \leq \lambda_N$ are of the form

$$\lambda + 2l\lambda_N, \qquad l = \pm 1, \pm 2, \ldots .$$

The selection of the sampling interval Δt is equivalent to selecting the Nyquist frequency. In fact the criterion for determining Δt in the design of a spectrum analysis is virtually always given in terms of conditions on the Nyquist frequency. An illustration of a poor choice of Δt is given in the following example.

Example 3.1 *An Illustration of the Effect of Sampling*

Suppose that a time series of ocean levels is sampled once each week to look for possible cycles in the data. Then, if time is measured in days, $\Delta t = 7$ days and the Nyquist folding frequency is

$$\lambda_N = \pi/7 \quad \text{rad/day}.$$

Because of the regular daily tides—which we will idealize as having a period of exactly one day—there will be a spectral line of substantial power at $\lambda = 2\pi$ rad/day. Now,

$$2\pi = 0 + \big(2\pi(7)/7\big) = 0 + 14\lambda_N$$

and, from the above theory, it follows that 2π is aliased with zero frequency. Thus, the power at one cycle per day will be added to the power at zero frequency and thus will simply appear as a contribution to the mean sea height. This can be seen intuitively, since if the weekly observations were always taken at the time of day, for example when high tide occurred, it would appear as though the mean sea level were elevated by an amount depending on the power in the daily cycle. It follows that this selection of Δt is a particularly poor one from the viewpoint of analyzing the daily cycle. This is a rather extreme example of what is known as the aliasing problem.

The Aliasing Problem and a Sampling Theorem

The *aliasing problem* arises when the spectrum of interest is that of the original, continuous-time series. To estimate this spectrum using digital methods it is necessary to sample the time series. However, the spectrum for which estimates will then be obtained is, unavoidably, the spectrum of the sampled series. The two spectra are related by expressions (3.6)–(3.8), but they can be quite different as is illustrated in Fig. 3.2. Consequently, even very accurate estimates of the sampled process spectrum may be poor estimates of the original spectrum.

This problem can be avoided or at least reduced to an acceptable level by an intelligent choice of Δt. By the natural high-frequency attenuation of physical processes and the limitations of recording equipment, for all practical purposes, the power in any real time series will be contained in a finite interval of frequencies. Consequently, realistic models will have spectra that tend to zero rapidly with increasing values of $|\lambda|$. More precisely, there will exist a frequency bound Λ such that if A is a set of frequencies which lies outside of the range $(-\Lambda, \Lambda)$, then $F(A) \cong 0$. Now, if Δt is taken small enough to ensure that

$$\lambda_N = \pi/\Delta t > \Lambda, \tag{3.10}$$

it will follow from (3.6) that

$$F_{\Delta t}(A) \cong F(A),$$

since each of the frequency sets $A + (2\pi l/\Delta t)$ for $l \neq 0$ will lie outside of $(-\Lambda, \Lambda)$. Similarly, if (3.10) is satisfied,

$$p_{\Delta t}(\lambda) \cong p(\lambda), \qquad -\Lambda \leq \lambda \leq \Lambda \tag{3.11}$$

and

$$f_{\Delta t}(\lambda) \cong f(\lambda), \qquad -\Lambda \leq \lambda \leq \Lambda. \tag{3.12}$$

Thus, over the frequency interval containing the major portion of the power, a proper choice of Δt will guarantee that the spectral functions and spectral density functions of the sampled process and continuous-time process are essentially identical. In this case, the aliasing problem does not arise. Note, for example, that the aliasing problem in Fig. 3.2 could have been avoided by selecting λ_N roughly three times larger than indicated—which corresponds to a sampling interval Δt one third as large.

The dependence of the sampling rate on the range of the spectrum is graphically demonstrated by the following theorem.

Sampling Theorem *If for some number Λ, the spectrum of $X(t)$ is zero outside of the frequency interval $-\Lambda \leq \lambda \leq \Lambda$, then the time series can be exactly reconstructed from its values at the time points*

$$\pi k/\Lambda, \qquad k = 0, \pm 1, \ldots.$$

More precisely,

$$X(t) = \sum_{k=-\infty}^{\infty} \frac{\sin \Lambda(t - (\pi k/\Lambda))}{\Lambda(t - (\pi k/\Lambda))} X\left(\frac{\pi k}{\Lambda}\right), \qquad -\infty < t < \infty. \tag{3.13}$$

The proof of this theorem depends on the Fourier series expansion of $e^{i\lambda t}$: Let $e_\Lambda^{i\lambda t}$ denote the function of λ which is equal to $e^{i\lambda t}$ for $-\Lambda < \lambda \leq \Lambda$ and is the periodic extension of this function (of period 2Λ) outside of this interval. The Fourier coefficients of $e_\Lambda^{i\lambda t}$ are easily computed from expression (1.25) and the Fourier series expansion (1.24) is

$$e_\Lambda^{i\lambda t} = \sum_{k=-\infty}^{\infty} \frac{\sin \Lambda(t - (\pi k/\Lambda))}{\Lambda(t - (\pi k/\Lambda))} e^{i\lambda(\pi k/\Lambda)}.$$

Now, by hypothesis, $F(d\lambda) = 0$, thus $Z(d\lambda) = 0$ for $|\lambda| > \Lambda$. Consequently,

$$X(t) = \int_{-\infty}^{\infty} e^{i\lambda t} Z(d\lambda) = \int_{-\Lambda}^{\Lambda} e^{i\lambda t} Z(d\lambda) \qquad \text{for every } t.$$

Finally,

$$X(t) = \int_{-\Lambda}^{\Lambda} e^{i\lambda t} Z(d\lambda) = \int_{-\Lambda}^{\Lambda} e_{\Lambda}^{i\lambda t} Z(d\lambda)$$

$$= \sum_{k=-\infty}^{\infty} \frac{\sin \Lambda(t - (\pi k/\Lambda))}{\Lambda(t - (\pi k/\Lambda))} \int_{-\Lambda}^{\Lambda} e^{i\lambda(\pi k/\Lambda)} Z(d\lambda)$$

$$= \sum_{k=-\infty}^{\infty} \frac{\sin \Lambda(t - (\pi k/\Lambda))}{\Lambda(t - (\pi k/\Lambda))} X\left(\frac{\pi k}{\Lambda}\right).$$

[For the justification of the mean-square convergence of this sum and the interchange of sum and integral see Rozanov (1967, pp. 26, 27).]

Expression (3.13), due to Shannon (1949), has been utilized extensively in applications, especially in engineering. From the viewpoint of spectrum analysis, this theorem implies that if the sampling interval is taken to satisfy the inequality

$$\Delta t \leq \pi/\Lambda,$$

which is equivalent to selecting the Nyquist folding frequency to satisfy

$$\lambda_N \geq \Lambda,$$

then

$$F_{\Delta t}(A) = F(A) \qquad \text{for all } A.$$

That is, no error is made by using $F_{\Delta t}(A)$ in place of $F(A)$. However, there are practical limitations to the usefulness of this result. It may not be feasible for one reason or another to sample at a rate sufficient to make $\lambda_N \geq \Lambda$. For example, if a long stretch of data must be sampled at a high rate it is rather easy to over-run the storage capacity of even the larger modern computers. It may be necessary to put up with a small amount of aliasing in order to fit the data into the machine. In Chapter 6 we will discuss the procedure of filtering and decimation by which the Nyquist folding frequency, thus the sampling rate, can be made smaller without significantly increasing the amount of aliasing. Once sampled, the number of data points can be reduced to any fraction of the initial number by this procedure. However, any aliasing introduced by the original sampling will remain.

In the case of a very broad-band spectrum (Λ large) it is possible to use an analog filter on the continuous-time data to decrease Λ before sampling in order to achieve the inequality $\lambda_N \geq \Lambda$ for an acceptable sampling rate. Even then, high-frequency noise is often reintroduced by the digital processing equipment. Consequently, in most practical applications it is necessary to strike a balance between the sampling rate and the amount of aliasing to be incurred. Details concerning the selection of the sampling rate to estimate spectra will be given in Chapter 9.

3.3 THE SPECTRAL MODEL FOR DISCRETE-TIME SERIES

Discrete time parameter, weakly stationary stochastic processes have a spectral theory which exactly parallels that given for continuous-time processes in Chapter 2. Although this theory can be derived quite independently, it necessarily agrees with the results for sampled processes derived in the last section. In fact, it is convenient to think of discrete processes as being sampled processes for which the sampling interval Δt is the "natural" unit of time. This is equivalent to setting $\Delta t = 1$ in the expressions of the last section. Let $X(k)$, $k = 0, \pm 1, \ldots$, denote a weakly stationary process with discrete time parameter. Then the *spectral representation of* $X(t)$ is

$$X(k) = \int_{-\pi}^{\pi} e^{i\lambda k} Z(d\lambda), \qquad k = 0, \pm 1, \ldots. \tag{3.14}$$

Although (3.4) and (3.5) have no real interpretation when an underlying continuous process does not exist, they still indicate an important property of the spectral measure $Z(A)$: It is easily seen from (3.5) that

$$Z(A + 2\pi) = Z(A). \tag{3.15}$$

Thus, with a slightly extended interpretation of the term, $Z(A)$ *is periodic of period* 2π.

Similarly, the *spectral distribution*

$$F(A) = E|Z(A)|^2$$

is periodic in the same sense,

$$F(A + 2\pi) = F(A). \tag{3.16}$$

The *spectral distribution function* is now given in terms of $F(A)$ by the expression

$$F(\lambda) = F((-\pi, \lambda]). \tag{3.17}$$

As before, the *spectral function* and *spectral density function* are, respectively,

$$p(\lambda) = F(\{\lambda\}) \tag{3.18}$$

and

$$f(\lambda) = \frac{dF(\lambda)}{d\lambda}. \tag{3.19}$$

It follows from (3.3) that $p(\lambda)$ and $f(\lambda)$ *are periodic functions of period* 2π in the usual sense. Thus, if we were to take $\Delta t = 1$ in Fig. 3.2, the graphs of $p(\lambda)$ and $f(\lambda)$ would extend periodically outside of the interval $(-\pi, \pi]$ as indicated by the dashed lines. Since periodic functions are completely determined by their values on a single cycle, it suffices to restrict attention to the interval $(-\pi, \pi]$.

The *autocovariance function*

$$C(k) = EX(j+k)X(j)$$

has the spectral representation

$$C(k) = \int_{-\pi}^{\pi} e^{i\lambda k} F(d\lambda) \tag{3.20}$$

as can be easily shown from (3.14). In the *case of a discrete spectrum* we have

$$C(k) = \sum_{j=-\infty}^{\infty} e^{i\lambda_j k} p(\lambda_j),$$

where the points of positive power λ_j satisfy the relations

$$\lambda_{-j} = -\lambda_j$$

for $\lambda_j \neq \pi$ and $-\pi < \lambda_j \leq \pi$. That is, if $\lambda_j = \pi$ is a point of positive power, then λ_{-j} is excluded from the interval.

Now from the theory of almost periodic functions in discrete time covered in Example 1.3, for a process with discrete spectrum the spectral function is determined from the autocovariance by the expression

$$p(\lambda) = \lim_{L\to\infty} \frac{1}{2L+1} \sum_{k=-L}^{L} e^{-i\lambda k} C(k) \qquad \text{for} \quad -\pi < \lambda \leq \pi. \tag{3.21}$$

When the *spectrum is continuous*, we have

$$C(k) = \int_{-\pi}^{\pi} e^{i\lambda k} f(\lambda)\, d\lambda. \tag{3.22}$$

If $\sum_{k=-\infty}^{\infty} |C(k)|^2 < \infty$, as is the case in most applications, then the numbers $C(k)/2\pi$ are simply the Fourier coefficients of the Fourier series expansion of the periodic function $f(\lambda)$,

$$f(\lambda) = \frac{1}{2\pi} \sum_{k=-\infty}^{\infty} e^{-i\lambda k} C(k). \tag{3.23}$$

(See Example 1.2.) Note that (3.23) also follows from the condition $\sum_{k=-\infty}^{\infty} |C(k)| < \infty$, since an absolutely summable series is automatically square summable. Expressions (3.22) and (3.23) are the discrete-time version of the *Wiener–Khintchine relations*.

As in the continuous-time case there is a general expression for obtaining the value of the spectral distribution F for every interval of the form $(\Lambda_1, \Lambda_2]$

for which $p(\Lambda_1) = p(\Lambda_2) = 0$. This determines $F(A)$ for every (measurable) set A and is valid for every type of spectrum—discrete, continuous, or mixed (Doob, 1953, p. 474). The expression is

$$F((\Lambda_1, \Lambda_2]) = \frac{1}{2\pi}\left\{(\Lambda_2 - \Lambda_1)C(0) + \lim_{L\to\infty} \sum_{k=-L,\, k\neq 0}^{L} \frac{e^{-i\Lambda_1 k} - e^{-i\Lambda_2 k}}{ik} C(k)\right\}. \tag{3.24}$$

Finally, the ergodic theorems in Sections 2.9 and 2.10 carry over word for word to discrete-time processes by the simple expedient of replacing every occurrence of the average

$$\lim_{T\to\infty} \frac{1}{2T}\int_{-T}^{T} \text{——}\, dt$$

by the average

$$\lim_{L\to\infty} \frac{1}{2L+1}\sum_{k=-L}^{L} \text{——}.$$

Thus, for example, when the process has zero mean and no added non-stochastic almost periodic term,

$$\lim_{L\to\infty} \frac{1}{2L+1}\sum_{k=-L}^{L} e^{-i\lambda k}X(k) = Z(\{\lambda\}). \tag{3.25}$$

If the process is Gaussian with continuous spectrum, we have

$$\lim_{L\to\infty} \frac{1}{2L+1}\sum_{k=-L}^{L} X(k+r)X(k) = C(r). \tag{3.26}$$

Both limits are limits in mean-square as before.

Example 3.2 *Discrete-Time Periodic Processes*
A discrete weakly stationary time series is periodic of period N if

$$X(k+N) = X(k), \qquad k = 0, \pm 1, \ldots.$$

By the mean ergodic theorem (3.25) and a change of variables argument, it can be shown that

$$EZ(\{\lambda\})\overline{Z(\{\mu\})} = \begin{cases} \lim\limits_{L\to\infty} \dfrac{1}{2L+1}\sum\limits_{r=-L}^{L} e^{-i\lambda r}C(r), & \lambda = \mu, \\ 0, & \lambda \neq \mu. \end{cases}$$

But $C(r)$ is seen to be periodic of period N, since $X(k)$ is. Thus (see Example 1.3),

$$E|Z(\{\lambda\})|^2 = \begin{cases} \dfrac{1}{N}\sum_{r=1}^{N} e^{-i\lambda r}C(r), & \lambda = \dfrac{2\pi v}{N}, \quad -\left[\dfrac{N-1}{2}\right] \le v \le \left[\dfrac{N}{2}\right], \\ 0, & \text{otherwise.} \end{cases}$$

Let $\lambda_v = 2\pi v/N$, $Z_v = Z(\{\lambda_v\})$, and $p_v = E|Z_v|^2$. We then have

$$C(r) = \sum_{v=-[(N-1)/2]}^{[N/2]} e^{i\lambda_v r} p_v,$$

which implies that all of the power in the time series is concentrated at the equally spaced frequencies λ_v. That is, the spectrum of the time series is discrete with spectral function

$$p(\lambda) = \begin{cases} p_v, & \lambda = \dfrac{2\pi v}{N}, \quad -\left[\dfrac{N-1}{2}\right] \le v \le \left[\dfrac{N}{2}\right], \\ 0, & \text{otherwise.} \end{cases}$$

The spectral representation of the process is

$$X(k) = \sum_{v=-[(N-1)/2]}^{[N/2]} e^{i\lambda_v k} Z_v, \qquad k = 0, \pm 1, \ldots \tag{3.27}$$

and (see Example 1.1)

$$Z_v = \frac{1}{N}\sum_{k=1}^{N} e^{-i\lambda_v k} X(k). \tag{3.28}$$

This is the *finite Fourier transform* of the discrete stochastic process. In Chapter 8, $X(1), X(2), \ldots, X(N)$ will represent the available observations from a discrete process or a sampled continuous-time process and a normalized version of the finite Fourier transform will be computed for this data. Then expression (3.27) will produce a periodic extension of the original process which repeats these N values over and over. A convenient relationship between the spectrum of the original time series and the spectrum of the induced periodic process will allow us to use the finite Fourier transform to estimate the spectrum of the underlying process. Moreover, the distribution theory for these spectral estimators will be especially simple and useful. Consequently, these expressions will play a fundamental role in the statistical theory of spectrum analysis.

Example 3.3 *White Noise*

Let $X(k) : k = 0, \pm 1, \ldots$ be a sequence of uncorrelated, zero-mean random variables with common variance σ^2. That is,

$$EX(k) = 0$$

and

$$EX(j)X(k) = \begin{cases} \sigma^2, & \text{if } j = k, \\ 0, & \text{if } j \neq k. \end{cases}$$

Such a sequence is easily seen to be a weakly stationary process with autocovariance function

$$C(k) = \begin{cases} \sigma^2, & k = 0, \\ 0, & k \neq 0. \end{cases}$$

Now, since this autocovariance satisfies the condition $C(k) \to 0$ as $k \to \infty$ (in a rather trivial sense), this time series has continuous spectrum and, in fact, (3.23) can be applied to obtain the spectral density

$$f(\lambda) = \sigma^2/2\pi, \qquad |\lambda| \leq \pi.$$

That is, the time series has a spectral density which weighs all frequencies equally. Because white light is popularly considered to be composed of a uniform mix of all light colors, a time series with a constant spectral density is called a *white noise process*. As we will see in Chapter 7, the more important discrete time series models are based on linear transformations of white noise processes.

CHAPTER

4

Linear Filters—General Properties with Applications to Continuous-Time Processes

4.1 INTRODUCTION

This chapter will be devoted to outlining the basic properties of linear filters. The various roles played by linear filters in both the theory and applications of time series analysis are among the most important features of the subject. Linear filters provide an important class of models for physical transformations. For example, to a good degree of approximation, the earth behaves like a linear filter to seismic waves and the ocean to ocean waves. To a more restricted extent such phenomena as economic systems and biological systems behave like linear filters for restricted lengths of time.

Linear filters transform time series into new time series where the term "time series" can be interpreted in the broadest sense as meaning any numerical function of time whether continuous or discrete, random or nonrandom. Because of this feature, an important topic is the construction of special purpose linear filters to modify data to meet particular objectives or to display specific features of the data. Methods for combining linear filters, by which a variety of special filters can be obtained, will be given in this chapter. We will show how high-pass, band-pass, and notch filters can be constructed from low-pass filters. Although much of the necessary filtering can be done electronically when the data is still in analog form, it is often more convenient to do the filtering after sampling the series. Thus, these constructions are also of considerable importance for the design of digital filters which will be considered in Chapter 6.

The importance of linear filters in the construction of stochastic time series models is due to the fact that they convert weakly stationary stochastic processes back into weakly stationary processes in an easily describable way. Because of this technical feature, they are used to provide the "dynamic" connecting links between time series in virtually all of the useful stochastic models employed in applications. Examples will be given in Chapter 5.

The important continuous-time linear filters will be defined in this chapter and their properties illustrated. Linear differential equations with stochastic inputs will be seen to be a special topic that can be dealt with under the more general theory of inverting linear filters.

4.2 LINEAR FILTERS

Because of the applicability of linear filters to a variety of different time series models, it is instructive to discuss them initially at a more or less intuitive level that isolates the features common to all models. A more precise discussion for the case of weakly stationary processes is given in the Appendix. The name "linear filter" is used in place of the more accurate (but more cumbersome) title "time-invariant linear transformation." The natural domain of a linear transformation is a vector space. Without being specific about the particular space in question, the vectors of the space will be (possibly) complex-valued time series $\{x(t)\}$, stochastic or nonstochastic, and in continuous or discrete time. The bracket notation will be used for a brief time to distinguish between the entire time series

$$\{x(t)\} = \{x(t) : -\infty < t < \infty\}$$

and its value at time t which is denoted by $x(t)$.

We define two time series to be *equal*,

$$\{x(t)\} = \{y(t)\},$$

if $x(t) = y(t)$ for all t.

The operations of scalar multiplication and addition are defined coordinatewise by the expressions

$$\alpha\{x(t)\} = \{\alpha x(t)\},$$

where α is a complex constant and

$$\{x(t)\} + \{y(t)\} = \{x(t) + y(t)\}.$$

That is, $\alpha\{x(t)\}$ is obtained by multiplying each component $x(t)$ by α and $\{x(t)\} + \{y(t)\}$ is the time series obtained by adding $x(t)$ and $y(t)$ at each time t.

The complex constant α will, in some cases, be a random quantity. "Constant" is interpreted to mean "not a function of t."

A linear filter L transforms a time series $\{x(t)\}$, *the input*, into an *output* time series $\{y(t)\}$,

$$\{y(t)\} = L(\{x(t)\}).$$

Linear, time-invariant transformations have three basic properties: (i) scale preservation, (ii) superposition, and (iii) time invariance. These properties can be expressed symbolically as follows:

(i) *scale preservation:*

$$L(\alpha\{x(t)\}) = \alpha L(\{x(t)\});$$

(ii) *the superposition principle:*

$$L(\{x(t)\} + \{y(t)\}) = L(\{x(t)\}) + L(\{y(t)\});$$

(iii) *time invariance:* If $L(\{x(t)\}) = \{y(t)\}$, then

$$L(\{x(t+h)\}) = \{y(t+h)\}$$

for every number h, where $\{x(t+h)\}$ and $\{y(t+h)\}$ are the time series whose values at time t are $x(t+h)$ and $y(t+h)$, i.e., the time series obtained from $\{x(t)\}$ and $\{y(t)\}$ by shifting the time origin by the amount h.

These properties can be described, in words, as follows: a transformation *preserves scale* if an amplification of the input by a given scale factor results in the amplification of the output by the same factor. The *superposition principle* states that if two time series are added together (in the sense that the amplitudes at each point are summed) and presented as the input to the filter, then the output will be the sum of the two time series which would have resulted from using the two initial series as inputs to the filter separately. The *time invariance property* requires that if two inputs to the filter are the same except for a relative displacement in time, then the outputs will also be the same except for the same time displacement. Intuitively, this means that no matter what time in history a given input is presented to the filter, it will always respond in the same way. Its "behavior" does not change with time.

The first two properties combine to express the *linearity* of L:

$$L(\alpha\{x(t)\} + \beta\{y(t)\}) = \alpha L(\{x(t)\}) + \beta L(\{y(t)\}).$$

By applying this expression inductively the linearity property can be extended to any finite number of time series $\{x_j(t)\}$ and complex constants α_j, $j = 1, 2, \ldots, N$,

$$L\left(\sum_{j=1}^{N} \alpha_j\{x_j(t)\}\right) = \sum_{j=1}^{N} \alpha_j L\{x_j(t)\}. \tag{4.1}$$

Moreover, under suitable conditions, the linearity property is preserved or, more properly, can be defined to hold "in the limit" as $N \to \infty$. The actual type of limit will depend on the particular model. In some cases the sum in (4.1) will become an infinite series and in others an integral. We will combine all possibilities symbolically by the expression

$$L\left(\sum_{\lambda} \alpha_\lambda \{x_\lambda(t)\}\right) = \sum_{\lambda} \alpha_\lambda L(\{x_\lambda(t)\}),$$

where α_λ and $\{x_\lambda(t)\}$ are, respectively, complex constants and time series belonging to infinite collections indexed by the parameter λ.

At this point, we will drop the bracket notation $L(\{x(t)\})$ in favor of the more commonly used notation $L(x(t))$. This notation will be subject to a certain amount of abuse in what follows. In some instances, $L(x(t))$ will denote the entire output series with input $\{x(t)\}$. In others, it will denote the value of the output series *at time* t. Thus, for example, the time invariance property can be written $L(x(t)) = y(t)$ implies $L(x(t+h)) = y(t+h)$, with the understanding that $L(x(t+h))$ is the response of L at time t to the entire input series whose value at time t is $x(t+h)$.

With the substitution $x_\lambda(t) = e^{i\lambda t}$, the above expression becomes

$$L\left(\sum_{\lambda} \alpha_\lambda e^{i\lambda t}\right) = \sum_{\lambda} \alpha_\lambda L(e^{i\lambda t}). \tag{4.2}$$

This is an extremely important relationship because it implies that *for any time series which can be represented in the form*

$$x(t) = \sum_{\lambda} \alpha_\lambda e^{i\lambda t}, \tag{4.3}$$

the action of a linear filter on $x(t)$ *is completely determined by what it does to the complex exponential functions* $e^{i\lambda t}$ *for all* λ. However, *every* type of time series we have considered in Chapters 1 and 2 has a representation of this form (although the ensuing results cannot in general be applied directly to the functions of Section 2.2). Almost periodic functions and functions possessing Fourier transforms have this form virtually by their definitions. Weakly stationary stochastic processes have the spectral representation given by (2.22). Thus, the description of the output of a linear filter L to any of these inputs depends only on the value of $L(e^{i\lambda t})$ for each λ. We will now show that properties (i)–(iii) of a linear time-invariant transformation completely determine the form of $L(e^{i\lambda t})$ in a simple and elegant way.

The Form of $L(e^{i\lambda t})$

Let L be any linear filter and for fixed λ, denote the response of L to $e^{i\lambda t}$ at time t by $\varphi_\lambda(t)$,

$$L(e^{i\lambda t}) = \varphi_\lambda(t).$$

By property (iii), for any h,

$$L(e^{i\lambda(t+h)}) = \varphi_\lambda(t+h).$$

However $e^{i\lambda(t+h)} = e^{i\lambda h}e^{i\lambda t}$. Consequently by property (i), since $e^{i\lambda h}$ is a complex number which does not depend on t,

$$\begin{aligned} L(e^{i\lambda(t+h)}) &= L(e^{i\lambda h}e^{i\lambda t}) \\ &= e^{i\lambda h}L(e^{i\lambda t}) = e^{i\lambda h}\varphi_\lambda(t). \end{aligned}$$

By the definition of equality for time series, these expressions yield the equation

$$\varphi_\lambda(t+h) = e^{i\lambda h}\varphi_\lambda(t) \qquad \text{for all } \lambda,\ t,\ h.$$

Now, set $t = 0$. This yields

$$\varphi_\lambda(h) = \varphi_\lambda(0)e^{i\lambda h}.$$

Then, since h is simply a value of the output time parameter, we can set $h = t$ to obtain the important result

$$L(e^{i\lambda t}) = B(\lambda)e^{i\lambda t}, \tag{4.4}$$

where $B(\lambda) = \varphi_\lambda(0)$. It follows that each linear filter L transforms $e^{i\lambda t}$ back into $e^{i\lambda t}$ multiplied by a factor $B(\lambda)$ called the *transfer function of the filter.* Moreover, this function is uniquely determined by the response of the filter to the input $e^{i\lambda t}$.

Combining (4.4) and (4.2) we obtain

$$L(x(t)) = \sum_\lambda \alpha_\lambda B(\lambda)e^{i\lambda t}. \tag{4.5}$$

This expression represents the output of the filter in the same form as the input with the amplitudes α_λ replaced by $\alpha_\lambda B(\lambda)$. It shows that *the action of L on $x(t)$ is completely determined by the transfer function of the filter.* The right-hand side will "make sense" and the interchange of L and sum in (4.2) will lead to a well-defined time series if whatever conditions required of α_λ to validate the representation (4.3) are also satisfied by $\alpha_\lambda B(\lambda)$. We will discuss this in the various particular cases of interest below. A better feeling for the effect of the filter on time series can be obtained by writing (4.3) in the form

$$x(t) = \sum_\lambda |\alpha_\lambda| e^{i(\lambda t + \psi(\lambda))},$$

where $|\alpha_\lambda|$ is the amplitude and $\psi(\lambda)$ is the phase angle of α_λ. Write $B(\lambda)$ in polar form,

$$B(\lambda) = |B(\lambda)| e^{i\vartheta(\lambda)}. \tag{4.6}$$

Then (4.5) becomes

$$L(x(t)) = \sum_{\lambda} |\alpha_\lambda| \, |B(\lambda)| \, e^{i(\lambda t + \psi(\lambda) + \vartheta(\lambda))}.$$

That is, *the amplitude at each frequency* λ *is multiplied by the factor* $|B(\lambda)|$ *and the phase is changed from* $\psi(\lambda)$ *to* $\psi(\lambda) + \vartheta(\lambda)$. *This completely specifies the action of the filter on the input.*

This property provides an important characterization of linear filters. Thus, for example, transformations for which the amplitudes and phases change as a function of time or for which power is transferred from one frequency to another are either nontime-invariant or nonlinear. In particular, a linear filter cannot convert a time series with a discrete spectrum into one with a continuous spectrum nor can the reverse conversion take place. Thus, *the discrete component of the input spectrum is always mapped into the discrete component of the output by a linear filter and the same is true of the continuous spectral component.*

The functions $|B(\lambda)|$ and $\vartheta(\lambda) = \arg B(\lambda)$ are called the *gain* and *phase(shift)* functions of the linear filter. In all cases of interest, $x(t)$ is a real-valued time series, which implies that $\alpha_{-\lambda} = \bar{\alpha}_\lambda$. Since $\alpha_\lambda B(\lambda)$ is the coefficient of $e^{i\lambda t}$ in the representation of $L(x(t))$, the filter output will be real-valued only if

$$B(-\lambda) = \overline{B(\lambda)}.$$

It follows that the gain and phase functions satisfy the relationships

$$|B(-\lambda)| = |B(\lambda)| \tag{4.7}$$

and

$$\vartheta(-\lambda) = -\vartheta(\lambda). \tag{4.8}$$

These functions provide an equivalent but somewhat more convenient representation of a linear filter than does the transfer function. Thus once these functions are known, the filter is completely specified.

Before looking at specific kinds of linear filters we will investigate the conditions under which (4.5) holds for the various time series models we have considered. These conditions are called *matching conditions*, since they match the filter to the input to produce a "valid" output.

The Matching Condition for a Nonstochastic Almost Periodic Function

An almost periodic function with finite power has the representation

$$x(t) = \sum_{k=-\infty}^{\infty} c_k \, e^{i\lambda_k t}, \qquad -\infty < t < \infty,$$

with $\sum_{k=-\infty}^{\infty} |c_k|^2 < \infty$. If L is a linear filter with transfer function $B(\lambda)$, it follows that

$$L(x(t)) = \sum_{k=-\infty}^{\infty} c_k B(\lambda_k) e^{i\lambda_k t}.$$

This output will be a well-defined almost periodic function with finite power if

$$\sum_{k=-\infty}^{\infty} |c_k|^2 |B(\lambda_k)|^2 < \infty. \tag{4.9}$$

This is the matching condition.

The Matching Condition for Nonstochastic Functions with Finite Energy

If $x(t)$ is a real-valued function for which

$$\int_{-\infty}^{\infty} x^2(t)\, dt < \infty.$$

then, as was seen in Example 1.4,

$$x(t) = \int_{-\infty}^{\infty} h(\lambda) e^{i\lambda t}\, d\lambda$$

with

$$\int_{-\infty}^{\infty} |h(\lambda)|^2\, d\lambda < \infty.$$

Now, if

$$\int_{-\infty}^{\infty} |h(\lambda)|^2 |B(\lambda)|^2\, d\lambda < \infty, \tag{4.10}$$

it follows that

$$L(x(t)) = \int_{-\infty}^{\infty} h(\lambda) B(\lambda) e^{i\lambda t}\, d\lambda$$

is a well-defined function with finite energy. Thus, the matching condition is (4.10).

The Matching Condition for Weakly Stationary Stochastic Processes

This is the case in which we will be most interested. If $X(t)$, $-\infty < t < \infty$, is weakly stationary (in continuous or discrete time), then the spectral representation theorem yields

$$X(t) = \int e^{i\lambda t} Z_X(d\lambda).$$

The limits of integration are $-\infty$, ∞ and $-\pi$, π in the continuous- and discrete-time cases, respectively. The condition that $X(t)$ has finite power is

$$\int F_X(d\lambda) = C(0) < \infty.$$

Now,

$$Y(t) = L(X(t)) = \int e^{i\lambda t} B(\lambda) Z_X(d\lambda)$$

represents the output of L as a weakly stationary process with spectral measure

$$Z_Y(d\lambda) = B(\lambda) Z_X(d\lambda). \tag{4.11}$$

Thus,

$$F_Y(d\lambda) = E|Z_Y(d\lambda)|^2 = |B(\lambda)|^2 F_X(d\lambda), \tag{4.12}$$

and the output will have finite power if $\int F_Y(d\lambda) < \infty$. Therefore, the matching condition is

$$\int |B(\lambda)|^2 F_X(d\lambda) < \infty. \tag{4.13}$$

Relationship between Input and Output Spectral Functions and Spectral Density Functions

By (4.12) *the spectral functions and spectral density functions of input and output are related by the expressions*

$$p_Y(\lambda) = |B(\lambda)|^2 p_X(\lambda), \tag{4.14}$$

$$f_Y(\lambda) = |B(\lambda)|^2 f_X(\lambda). \tag{4.15}$$

The simplicity of these relationships is one of the most important reasons for preferring the frequency domain representation of time series to the equivalent time domain representation.

It is important to note that even though the spectral representation of the time series defined in Section 2.2 cannot be used directly to compute spectra, the properties of the spectra under linear transformation are precisely the same as those for weakly stationary processes. Thus, if the matching condition (4.13) is satisfied, where $F_X(\lambda)$ now represents the Wiener spectral distribution function of the input, then the output spectrum is again given by the first and last expressions of (4.12) and by (4.14) and (4.15). Consequently, from the viewpoint of computing spectra for time series models, if a stochastic or nonstochastic function of finite power is included in the model, it can be replaced, operationally, with a weakly stationary process and the resulting calculations will be correct.

Specific Types of Linear Filters for Continuous-Time Processes

We now look at some important types of linear filters for continuous-time processes. First, observe the following, easy-to-remember *principle for computing the transfer functions of linear filters* based on expression (4.5). *Apply the transformation to the time series* $x(t) = e^{i\lambda t}$. *The coefficient of* $e^{i\lambda t}$ *in the resulting output expression is the transfer function* $B(\lambda)$.

The Derivative and High-Pass Filters

We look at the derivative defined in Example 2.2 from the viewpoint of linear filtering. From the discussion given above, the derivative can be defined for a variety of continuous-time series. Using the principle for deriving the transfer function just given,

$$L(e^{i\lambda t}) = \frac{de^{i\lambda t}}{dt} = i\lambda e^{i\lambda t}.$$

From this we have

$$B(\lambda) = i\lambda.$$

Thus, the gain function is

$$|B(\lambda)| = |\lambda|,$$

and the phase shift is

$$\vartheta(\lambda) = \arg B(\lambda) = \begin{cases} \pi/2, & \lambda \geq 0, \\ -\pi/2, & \lambda < 0. \end{cases}$$

It is customary (though not universal) to graph the square of the gain function $|B(\lambda)|^2$ rather than the gain function itself since this is the factor which multiplies the input spectrum to obtain the spectrum of the output. This function is given for the derivative in Fig. 4.1. Note that the power in small frequencies is attenuated or decreased while the power in higher frequencies is magnified.

A filter which passes high frequencies and cuts out low frequencies is called a *high-pass filter*. The gain function of an ideal high-pass filter is given in Fig. 4.2. Note that the ideal is to completely eliminate the power below a

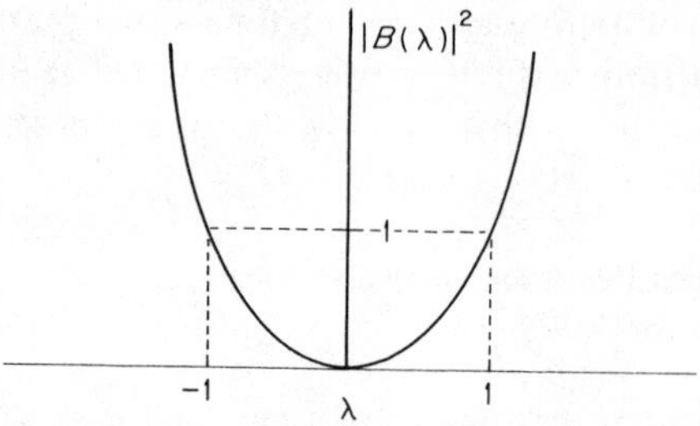

Fig. 4.1 *Graph of the squared gain function for the derivative.*

given frequency λ_0, called the *cutoff frequency*, and to pass unchanged the power above this value.

Clearly, the derivative is anything but ideal as a high-pass filter. It magnifies the power in the high frequencies to such an extent that for some time series the input cannot be matched to the derivative. Intuitively, the derivative magnifies the power in the higher frequencies in such a manner that the output has infinite power. For example, a time series with continuous spectrum defined by the spectral density function

$$f(\lambda) = \begin{cases} 1, & |\lambda| < 1, \\ 1/\lambda^2, & |\lambda| \geq 1, \end{cases}$$

would not have a derivative. Now, it is an important fact from the theory of stochastic processes that if a valid spectrum (or autocovariance function) is specified, then there exists a stationary Gaussian process with that spectrum. Thus, models of time series which cannot be differentiated are easy to construct, in theory. These models will not apply to time series encountered in "real-life" since the power in real time series is effectively zero for large

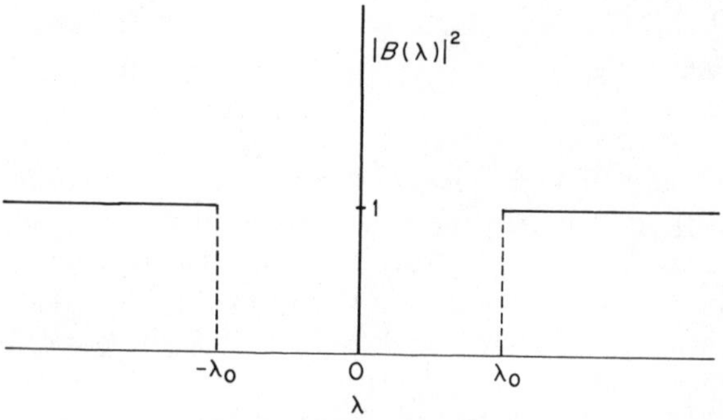

Fig. 4.2 *Squared gain function of ideal high-pass filter which passes the frequencies* $|\lambda| \geq \lambda_0$.

enough values of $|\lambda|$. However, they do crop up as theoretical models for such phenomena as turbulence.

The Convolution Integral

One of the most important time domain representations of a linear filter is the convolution integral

$$L(x(t)) = \int_{-\infty}^{\infty} h(u)x(t-u)\,du, \qquad -\infty < t < \infty. \tag{4.16}$$

It is easy to check that the three properties of a linear filter are satisfied by this type of operation on time series. The properties of the filter are completely determined by the function $h(u)$ called the *impulse response function.*

Although we will make little use of generalized functions in this book, it is instructive to introduce the usual model for an impulse, the Dirac delta function $\delta(t)$, in order to understand the reason for the term "impulse response." The delta function has the property that for every "reasonable" function $g(t)$,

$$\int_{-\infty}^{\infty} g(t)\delta(t)\,dt = g(0).$$

That is, the delta function has a spike at $t = 0$ which picks out the value of the function $g(t)$ at that point.

Now, if the delta function is used as the input to the convolution integral, we obtain

$$L(\delta(t)) = \int_{-\infty}^{\infty} h(u)\delta(t-u)\,du = h(t).$$

That is, $h(t)$ is the response of the filter to the impulse function input.

Some condition must be imposed on the impulse response function to guarantee a proper match of the filter to various inputs. Note, that by the principle for obtaining the transfer function of the filter,

$$L(e^{i\lambda t}) = \int_{-\infty}^{\infty} h(u)e^{i\lambda(t-u)}\,du = \left(\int_{-\infty}^{\infty} h(u)e^{-i\lambda u}\,du\right)e^{i\lambda t}.$$

Thus, *the transfer function is the Fourier transform of the impulse response function,*

$$B(\lambda) = \int_{-\infty}^{\infty} h(u)e^{-i\lambda u}\,du. \tag{4.17}$$

In Example 1.4 we saw that if

$$\int_{-\infty}^{\infty} |h(u)|\,du < \infty, \tag{4.18}$$

then $B(\lambda)$ is a bounded function. It is easily checked that the filter will then match any of the time series models given above.

A linear filter is said to be *stable* if whenever the input is bounded, the output is also bounded. Consider a convolution filter with impulse reponse function $h(u)$ and let the input satisfy the inequality

$$|x(t)| \leq M, \qquad -\infty < t < \infty.$$

Then, since

$$|y(t)| \leq \int_{-\infty}^{\infty} |h(u)|\,|x(t-u)|\; du \leq M \int_{-\infty}^{\infty} |h(u)|\; du,$$

it follows that (4.18) is a sufficient condition for the filter to be stable. If $h(u) = 0$ for $u < 0$, then the convolution integral becomes $y(t) = L(x(t)) = \int_0^\infty h(u)x(t-u)\,du$, and the output at time t is seen to depend on the input only for times $s \leq t$, i.e., on the present and past of $x(t)$. Filters with this property are called *realizable*.

When the input is a weakly stationary process, the usual condition imposed on the impulse response function is

$$\int_{-\infty}^{\infty} h^2(u)\, du < \infty. \tag{4.19}$$

In this case, by the Parseval relation, we have

$$\int_{-\infty}^{\infty} |B(\lambda)|^2\, d\lambda < \infty.$$

If the input process has a continuous spectrum, then, because of expression (4.15), the matching condition will be satisfied if the spectral density is bounded. In fact, the spectral density need not even have a finite integral. This makes it possible to consider generalized stochastic processes with bounded spectral densities but possibly infinite power as the inputs to such filters with perfectly well-defined, finite-power outputs.

For example, a generalized weakly stationary stochastic process which is often used as the input to convolution filters is the *continuous-time white noise process* $\varepsilon(t)$, $-\infty < t < \infty$. This is a weakly stationary process with zero mean and autocovariance function

$$C(\tau) = \begin{cases} \sigma^2 > 0, & \tau = 0, \\ 0, & \tau \neq 0. \end{cases}$$

Since this function is not continuous at $\tau = 0$, the sample functions of the process are exceedingly jumbled. However, the process has a continuous spectrum and the spectral density is

$$f(\lambda) = \sigma^2/2\pi, \qquad -\infty < \lambda < \infty.$$

This function has infinite integral, thus the total power of the process is infinite. However, the spectral density is bounded and the output $Y(t)$ of any convolution filter with input $\varepsilon(t)$ will have spectral density function

$$f_Y(\lambda) = (\sigma^2/2\pi)|B(\lambda)|^2, \qquad -\infty < \lambda < \infty.$$

Since this function is integrable as a consequence of (4.19), the output has finite power. A standard application of the continuous white noise process is given in the following example.

Example 4.1 *First-Order Linear Differential Equations with White Noise Inputs*

A simple physical system subjected to a random input might be described in terms of the differential equation

$$\frac{dX(t)}{dt} + aX(t) = \varepsilon(t), \qquad -\infty < t < \infty.$$

The solution of this equation over a time interval $-b \le t \le \infty$ is

$$\begin{aligned} X(t) &= ce^{-a(t+b)} + e^{-at}\int_{-b}^{t} e^{au}\varepsilon(u)\,du \\ &= ce^{-a(t+b)} + \int_0^{b+t} e^{-av}\varepsilon(t-v)\,dv. \end{aligned}$$

If $a > 0$, the transient term $c\exp(-a(t+b))$ "washes out" as $b \to \infty$ and the *steady state* solution for the time interval $-\infty < t < \infty$ is

$$X(t) = \int_0^{\infty} e^{-av}\varepsilon(t-v)\,dv.$$

Thus the solution is the output of a convolution filter with impulse response function

$$h(u) = \begin{cases} e^{-au}, & u \ge 0, \\ 0, & u < 0. \end{cases}$$

This function clearly satisfies (4.19). Note also that this filter is realizable. This is an important property, since it is implicitly assumed that the filter is operating in *real time*, i.e., on the present and past of the input. If $a \le 0$, then a real-time stationary solution does not exist.

Now, by (4.17),

$$B(\lambda) = \int_0^{\infty} e^{-au}e^{-i\lambda u}\,du = 1/(a+i\lambda) = (a-i\lambda)/(a^2+\lambda^2)$$

The squared gain is

$$|B(\lambda)|^2 = 1/(a^2 + \lambda^2).$$

Thus, the spectral density function of the process $X(t)$ is

$$f_X(\lambda) = \sigma^2/2\pi(a^2 + \lambda^2).$$

Low-Pass Filters

Intuitively the filter in Example 4.1 attenuates the high-frequency power of the input to such an extent that the remaining power is finite. A filter that attenuates high frequencies and passes low frequencies relatively unchanged is called a *low-pass filter*. An *ideal low-pass filter* would cut off all of the power above a prescribed frequency λ_1, and leave untouched the frequencies below that point. The squared gain function of such a filter is shown in Fig. 4.3.

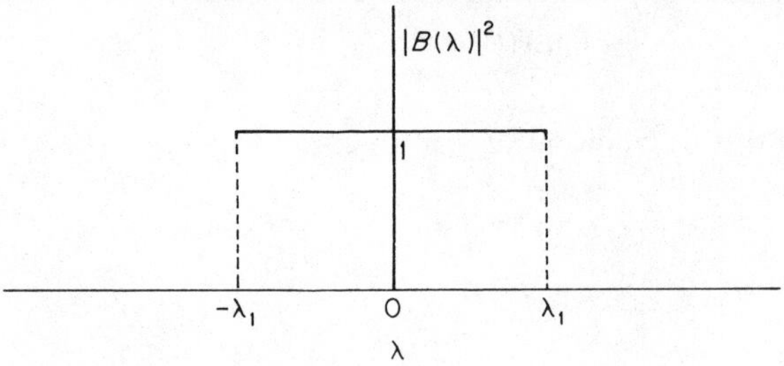

Fig. 4.3 *Gain function of ideal low-pass filter which passes frequencies* $|\lambda| \leq \lambda_1$.

The filter of Example 4.1 is a good illustration of a low-pass filter that can be (nearly) realized in practice. The recursive exponential filter to be discussed in Chapter 6 is the discrete-time equivalent of this filter with comparable transfer function characteristics. See Fig. 6.3 for the gain function of the exponential filter.

The attenuation of high-frequency power leads to time series with smoother, less erratic appearance. For this reason, low-pass filters are often called *smoothing filters*. An example in which the smoothing is a natural and desirable operation is the following:

Example 4.2 *Accumulated Processes and Their Aliased Spectra*

In many instances, discrete time series are obtained as the accumulation of "material" during equally spaced intervals of time. Thus, for example, the thicknesses of tree rings are the yearly accumulations of wood produced by the continuous growth process of the tree. The daily receipts of a business are the total of funds accumulated on a minute by minute basis during the business day.

A mathematical model of an accumulated series is the following. Let $X(t)$ be a weakly stationary time series which represents the instantaneous activity of the underlying process and let $Y(t)$ denote the accumulated process. Then, in a time interval of length du centered at $t = u$, the contribution to $Y(t)$ is $X(u)\,du$. If the length of time over which the accumulation takes place is T, then we can represent $Y(t)$ as a function of $X(t)$ by the expression

$$Y(t) = \int_{t-T}^{t} X(u)\,du.$$

Next, a discrete time series is formed by sampling $Y(t)$, most commonly at the sampling interval $\Delta t = T$. Thus, indirectly, the variations in the discrete process $Y_T(k) = Y(kT)$ are due to variations in the $X(t)$ process and this dependence can be easily understood by obtaining the transfer function of the transformation of $X(t)$ into $Y(t)$.

For simplicity, rescale $X(u)$ by the factor $1/T$. Then, by the change of variables $v = t - u$, the above representation can be written in the form

$$Y(t) = \frac{1}{T}\int_{0}^{T} X(t - v)\,dv.$$

This is a convolution filter with impulse response function

$$h(v) = \begin{cases} 1/T, & 0 \leq v \leq T, \\ 0, & \text{otherwise.} \end{cases}$$

The transfer function of this filter is

$$B(\lambda) = \frac{1}{T}\int_{0}^{T} e^{-i\lambda v}\,dv = e^{-i\lambda T/2}\,\frac{\sin(\lambda T/2)}{\lambda T/2}.$$

Consequently, the gain function is

$$|B(\lambda)| = \left|\frac{\sin(\lambda T/2)}{\lambda T/2}\right|.$$

A graph of the squared gain function is given in Fig. 4.4.

Now, since $\Delta t = T$, the Nyquist folding frequency is

$$\lambda_N = \pi/T.$$

Thus, from Fig. 4.4 it is seen that the power from frequencies $\pi/T < |\lambda| \leq 2\pi/T$ is folded back into the range $(-\lambda_N, \lambda_N]$ but essentially no power from other frequencies contributes to the aliasing problem. Moreover, the low-frequency part of the aliased spectrum will be virtually identical to the spectrum of $X(t)$ over the same range. Since it is generally the low end of the spectrum that is of

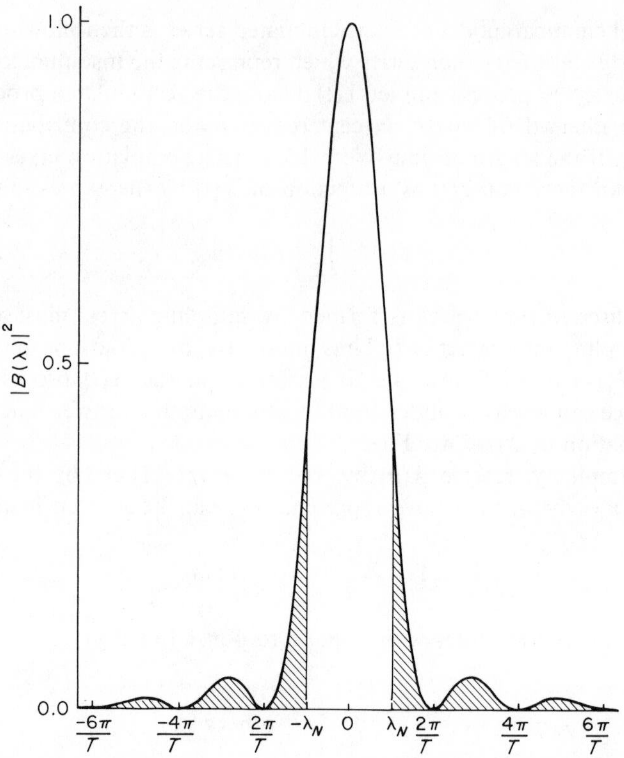

Fig. 4.4 *Squared gain function of the accumulation filter with aliased power range cross-hatched.*

interest and since it is characteristic of physical time series that the high-frequency power of the $X(t)$ series is significantly smaller in magnitude than the power in the lower frequencies, aliasing will generally not be a problem. Thus, for example, an 11-year period which might be discovered in (yearly) tree ring data could be expected to be due to an 11-year cycle in the growth mechanism of the tree rather than to its closest alias, an 11/12-year or 334^+-day cycle.

The following example illustrates the ideal version of a type of linear filter of practical interest.

Example 4.3 *The Ideal Amplifier*

Electronic amplifiers are complex and expensive pieces of equipment involving both linear and highly nonlinear components. The sum purpose of these devices (aside from suppression of extraneous noise) is to convert a weak signal into a strong version of the same signal with as little distortion as possible. The more closely the amplifier approaches the ideal of no distortion over a wide band of frequencies, the more complicated and expensive it must be.

As will be argued shortly, since amplifiers are constrained to operate in real time, they will necessarily shift the phases of the incoming signal to some extent. To avoid distortion, the ideal would be to shift all phases in such a manner that the frequency components of the output are displaced in time relative to those of the input by the same amount τ. In this way, the entire input would be simply displaced by τ time units. Thus, the mathematical model for a *realizable ideal amplifier* could be described by the expression

$$A\big(x(t)\big) = cx(t - \tau),$$

where the positive constant c is the amplification factor and τ is the time delay imposed on the input $x(t)$. However, this is easily seen to be a linear filter. Moreover, by applying the filter to $e^{i\lambda t}$, it can be checked that the transfer function is

$$B(\lambda) = ce^{-i\lambda\tau}.$$

Thus, the gain and phase functions are

$$|B(\lambda)| = c \qquad \text{and} \qquad \vartheta(\lambda) = -\tau\lambda.$$

That is, the response (gain function) of an ideal amplifier is "flat" and the phase shift is a linear function of λ.

Modern high-fidelity amplifiers come very close to achieving the ideal over rather wide frequency ranges. This has made possible the accurate amplification of signals previously recorded with low fidelity or not recorded at all. Thus, for example, the detailed analysis of such data as electroencephalograms, electrocardiograms, seismograms, and cosmic ray records has been made a reality in large part by the development of modern recording and amplifying equipment.

The Time Shift Parameter

Example 4.3 suggests the definition of a new parameter equivalent to the phase shift. Whereas $\vartheta(\lambda)$ measures the angular displacement of the output phase relative to the input at frequency λ, the parameter

$$\tau(\lambda) = \vartheta(\lambda)/\lambda, \qquad \lambda \neq 0, \tag{4.20}$$

which we will call the *time shift function*, measures the phase shift in time units. Thus, if $|B(\lambda)|$ and $\vartheta(\lambda)$ are the gain and phase shift of a linear filter L, respectively, we will have

$$L(e^{i\lambda t}) = |B(\lambda)|\, e^{i\lambda(t+\tau(\lambda))}.$$

Consequently, L displaces the time origin of the harmonic of frequency λ by $\tau(\lambda)$ time units. Note that since both $\vartheta(\lambda)$ and λ are odd functions, $\tau(\lambda)$ is an even function of λ.

In the example of the ideal amplifier,

$$\tau(\lambda) = -\tau, \qquad -\infty < \lambda < \infty.$$

That is, all harmonics were displaced back in time by the same amount. When this is not the case, *phase distortion* takes place. Thus, for example, the derivative, for which

$$\tau(\lambda) = \pi/2|\lambda|, \qquad \lambda \neq 0,$$

suffers greatly from phase distortion when $|\lambda|$ is small but less so as $|\lambda|$ becomes larger. Since the lower-frequency components are greatly attenuated relative to those with higher frequencies $[|B(\lambda)|^2 = |\lambda|^2]$, relatively little phase distortion would actually be observed in the output.

Symmetric Filters

A convolution filter for which $\tau(\lambda) = \vartheta(\lambda) = 0$ for all λ will necessarily have a symmetric (even) impulse response function and, for this reason, is called a *symmetric filter*.

To see why this is true, observe that $\vartheta(\lambda) = 0$ for all λ only if $B(\lambda)$ is real-valued. Moreover, this function is the Fourier transform of the impulse response function by (4.17). However, as is easily argued, the Fourier transform of a real-valued function can be real-valued only if the function is even. Note that since symmetric filters operate simultaneously on the past and future of the input series, they cannot be realizable unless they are instantaneous, i.e., unless $L(x(t)) = cx(t)$ for some constant c. It follows that noninstantaneous realizable filters will always have $\vartheta(\lambda) \neq 0$ for some (usually most) values of λ. This was noted in our last example. Symmetry of the filter is necessary but not sufficient for $\vartheta(\lambda) = 0$ for all λ. The condition a filter must satisfy to guarantee this property is discussed in Section A6.2.

4.3 COMBINING LINEAR FILTERS

Two important methods of combining linear filters to produce new filters are the following:

(i) *Linear combinations of filters: If L_1 and L_2 are linear filters with transfer functions $B_1(\lambda)$ and $B_2(\lambda)$, both matched with the input $x(t)$, then the transformation $L_3 = aL_1 + bL_2$, defined by*

$$L_3(x(t)) = aL_1(x(t)) + bL_2(x(t)),$$

is again a linear filter, matched with $x(t)$ for any pair of complex numbers a and b. Moreover, the transfer function of this filter is

$$B_3(\lambda) = aB_1(\lambda) + bB_2(\lambda).$$

(ii) *Sequential application of filters: If L_1 and L_2 are linear filters with transfer functions $B_1(\lambda)$ and $B_2(\lambda)$ such that L_1 is matched with $x(t)$ and L_2 is matched with $L_1(x(t))$, then the transformation $L_4 = L_2 L_1$, defined by*

$$L_4(x(t)) = L_2(L_1(x(t)),$$

is a linear filter matched with $x(t)$ with transfer function

$$B_4(\lambda) = B_2(\lambda)B_1(\lambda).$$

The inputs can be of any type for which a spectral representation is defined and a matching condition is satisfied. However, to simplify the ensuing discussion we will restrict attention to weakly stationary processes. The theory can be applied to processes in either continuous or discrete time by the appropriate interpretation of the transfer functions. This fact will be utilized in Chapter 6.

That the linear combination filter L_3 matches $X(t)$ if L_1 and L_2 do is a simple consequence of the Minkowski inequality (Section 1.3). The order in which the terms appear is quite immaterial. That is, $aL_1 + bL_2 = bL_2 + aL_1$. This is not the case with the sequential filter, however. Although the transfer function of L_4 is the product $B_1(\lambda)B_2(\lambda)$ and this product commutes, the order in which the matching conditions are to be satisfied does not. The matching conditions for the filter $L_2 L_1$ are

$$\int |B_1(\lambda)|^2 F(d\lambda) < \infty \qquad \text{and} \qquad \int |B_1(\lambda)B_2(\lambda)|^2 F(d\lambda) < \infty. \tag{4.21}$$

The second condition is a consequence of the fact that the spectral distribution of the process $L_1(X(t))$ is $|B_1(\lambda)|^2 F(d\lambda)$ and this time series is the input to L_2. It is possible for both of these conditions to be satisfied but for $\int |B_2(\lambda)|^2 F(d\lambda)$ to be infinite. Then, L_2 does not match $X(t)$ and the filter $L_1 L_2$ would not be defined. However, if, in addition to the above conditions,

$$\int |B_2(\lambda)|^2 F(d\lambda) < \infty, \tag{4.22}$$

then $L_4 = L_2 L_1 = L_1 L_2$.

The gain and phase functions of the sequential filter are related to those of L_1 and L_2 in a simple way;

$$|B_4(\lambda)| = |B_1(\lambda)|\,|B_2(\lambda)|, \tag{4.23}$$

$$\vartheta_4(\lambda) = \vartheta_1(\lambda) + \vartheta_2(\lambda). \tag{4.24}$$

Explicit expressions for the gain and phase shift functions of L_3 must be determined in each case from the real and imaginary components of $B_3(\lambda)$. No simple general expression in terms of the gain and phase shift functions of L_1 and L_2 exists. We now give some applications of these methods for combining filters.

Construction of High-Pass Filters from Low-Pass Filters

The *identity filter*, or *do-nothing filter*, as it is sometimes called, is defined by

$$I(X(t)) = X(t).$$

The transfer function is easily seen to be

$$B(\lambda) = 1.$$

This filter matches any input. Let L be any low-pass filter matched to the input $X(t)$. For convenience take L to be the ideal symmetric low-pass filter with transfer function

$$B(\lambda) = |B(\lambda)| = \begin{cases} 1, & |\lambda| \leq \lambda_1, \\ 0, & \text{otherwise.} \end{cases}$$

Now pass $X(t)$ through L and subtract the result from $X(t)$. Call this series $Y(t)$;

$$Y(t) = X(t) - L(X(t)).$$

These operations define the linear combination filter

$$L_3 = I - L$$

which has the transfer function

$$\begin{aligned} B_3(\lambda) &= 1 - B(\lambda) \\ &= \begin{cases} 0, & |\lambda| \leq \lambda_1, \\ 1, & \text{otherwise.} \end{cases} \end{aligned}$$

This is the ideal symmetric high-pass filter with transfer function graphed in Fig. 4.2 with cut-off frequency $\lambda_0 = \lambda_1$.

This example illustrates an important general rule for constructing high-pass filters from low-pass filters;

$$\text{high-pass}(X(t)) = X(t) - \text{low-pass}(X(t)).$$

The transfer function of the high-pass filter $I - L$ is always $1 - B(\lambda)$, where $B(\lambda)$ is the transfer function of the low-pass filter L. The gain and phase shift of $I - L$ can be obtained from the expressions

$$|1 - B(\lambda)| = [(1 - \text{Re}\, B(\lambda))^2 + (\text{Im}\, B(\lambda))^2]^{1/2},$$
$$\arg(1 - B(\lambda)) = \text{Arctan}[-\text{Im}\, B(\lambda)/(1 - \text{Re}\, B(\lambda))].$$

Band-Pass Filters

A high-pass and low-pass filter can be combined sequentially to produce a filter that removes or attenuates all but a band of frequencies from the input. Again, for illustrative purposes, we will consider the ideal symmetric high- and low-pass filters of Section 4.2. Let L_1 be the low-pass filter with transfer function

$$B_1(\lambda) = \begin{cases} 1, & |\lambda| \le \lambda_1, \\ 0, & \text{otherwise,} \end{cases}$$

and L_2 the high-pass filter with transfer function

$$B_2(\lambda) = \begin{cases} 0, & |\lambda| < \lambda_0, \\ 1, & \text{otherwise,} \end{cases}$$

where $0 < \lambda_0 < \lambda_1$. Since both filters match any input of finite power and $B_1(\lambda)B_2(\lambda) \le B_1(\lambda)$, both (4.21) and (4.22) are satisfied and the order in which the filters are applied is immaterial. Thus $L_4 = L_1 L_2 = L_2 L_1$ has transfer function

$$B_4(\lambda) = B_1(\lambda)B_2(\lambda) = \begin{cases} 1, & \lambda_0 \le |\lambda| \le \lambda_1, \\ 0, & \text{otherwise.} \end{cases}$$

This is simultaneously the transfer function and gain function of an ideal symmetric band-pass filter. See Fig. 4.5 for a graph of this function. The interval $[\lambda_0, \lambda_1]$ is called the *pass-band* of the filter.

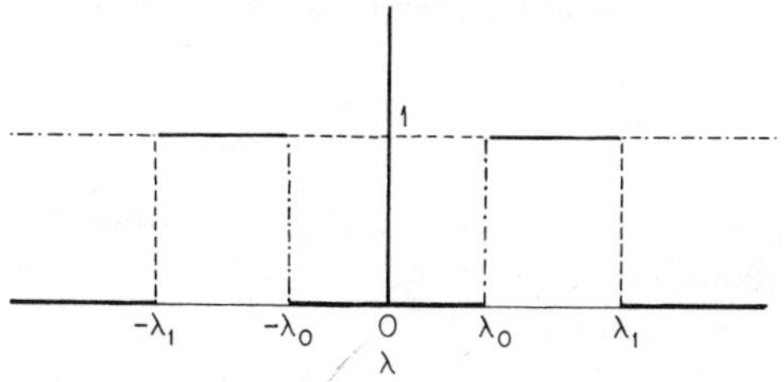

Fig. 4.5 *Graph of the transfer functions of an ideal symmetric band-pass filter and the high- and low-pass filters from which it is obtained:* – – –, $B_1(\lambda)$; – · –, $B_2(\lambda)$; —, $B_4(\lambda)$.

As a general principle, any properly overlapping high- and low-pass filters can be used to obtain a band-pass filter;

$$\begin{aligned}\text{band-pass}(X(t)) &= \text{high-pass}(\text{low-pass}(X(t)))\\ &= \text{low-pass}(\text{high-pass}(X(t))).\end{aligned}$$

Band-pass filters are used, for example, when the properties of the time series attributable to a prescribed range of frequencies are to be studied. One of the oldest methods of spectrum analysis, used primarily on analog data but with recent digital applications, is to estimate the power in a given frequency band Λ by filtering the time series with a band-pass filter having a pass-band as close as possible to Λ then measure the power of the filtered series. The filtering and power measurement operations are carried out electronically in the analog situation.

By taking a sequence of such filters with pass-bands covering the entire frequency range, an overall view of the spectrum can be obtained. For multichannel discrete time series data with a large number of channels this technique has also been used by employing digital band-pass filters. This method does not yield the degree of spectral resolution obtainable by most of the techniques we will discuss in Chapters 8 and 9. However, it has the advantage that large quantities of data can be processed with reasonable dispatch. Moreover, it permits real-time estimation of the spectrum which has advantages if the time series is nonstationary. For then, "local" estimates of the spectrum can be obtained as time progresses.

Notch Filters

By the operation

$$L(X(t)) = X(t) - \text{band-pass}(X(t)),$$

a *notch filter* is formed which deletes a band of frequencies from a time series. If the ideal symmetric band-pass filter given above were used, the transfer function of the corresponding ideal symmetric notch filter would be

$$B(\lambda) = \begin{cases} 0, & \lambda_0 \le |\lambda| \le \lambda_1, \\ 1, & \text{otherwise.} \end{cases}$$

If several band-pass filters $L_1, L_2, \ldots, L_n$ with nonoverlapping pass-bands $\Lambda_1, \Lambda_2, \ldots, \Lambda_n$ are applied to $X(t)$ one at a time and the filtered series subtracted from $X(t)$, the result,

$$L = I - (L_1 + L_2 + \cdots + L_n),$$

is a notch filter with *notches* $\Lambda_1, \Lambda_2, \ldots, \Lambda_n$. A graph of the squared gain function of such a filter based on ideal band-pass filters is given in Fig. 4.6.

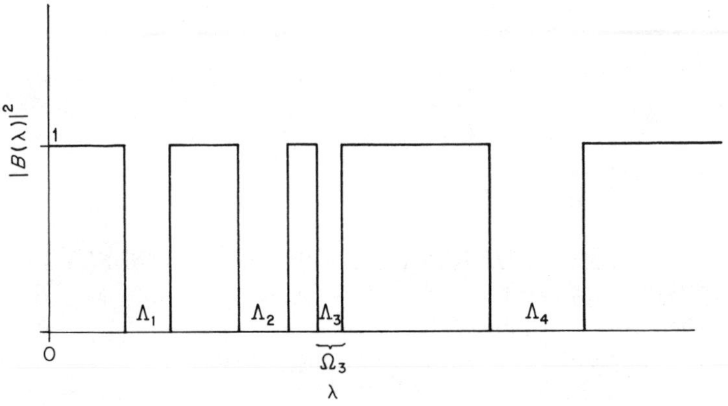

Fig. 4.6 *Squared gain function of an ideal notch filter with four notches.*

Since gain functions are always symmetric about $\lambda = 0$, we will, hereafter, graph them only for $\lambda \geq 0$.

In theory, notch filters can be used to eliminate lines from the spectrum. A narrow notch is simply located over each line. Then, since the output spectral function is related to that of the input by

$$p_{\text{out}}(\lambda) = |B(\lambda)|^2 p_{\text{in}}(\lambda),$$

any line occurring at a frequency λ for which $B(\lambda) = 0$ will be removed. This will also affect the continuous part of the spectrum, however, since

$$f_{\text{out}}(\lambda) = |B(\lambda)|^2 f_{\text{in}}(\lambda).$$

Realizable notch and band-pass filters will not have the sharp "cutoff" features of the ideal filters. It is not possible, for example, to construct a notch filter for which $|B(\lambda)| = 0$ for intervals of frequencies. Thus, spectral lines, which actually appear as narrow peaks in the spectrum for reasons which will be given in Section A8.2, can be largely but usually not entirely removed with notch filters.

It is interesting to reflect on the fact that *all of the above filters were constructed from low-pass filters by the two basic operations for combining linear filters*. Thus, with a single computer package of low-pass filters with various cutoff frequencies, a large variety of special purpose filters can be designed for processing time series. This important feature will be exploited in the design of digital filters in Chapter 6.

The sequential application of filters also admits the possibility of constructing filters by repeating the same linear filter several times. This topic is considered next.

Repeating Linear Filters

If a linear filter L with transfer function $B(\lambda)$ is applied n times to a weakly stationary time series $X(t)$ with spectral distribution $F(A)$, the resulting operator L^n will be a linear filter matched with $X(t)$ provided

$$\int |B(\lambda)|^{2k} F(d\lambda) < \infty \qquad \text{for} \quad k = 1, 2, \ldots, n.$$

The transfer function of L^n is then $B^n(\lambda)$, the nth power of $B(\lambda)$. Moreover, from (4.23) and (4.24), the gain and phase shift functions are $|B(\lambda)|^n$ and $n\vartheta(\lambda)$, where $\vartheta(\lambda) = \arg B(\lambda)$.

As an example of a family of low-pass filters generated by repeating a given low-pass filter, consider the filter L of Example 4.1 which has squared gain function

$$|B(\lambda)|^2 = 1/(a^2 + \lambda^2).$$

Then, taking $a = 1$, the squared gain of L^n is

$$|B(\lambda)|^{2n} = 1/(1 + \lambda^2)^n.$$

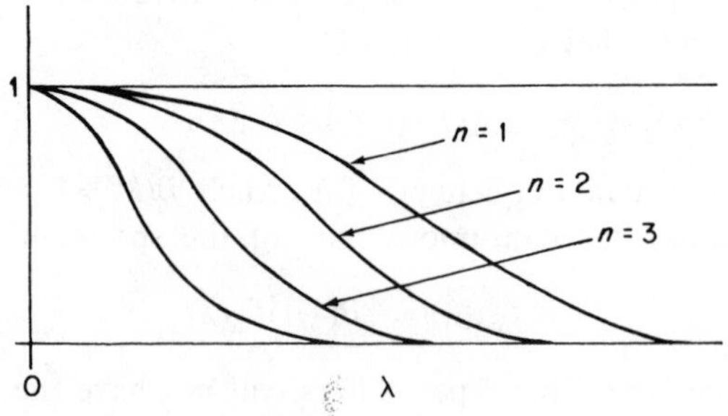

Fig. 4.7 *Squared gain functions of low-pass filters generated by repeating a given low-pass filter.*

The squared gain functions are plotted in Fig. 4.7. Note the decreasing length of the pass-bands as well as the increasing sharpness of the "cutoff" with increasing n. Repeated linear filters have other important applications as is illustrated by the following example.

Example 4.4 *Higher-Order Derivatives and Differential Operators*

It is convenient to change the notation for the derivative to conform to our filter notation. Thus hereafter D will denote

$$D(X(t)) = \frac{dX(t)}{dt}.$$

In Section 4.2, the transfer function for D was shown to be

$$B(\lambda) = i\lambda.$$

Now, D^2, the second derivative, is defined by $D^2(X(t)) = D(D(X(t)))$ and, in general, the nth derivative is the repeated filter D^n. Consequently, if

$$\int \lambda^{2k} F(d\lambda) < \infty \qquad \text{for} \quad k = 1, 2, \ldots, n,$$

the time series possesses n derivatives with transfer functions $(i\lambda)^k$, $k = 1, 2, \ldots, n$.

Now, an *nth order linear, constant coefficient differential operator* L is a linear combination of these derivatives

$$L = a_n D^n + a_{n-1} D^{n-1} + \cdots + a_1 D + a_0 I,$$

where the a_j's are real-valued coefficients and $a_n \neq 0$. This is a linear filter which matches an input $X(t)$ provided all indicated derivatives of $X(t)$ exist. The transfer function of L is

$$B(\lambda) = a_n(i\lambda)^n + a_{n-1}(i\lambda)^{n-1} + \cdots + a_1(i\lambda) + a_0 .$$

The *characteristic polynomial* of L is defined to be

$$P(z) = a_n z^n + a_{n-1} z^{n-1} + \cdots + a_1 z + a_0 ,$$

where z is a complex variable. It follows that the transfer function is the restriction of $P(z)$ to the imaginary axis,

$$B(\lambda) = P(i\lambda).$$

The characteristic polynomial plays an important role in determining the solutions of linear constant coefficient differential equations, where a solution is a function $x(t)$ which satisfies the equation

$$L(x(t)) = y(t)$$

over a specified range of t values for a given *forcing function* $y(t)$. We will discuss the solution of such equations when $y(t)$ is a weakly stationary stochastic process in the next section.

Combined and Repeated Convolution Filters

If L_1 and L_2 are convolution filters with impulse response functions $h_1(u)$ and $h_2(u)$ which are bounded and satisfy (4.18), then the operator $L = L_2 L_1 = L_1 L_2$ is again a convolution filter which matches any input with finite power. The impulse response function of L can be calculated as follows: Since

$$L_1(X(t)) = \int_{-\infty}^{\infty} h_1(v) X(t - v)\, dv,$$

we have

$$L(X(t)) = L_2(L_1(X(t))) = \int_{-\infty}^{\infty} h_2(u)\left[\int_{-\infty}^{\infty} h_1(v)X(t-u-v)\,dv\right] du.$$

By the change of variables $w = u + v$ this becomes

$$L(X(t)) = \int_{-\infty}^{\infty} h_2(u)\left[\int_{-\infty}^{\infty} h_1(w-u)X(t-w)\,dw\right]$$
$$= \int_{-\infty}^{\infty} h(w)X(t-w)\,dw,$$

where

$$h(w) = \int_{-\infty}^{\infty} h_2(u)h_1(w-u)\,du.$$

This expression is the impulse response function of L. The exchange of integrals is justifiable and the operations can be carried out in either order. It follows that

$$\int_{-\infty}^{\infty} h_2(u)h_1(w-u)\,du = \int_{-\infty}^{\infty} h_1(u)h_2(w-u)\,du.$$

Moreover, L is a *stable* operator, since condition (4.18) and the boundedness of the impulse response functions imply

$$\int_{-\infty}^{\infty} |h(w)|\,dw \le M \int_{-\infty}^{\infty} |h_2(u)|\,du < \infty,$$

where we have taken $|h_1(u)| \le M$. It is also easily seen that *if both L_1 and L_2 are realizable, then L is realizable.*

Condition (4.18) and the boundedness condition can be replaced by (4.19). If both $h_1(u)$ and $h_2(u)$ satisfy (4.19), then $h(u)$ is integrable by the Schwarz inequality. Thus, the convolution filter will match any input with finite power. If, as is often the case, $h(u)$ also satisfies (4.19), then the convolution filter will also match inputs with continuous, bounded spectra.

Since the impulse response function and transfer function are related by Fourier transformation, if $B_1(\lambda)B_2(\lambda)$ is integrable or $h(w)$ satisfies (4.19), we can obtain the impulse response function of $L = L_1L_2$ by the expression

$$h(w) = \frac{1}{2\pi}\int_{-\infty}^{\infty} B_1(\lambda)B_2(\lambda)e^{i\lambda w}\,d\lambda.$$

These results immediately generalize to repetitions of the same convolution operator. Thus, if L has impulse response function $h(u)$, L^n will have impulse response function $h^{(n)}(u)$ determined iteratively by the expression

$$h^{(k)}(u) = \int_{-\infty}^{\infty} h^{(k-1)}(v)h(u-v)\,dv$$
$$= \int_{-\infty}^{\infty} h(v)h^{(k-1)}(u-v)\,dv, \qquad k = 2, 3, \ldots, n,$$

where $h^{(1)}(u) = h(u)$, or directly by Fourier transformation from

$$h^{(n)}(u) = \frac{1}{2\pi}\int_{-\infty}^{\infty} B^n(\lambda)e^{i\lambda u}\,d\lambda$$

whenever $B^n(\lambda)$ is integrable or square integrable.

4.4 INVERTING LINEAR FILTERS

A commonly occurring question in time series analysis is the following: A linear filter L with transfer function $B(\lambda)$ has been applied to a time series $X(t)$, to which it was matched, resulting in an output $Y(t)$,

$$L(X(t)) = Y(t). \tag{4.25}$$

When is it possible to apply another linear transformation, say L^*, to $Y(t)$ and exactly reproduce $X(t)$? This question has a simple answer with important consequences. Again we restrict attention to weakly stationary time series.

Note that if

$$B(\lambda) \neq 0 \qquad \text{for all } \lambda,$$

then, taking L^* to be the filter with transfer function $B^*(\lambda) = 1/B(\lambda)$, it follows that

$$\int |B(\lambda)|^2 F_X(d\lambda) < \infty,$$

and

$$\int |B^*(\lambda)B(\lambda)|^2 F_X(d\lambda) = \int F_X(d\lambda) < \infty.$$

Consequently, the filter L^*L is well defined and is identical to the do-nothing filter I which has transfer function identically equal to 1. By (4.21) it follows that L^* is matched to $Y(t)$ and

$$X(t) = L^*L(X(t)) = L^*(Y(t)).$$

The spectrum of $X(t)$ can be recovered from that of $Y(t)$ by the expression

$$F_X(d\lambda) = (1/|B(\lambda)|^2)F_Y(d\lambda). \tag{4.26}$$

Thus, the invertibility of L follows immediately from the condition $B(\lambda) \neq 0$ *for all* λ.

If $B(\lambda) = 0$ on a set of frequencies A, then, necessarily, the output spectrum will have zero power on A, since

$$F_Y(A) = \int_A |B(\lambda)|^2 F_X(d\lambda).$$

Thus, any power the $X(t)$ process might have had for frequencies in A will be lost. Once lost, this power can never be recovered by an inverting filter, since a linear filter cannot create power or transfer it from another frequency range. However, if L^* is taken to be the linear filter with transfer function

$$B^*(\lambda) = \begin{cases} 1/B(\lambda), & \lambda \notin A, \\ 0, & \lambda \in A, \end{cases}$$

say, then $L^*(Y(t))$ will be well defined and the spectrum of $L^*(Y(t))$ will agree with that of $X(t)$ on A^c.

There is an important application of these results to the statistical estimation of spectra. As mentioned before, it is often necessary to prefilter a time series before estimating the spectrum. The prefiltering can be carried out by a (compound) linear filter L and the transfer function $B(\lambda)$ of L can be computed by the techniques described above. The spectrum of interest is that of the input $X(t)$. However, estimates will actually be obtained for the output spectrum. If $f_X(\lambda)$ and $f_Y(\lambda)$ denote the spectral density functions of input and output, and $\hat{f}_X(\lambda)$ and $\hat{f}_Y(\lambda)$ denote statistical estimates of these parameters, then it is reasonable to take

$$\hat{f}_X(\lambda) = (1/|B(\lambda)|^2)\hat{f}_Y(\lambda). \tag{4.27}$$

That is, relation (4.26) is used to define a procedure for correcting the estimates for *filter bias*. This method is quite successful except at frequencies for which $|B(\lambda)|^2$ is near zero. For such frequencies very small variations in $\hat{f}_Y(\lambda)$, due, for example, to noise or the inherent variability of the estimation process, are greatly magnified leading to essentially useless estimates.

A second question, similar to the one stated above, but not to be confused with it, can be posed as the following problem: Given a linear filter L with transfer function $B(\lambda)$ and a time series $X(t)$, find a linear filter L^* *which matches* $X(t)$ such that the time series

$$Y(t) = L^*(X(t)) \tag{4.28}$$

satisfies the equation

$$X(t) = L(Y(t)). \tag{4.29}$$

This is a stochastic functional equation which is to be solved for $Y(t)$. There can be a variety of solutions or no solution depending on the interval of time for which the equation is to be satisfied, the properties of L and of $X(t)$. We will be concerned here only with weakly stationary (*steady state*) solutions which are defined on the interval $(-\infty, \infty)$. We now indicate the conditions under which (4.28) will provide such a solution to the equation.

Note that there is only one function $B^*(\lambda)$ which can qualify as the transfer function of L^* and produce the result

$$L(Y(t)) = L(L^*(X(t))) = X(t).$$

This function is

$$B^*(\lambda) = 1/B(\lambda).$$

Thus, the *key condition for the existence of the solution to* (4.29) *is the matching condition for* L^* *and* $X(t)$;

$$\int_{-\infty}^{\infty} (1/|B(\lambda)|^2)F_X(d\lambda) < \infty. \tag{4.30}$$

For integrals, we assume that the convention $0/0 = 0$ holds. *Thus, this condition can be satisfied even when* $B(\lambda) = 0$ *on a set* A *provided it is also true that* $F_X(A) = 0$. If this condition is satisfied, the solution to (4.29) is given by

$$Y(t) = \int_{-\infty}^{\infty} (1/B(\lambda))e^{i\lambda t}Z_X(d\lambda). \tag{4.31}$$

To see this, suppose $A = \{\lambda\colon B(\lambda) = 0\}$. Then, since we have assumed that $F_X(A) = 0$, we will have $E|Z_X(d\lambda)|^2 = F_X(d\lambda) = 0$, thus $Z_X(d\lambda) = 0$ for all $\lambda \in A$. It follows that

$$Y(t) = \int_{A^c} (1/B(\lambda))e^{i\lambda t}Z_X(d\lambda).$$

Then, as before, L matches $Y(t)$ and

$$\begin{aligned} L(Y(t)) &= \int_{A^c} B(\lambda)(1/B(\lambda))e^{i\lambda t}Z_X(d\lambda) \\ &= \int_{A^c} e^{i\lambda t}Z_X(d\lambda) \\ &= \int_{-\infty}^{\infty} e^{i\lambda t}Z_X(d\lambda) = X(t). \end{aligned}$$

If $X(t)$ has continuous spectrum and the spectral density function $f_X(\lambda)$ is bounded, then (4.30) will be satisfied if the following condition holds;

$$\int_{-\infty}^{\infty} (1/|B(\lambda)|^2)\, d\lambda < \infty.$$

The solution can then be put in the form of a convolution integral, since $1/B(\lambda)$ has a square integrable Fourier transform $h(u)$;

$$\frac{1}{B(\lambda)} = \int_{-\infty}^{\infty} h(u)e^{-i\lambda u}\,du.$$

It follows that

$$\begin{aligned} Y(t) &= \int_{-\infty}^{\infty}\left(\int_{-\infty}^{\infty} h(u)e^{-i\lambda u}\,du\right)e^{i\lambda t}Z_X(d\lambda) \\ &= \int_{-\infty}^{\infty} h(u)\left(\int_{-\infty}^{\infty} e^{i\lambda(t-u)}Z_X(d\lambda)\right)du \\ &= \int_{-\infty}^{\infty} h(u)X(t-u)\,du. \end{aligned} \tag{4.32}$$

This yields an explicit time domain representation for the solution.

If $Y(t)$ is to depend only on the present and past of the $X(t)$ process, i.e., if the filter is to be *realizable*, additional conditions will have to be imposed on L. We will see an example of this shortly. Solution (4.31) is unique for every forcing function $X(t)$ satisfying (4.30) if $B(\lambda) \neq 0$ for all λ.

If $B(\lambda)$ is zero on a set A of positive measure, then many solutions can exist. For example, suppose that $X(t)$ has a continuous spectrum with spectral density $f_X(\lambda)$ and let $U(t)$ be any weakly stationary process with continuous spectrum for which the spectral density satisfies

$$f_U(\lambda) = 0 \qquad \text{on} \quad A^c.$$

By the Schwarz inequality, $|EZ_X(d\lambda)\overline{Z_U(d\lambda)}|^2 \le f_X(\lambda)f_U(\lambda)(d\lambda)^2 = 0$, since we must have $f_X(\lambda) = 0$ on A in order to satisfy (4.30). Thus, $U(t)$ and $X(t)$ are uncorrelated processes. Let

$$Y(t) = \int_{A^c} (1/B(\lambda))e^{i\lambda t}Z_X(d\lambda) + U(t).$$

This process has spectral density

$$f_Y(\lambda) = (1/|B(\lambda)|^2)f_X(\lambda) + f_U(\lambda)$$

and, since $|B(\lambda)|^2 f_U(\lambda) = 0$, we obtain

$$|B(\lambda)|^2 f_Y(\lambda) = f_X(\lambda).$$

It follows that $Y(t)$ matches L. Moreover, since $B(\lambda)Z_U(d\lambda) = 0$ for all λ,

$$\begin{aligned} L(Y(t)) &= \int_{A^c} e^{i\lambda t}Z_X(d\lambda) + \int_{-\infty}^{\infty} B(\lambda)e^{i\lambda t}Z_U(d\lambda) \\ &= \int_{-\infty}^{\infty} e^{i\lambda t}Z_X(d\lambda) = X(t). \end{aligned}$$

Thus, $Y(t)$ is a weakly stationary solution of (4.29). Needless to say, the important situation is the one in which the solution is unique and can be expressed as a convolution integral. We give an illustration of the application of these results in an important special case.

Example 4.5 *Linear Constant Coefficient Differential Equations*

In the notation of Example 4.4 an nth order, linear, constant coefficient differential equation with forcing function $X(t)$ can be written in the form

$$a_n D^n(Y(t)) + a_{n-1} D^{n-1}(Y(t)) + \cdots + a_1 D(Y(t)) + a_0 Y(t) = X(t). \quad (4.33)$$

We assume that $X(t)$ has continuous spectrum with bounded, nonzero spectral density function. In particular $X(t)$ could be taken to be the continuous-time white noise process $\varepsilon(t)$. The condition under which a unique weakly stationary solution $Y(t)$ exists is extremely simple and pleasant as we now show.

The *characteristic equation* of (4.33) is

$$P(z) = 0,$$

where $P(z)$ is the characteristic polynomial defined in Example 4.4. *A unique solution of the differential equation exists if none of the roots of the characteristic equation lie on the imaginary axis (i.e., have zero real parts). The solution can always be expressed as a convolution filter operating on $X(t)$,*

$$Y(t) = \int_{-\infty}^{\infty} h(u) X(t-u)\, du,$$

and the filter is always stable. Moreover, if all of the roots of the characteristic equation have negative real parts, the solution only depends on the past and present of the $X(t)$ process, i.e., the convolution filter is realizable.

The first two statements follow immediately from the fact that if $P(z) \neq 0$ for z on the imaginary axis, then the transfer function of the differential operator, $B(\lambda) = P(i\lambda)$, is bounded away from zero for all λ. Thus, $1/|B(\lambda)|^2$ is bounded from above and goes to zero as $|\lambda| \to \infty$ at least as fast as $1/\lambda^2$. It follows that

$$\int_{-\infty}^{\infty} (1/|B(\lambda)|^2)\, d\lambda < \infty,$$

and (4.31) and (4.32) apply. Note that if the real part of any root of $P(z) = 0$ is zero, then (4.30) cannot hold since we have assumed $f_X(\lambda)$ to be nonzero. Thus, the condition that the real parts of all roots be nonzero is necessary as well as sufficient for the existence of a solution of the differential equation.

The argument demonstrating the realizability and stability of the solution will be sketched for the second-order equation ($n = 2$). This derivation contains all of the salient features of the general case.

By dividing both sides of the differential equation by the coefficient of $D^2(Y(t))$, we obtain an (equivalent) equation of the form

$$D^2(Y(t)) + aD(Y(t)) + bY(t) = X(t)$$

with characteristic equation

$$P(z) = z^2 + az + b = 0.$$

Let $z_j = \alpha_j + i\beta_j$, $j = 1, 2$, denote the two roots of this equation. The characteristic polynomial can also be written in factored form as $P(z) = (z - z_1)(z - z_2)$.

Suppose, first, that $z_1 \neq z_2$. Then, by a partial fraction expansion,

$$\frac{1}{P(z)} = \frac{1}{(z - z_1)(z - z_2)} = \frac{A_1}{z - z_1} + \frac{A_2}{z - z_2}.$$

The constants are easily computed by putting both sides of this equation over a common denominator and equating the coefficients of like powers of z. In this case it can be shown that $A_2 = -A_1$. Then,

$$\frac{1}{B(\lambda)} = \frac{1}{P(i\lambda)} = \frac{A_1}{i(\lambda - \beta_1) - \alpha_1} + \frac{A_2}{i(\lambda - \beta_2) - \alpha_2}.$$

This is the transfer function $B^*(\lambda)$ of L^*.

Let

$$B_j(\lambda) = 1/(i(\lambda - \beta_j) - \alpha_j), \qquad j = 1, 2.$$

Then, the impulse response function $h(u)$ of L^* will be

$$h(u) = A_1 h_1(u) + A_2 h_2(u),$$

where

$$B_j(\lambda) = \int_{-\infty}^{\infty} h_j(u) e^{-i\lambda u}\, du.$$

Now, a simple integration establishes that

$$\begin{aligned} h_j(u) &= \begin{cases} e^{(\alpha_j + i\beta_j)u}, & u > 0, \\ 0, & u \leq 0, \end{cases} \quad \text{if} \quad \alpha_j < 0, \\ &= \begin{cases} 0, & u \geq 0, \\ -e^{(\alpha_j + i\beta_j)u}, & u < 0, \end{cases} \quad \text{if} \quad \alpha_j > 0. \end{aligned} \tag{4.34}$$

It is immediate from this and the expression for $h(u)$ that the filter will be realizable, i.e., $h(u) = 0$ for $u \leq 0$, if and only if the real parts α_1 and α_2 are negative. If one root has positive real part and the other negative real part (which can only happen if both roots are real), $h(u)$ will be nonzero for all u and the convolution filter will depend on both the future and the past of the process. If both real parts are positive, the filter will depend only on the future.

Note that although $h_1(u)$ and $h_2(u)$ are complex-valued, since $A_2 = -A_1$ and complex roots appear in conjugate pairs for algebraic equations with real coefficients, $h(u)$ will always be real-valued. Moreover, since $\alpha_j < 0$ if $u > 0$ and $\alpha_j > 0$ if $u < 0$ in (4.34), we always have

$$\int_{-\infty}^{\infty} |h(u)| \, du < \infty.$$

Thus, in all cases, the convolution filter is stable.

In the case of a repeated root ($z_1 = z_2$), necessarily $\beta_1 = \beta_2 = 0$. Then, $B^*(\lambda)$ is simply

$$1/B(\lambda) = 1/(i\lambda - \alpha_1)^2.$$

This is the transfer function of two applications of the convolution filter with impulse response function (4.34). Since this is a convolution filter which is realizable when the component filters are, it follows that condition $\alpha_1 < 0$ is again necessary and sufficient for realizability. The filter is also stable since the impulse response function (4.34) is both bounded and integrable.

For a general, nth order differential equation, the partial fraction expansions are longer and may contain higher powers of the terms $B_j(\lambda)$, but otherwise the argument is the same. The impulse response function of the convolution filter will always be composed of linear combinations of the functions $h_j(u)$ given by (4.34) and higher-order convolutions $h_j^{(k)}(u)$ as defined in the last section. Thus, it is possible to obtain an explicit time-domain solution to any linear, constant coefficient differential equation excited by a weakly stationary forcing function which satisfies the conditions given above.

If the forcing function is a continuous-time white noise process and the zeros of $P(z)$ have negative real parts, the resulting weakly stationary process is called a *continuous-time autoregression* for reasons that will become clear in Chapter 7. These processes depend on only a finite number of parameters and have been the subject of a number of interesting applications of spline functions to time series. An expository paper which discusses these applications and provides a good bibliography is by Davis (1972).

4.5 NONSTATIONARY PROCESSES GENERATED BY TIME VARYING LINEAR FILTERS

Two important properties of models for real phenomena are that they should be both broadly applicable and computationally tractable. We have discussed one important nonstationary model which is both easy to handle mathematically and which fits a large number of physical problems; the signal-plus-noise process with weakly stationary noise and (more or less) arbitrary signal. A second type of model which also has both of these virtues is obtained as follows: Recall that a linear, time invariant transformation is determined by a transfer function $B(\lambda)$. The properties of the filter could be made to change with time if the transfer function were allowed to depend on the time parameter t as well as on λ. If $B(\lambda, t)$ denotes the transfer function of the time varying filter and if $X(t)$ is a matched weakly stationary stochastic process with random spectral measure $Z(d\lambda)$, then the output of the filter would be

$$Y(t) = \int B(\lambda, t)e^{i\lambda t}Z(d\lambda). \tag{4.35}$$

The $Y(t)$ process is nonstationary, and an easy computation yields

$$\operatorname{Cov}(Y(t), Y(s)) = \int B(\lambda, t)\overline{B(\lambda, s)}e^{i\lambda(t-s)}F(d\lambda),$$

where $F(A)$ is the spectral distribution of the $X(t)$ process. Thus, $EY(t) = 0$ and the variance of $Y(t)$ is

$$E(Y^2(t)) = \int |B(\lambda, t)|^2 F(d\lambda). \tag{4.36}$$

Now, a stochastic process $U(t)$ is said to be a *second-order process* if $E(U^2(t)) < \infty$ for all t. It follows that the matching condition which makes $Y(t)$ a second-order process is the finiteness of the integral (4.36) for all t.

Example 4.6 *Linear Differential Equations with Time-Dependent Coefficients*
If the coefficients in the differential operator of (4.33) are allowed to depend on t, we obtain the equation

$$a_n(t)D^n(Y(t)) + a_{n-1}(t))D^{n-1}(Y(t)) + \cdots + a_1(t)D(Y(t)) + a_0(t)Y(t) = X(t),$$

where we assume that the forcing function $X(t)$ remains as before. The solution $Y(t)$ will now be a nonstationary process of the type defined by (4.35). In fact, by analogy with the development in the last section, the time varying transfer function is

$$B(\lambda, t) = [a_n(t)(i\lambda)^n + \cdots + a_1(t)(i\lambda) + a_0(t)]^{-1}.$$

If the zeros of the characteristic equation

$$a_n(t)z^n + a_{n-1}(t)z^{n-1} + \cdots + a_1(t)z + a_0(t) = 0$$

are bounded away from the imaginary axis, the theory of the last section can be carried over to obtain an explicit time domain representation of $Y(t)$ in terms of $X(t)$. This will no longer be a simple convolution but rather will have the form

$$Y(t) = \int_{-\infty}^{\infty} h(u, t)X(t-u)\, du,$$

where

$$B(\lambda, t) = \int_{-\infty}^{\infty} h(u, t)e^{-i\lambda u}\, du \qquad \text{for each } t.$$

The partial fraction expansion method can again be used to obtain $h(u, t)$ in terms of the roots of the characteristic equation. The stability and realizability of the operator can be described in terms of these roots or, alternatively, in terms of conditions on the coefficients.

Differential equations of this type arise in a number of physical situations. For example, the characteristics of a simple structure subjected to random vibrations can change due to fatigue damage. The response of the system would be described by such a differential equation where the time varying coefficients depend on the "stress history" of the structure.

Example 4.7 *A Model for Stochastic Transients*

A model for the (potential) accelerograms of strong motion earthquakes used in earthquake engineering [see Wirshing and Yao (1971)] is the following: A "broad-band" stochastic process $X(t)$, for example, a white noise process, is passed through a band-pass filter with transfer function $B(\lambda)$. The resulting time series is then tapered by a weight function $G(t)$ for which $\int_{-\infty}^{\infty} G^2(t)\, dt < \infty$. A tapering function used in the above reference is

$$G(t) = \begin{cases} (e^{-\alpha t} - e^{-\beta t}), & t \geq 0, \\ 0, & \text{otherwise,} \end{cases}$$

where $0 < \alpha < \beta$. By proper selection of α and β and the pass-band of $B(\lambda)$, the sample functions of the process can be made to closely resemble actual accelerogram records. The output of the filtering and tapering operations is

$$Y(t) = G(t) \int B(\lambda)e^{i\lambda t} Z(d\lambda),$$

where $Z(d\lambda)$ is the spectral measure of the input process. This nonstationary process is of the above form with

$$B(\lambda, t) = G(t)B(\lambda)$$

and the matching condition is satisfied since

$$\int_{-\infty}^{\infty} |B(\lambda, t)|^2 F(d\lambda) = G^2(t) \int_{-\infty}^{\infty} |B(\lambda)|^2 F(d\lambda) < \infty.$$

Moreover, it can be argued that

$$E\int_{-\infty}^{\infty} Y^2(t)\,dt = \int_{-\infty}^{\infty} EY^2(t)\,dt$$
$$= \int_{-\infty}^{\infty} G^2(t)\,dt \int_{-\infty}^{\infty} |B(\lambda)|^2 F(d\lambda) < \infty.$$

It follows that the energy of the output process,

$$\int_{-\infty}^{\infty} Y^2(t)\,dt,$$

is a well-defined random variable with finite mean. Thus, $Y(t)$ can be viewed as a stochastic process with finite energy and, as such, provides a good model for many kinds of transient random phenomena. Bounds on level crossing probabilities for Gaussian processes of this type are given by Koopmans *et al.* (1973) in a special case and by Koopmans and Qualls (1972) in the general case. Several examples of models covered by the general representation (4.35) are given by Granger and Hatanaka (1964). Other work on models of this variety has been carried out by Priestley (1965).

APPENDIX TO CHAPTER 4

A4.1 Linear Filters for Weakly Stationary Processes

The spectral representation theorem given in Chapter 2 and its appendix provides the basis for a mathematically satisfactory discussion of linear filters for weakly stationary processes. Let $X(t)$, $-\infty < t < \infty$, be a zero-mean, weakly stationary process with spectral distribution F and random spectral measure Z. Recall that the mapping

$$g(\lambda) \leftrightarrow \int g(\lambda) Z(d\lambda)$$

is an inner product preserving isomorphism of $L_2(F)$ with $\mathcal{M}^X$, the linear subspace generated by the process.

We define a family of linear transformations U_t, $-\infty < t < \infty$, as follows: First, for fixed t, U_t is defined on the generators of $\mathcal{M}^X$ by

$$U_t\big(X(s)\big) = X(s+t), \qquad -\infty < s < \infty.$$

Note that U_t preserves inner products on this set of elements, since

$$\langle U_t\big(X(s)\big), U_t\big(X(u)\big)\rangle = \langle X(s+t), X(u+t)\rangle = C(s-u)$$
$$= \langle X(s), X(u)\rangle,$$

where $C(\tau)$ is the autocovariance function. This property is preserved in the following extensions: U_t is extended by linearity to finite linear combinations,

$$U_t\left(\sum_j \alpha_j X(t_j)\right) = \sum_j \alpha_j\, U_t(X(t_j)),$$

and by continuity to limits of Cauchy sequences of linear combinations,

$$U_t\left(\lim_n X_n\right) = \lim_n U_t(X_n).$$

As a result, U_t is a well-defined, inner product preserving, linear transformation on $\mathcal{M}^X$ to $\mathcal{M}^X$. Although this will not be used here U_t, $-\infty < t < \infty$ is a group of unitary transformations under the operation of composition; $U_t \cdot U_s = U_{t+s}$, where U_0 is the identity transformation and $U_t^{-1} = U_{-t}$. This fact has important consequences in the study of the mathematical structure of weakly stationary processes.

Now, a *linear filter* is defined to be a linear transformation L with domain $\mathcal{D}(L) \subset \mathcal{M}^X$ and values in $\mathcal{M}^X$ with the property

$$LU_t = U_t L \qquad \text{for} \quad -\infty < t < \infty.$$

This is the time invariance property.

We will now sketch a proof of the following basic characterization of linear filters: *Linear filters are completely determined by their values at the random variable $X(0)$. The class of linear filters is in one-to-one correspondence with $L_2(F)$. The correspondence is $L \leftrightarrow B(\lambda)$ if*

$$L(X(0)) = \int B(\lambda) Z(d\lambda), \tag{A4.1}$$

where the function $B(\lambda) \in L_2(F)$ is the transfer function of the filter.

The action of L on $\mathcal{M}^X$ can be described as follows: *Let $U \in \mathcal{M}^X$ and let $U \leftrightarrow g(\lambda)$ for $g \in L_2(F)$. Now, $U \in \mathcal{D}(L)$ if, in addition to the condition $\int |g(\lambda)|^2 F(d\lambda) < \infty$, we have*

$$\int |g(\lambda)|^2 |B(\lambda)|^2 F(d\lambda) < \infty.$$

Then

$$L(U) = \int g(\lambda) B(\lambda) Z(d\lambda). \tag{A4.2}$$

In particular, since $X(t) \leftrightarrow e^{i\lambda t}$, we have

$$L(X(t)) = \int e^{i\lambda t} B(\lambda) Z(d\lambda).$$

We indicate briefly how (A4.1) leads to (A4.2). Since $B(\lambda) \in L_2(F)$, $B(\lambda) = \lim_n B_n(\lambda)$, where $B_n(\lambda)$ is a Cauchy sequence of finite linear combinations of complex exponentials,

$$B_n(\lambda) = \sum_{j \in J_n} a_{n,j} e^{-i\lambda t_{n,j}}.$$

Thus,

$$\begin{aligned} L(X(0)) = \lim_n \int B_n(\lambda) Z(d\lambda) &= \lim_n \sum_{j \in J_n} a_{n,j} \int e^{-i\lambda t_{n,j}} Z(d\lambda) \\ &= \lim_n \sum_{j \in J_n} a_{n,j} X(-t_{n,j}), \end{aligned}$$

by the spectral representation of the process. However, time invariance and the properties of U_t imply

$$\begin{aligned} L(X(t)) = LU_t(X(0)) = U_t L(X(0)) & \\ &= \lim_n \sum_{j \in J_n} a_{n,j} U_t(X(-t_{n,j})) \\ &= \lim_n \sum_{j \in J_n} a_{n,j} X(t - t_{n,j}). \end{aligned} \tag{A4.3}$$

Again the spectral representation yields

$$\begin{aligned} L(X(t)) &= \lim_n \sum_{j \in J_n} a_{n,j} \int e^{i\lambda(t - t_{n,j})} Z(d\lambda) \\ &= \lim_n \int e^{i\lambda t} B_n(\lambda) Z(d\lambda) \\ &= \int e^{i\lambda t} B(\lambda) Z(d\lambda). \end{aligned}$$

Now, suppose that $U = \int g(z) Z(d\lambda)$ with

$$\int |g(\lambda)|^2 F(d\lambda) < \infty,$$

and assume in addition that

$$\int |g(\lambda)|^2 |B(\lambda)|^2 F(d\lambda) < \infty. \tag{A4.4}$$

That is, both $g(\lambda)$ and $g(\lambda)B(\lambda)$ are in $L_2(F)$. Then it can be shown [e.g., Koopmans (1964a)] that $g(\lambda) = \lim g_n(\lambda)$, where $g_n(\lambda)$ is a sequence of finite linear combinations of complex exponentials,

$$g_n(\lambda) = \sum_{k \in K_n} b_{n,k} e^{i\lambda t_{n,k}},$$

such that both $g_n(\lambda)$ and $g_n(\lambda)B(\lambda)$ are Cauchy sequences in $L_2(F)$. It follows that

$$U = \lim_n \int g_n(\lambda)Z(d\lambda) = \lim_n \sum_{k \in K_n} b_{n,k} X(t_{n,k}).$$

Then, by the linearity and continuity properties of L we have

$$\begin{aligned} L(U) &= \lim_n \sum_{k \in K_n} b_{n,k} L(X(t_{n,k})) \\ &= \lim_n \sum_{k \in K_n} b_{n,k} \int e^{i\lambda t_{n,k}} B(\lambda)Z(d\lambda) \\ &= \lim_n \int g_n(\lambda)B(\lambda)Z(d\lambda) \\ &= \int g(\lambda)B(\lambda)Z(d\lambda). \end{aligned}$$

It follows that (A4.4) is the condition that $U \in \mathscr{D}(L)$.

Now, starting with any $B(\lambda) \in L_2(F)$ and defining $L(X(0))$ by (A4.1), a linear filter can be constructed by this process. Moreover, every linear filter (with respect to the given process) is of this type. Consequently, the class of linear filters which match this process is in one-to-one correspondence with $L_2(F)$.

In theory, then, linear filters can be constructed to perform a great variety of operations on time series. It is only necessary to specify the desired transfer function and the filter is completely determined. In practice, only restricted classes of filters are used. The convolution filters and linear combinations of derivatives predominate. In general these filters constitute a rather small subclass of the collection of possible filters, but, fortunately, most of the important practical filters are in this class or can be closely approximated by members of the class. For example, most filters of interest have bounded transfer functions $|B(\lambda)| \le M$. The various ideal filters given in the text have this property. If $X(t)$ is a process in continuous time which has a continuous spectrum with bounded spectral density, then, as was seen in the text, $\mathscr{L}_2 \subset L_2(F)$, where $\mathscr{L}_2$ is the class of functions $g(\lambda)$ such that $\int |g(\lambda)|^2 \, d\lambda < \infty$. However, every element of $\mathscr{L}_2$ corresponds to a convolution filter, since if $D(\lambda) \in \mathscr{L}_2$, then

$$D(\lambda) = \int e^{-i\lambda u} h(u) \, du,$$

with $\int h^2(u) \, du < \infty$. Thus, if $L \leftrightarrow D(\lambda)$,

$$\begin{aligned} L(X(t)) &= \int e^{it\lambda} D(\lambda)Z(d\lambda) \\ &= \int h(u)X(t-u) \, du. \end{aligned}$$

Now, if $B(\lambda)$ is a bounded transfer function corresponding to a filter L, taking

$$B_K(\lambda) = \begin{cases} B(\lambda), & |\lambda| \le K, \\ 0, & |\lambda| > K, \end{cases}$$

K can be made large enough so that

$$\int |B(\lambda) - B_K(\lambda)|^2 F(d\lambda) < \varepsilon$$

for any prespecified $\varepsilon > 0$. However, $B_K(\lambda) \in \mathscr{L}_2$ and the corresponding convolution filter

$$L_K(X(t)) = \int h_K(u) X(t-u)\, du$$

then has the property

$$E(L_K(X(t)) - L(X(t)))^2 < \varepsilon \qquad \text{for} \quad \text{all } t.$$

Consequently, in this sense, convolution filters can be made to approximate a large class of filters of practical interest when the input is a weakly stationary process. This accounts for the widespread use of convolution filters in practice. As we will see in Chapter 6, the discrete analogs of convolution filters play an even larger role in filtering discrete-time series.

CHAPTER

5

Multivariate Spectral Models and Their Applications

5.1 INTRODUCTION

It is more the rule than the exception that real, time-varying physical processes require more than one measurement to adequately describe their behavior. Thus, the position or state of the process at each instant of time will be represented by a vector of time-dependent measurements

$$\mathbf{x}(t) = \begin{pmatrix} x_1(t) \\ x_2(t) \\ \vdots \\ x_p(t) \end{pmatrix},$$

called a *multivariate* (*vector*, *multidimensional*) *time series*. We will consider only those multivariate time series models for which the components are univariate time series, either all in continuous time or all in discrete time, possessing power spectra. As in Chapter 2, will will deal most extensively with stochastic models for which the components are stationary processes.

These models apply quite well to a variety of real phenomena. In geophysics they are used, for example, to describe wind velocities (two components of velocity at each of, say, n levels leads to a $2n$-dimensional series), the oscillation of the earth at a given location (the vertical and two components of horizontal motion produce a three-dimensional time series) and sea state over a given region (sea heights at p locations in the region generate a p-dimensional series). Economic systems characteristically require several descriptive variables such as price, available supply, and demand among others. Under stable conditions, the "stationary" models are reasonably accurate. Even when conditions are not stable, these models provide useful results over restricted time periods. Early applications of multivariate time series analysis were made in engineering, particularly in the areas of communication and control theory. Since then, applications of these methods based on the models of this chapter have pervaded almost all of the physical and engineering sciences. A bibliography of applications of both the univariate and multivariate theory to the physical sciences and engineering is given by Tukey (1959).

More recently, these models have been applied to problems in the social, biological, and medical sciences as quantitative measurement techniques have been developed. Time series with large numbers of components are common in these areas. For example, the recording of EEG data for the study of brain function [see Walter *et al.* (1966)] often requires in excess of ten recording channels and results in a time series of as many dimensions. The future of time series analysis in these disciplines appears especially promising.

Time series models can be constructed with varying degrees of complexity depending on the purpose to which they are to be put and the knowledge available about the physical mechanism generating the data. At the purely descriptive level, a simple model can be fitted to the data much as polynomials are fitted to regression curves in statistics. The object is to fit the model to the data as closely as possible without trying to "understand" the underlying generating mechanism. For reasons that will be discussed in Chapter 7, the finite-parameter models defined therein and their multivariate counterparts are the ones most widely used for this purpose. Many important applications of time series analysis can be effectively made at this level. This is amply demonstrated by Box and Jenkins (1970), for example.

However, one of the more important uses of time series analysis is to improve our understanding, either qualitatively or quantitatively, of various properties of the generating mechanism. For this purpose, it is convenient to have the natural parameters of the time series model related in a simple way to the physical characteristics of the mechanism—or more precisely—to the parameters of a mathematical model of the mechanism. The spectra of the models to be discussed in this chapter will be seen to have this property in many applications. Since spectra can be readily estimated by the methods to be covered in Chapter 8, the validation of physical models by comparing predicted and measured spectra and the estimation of model parameters can be readily carried out.

We will see that by a simple application of the rules for operating with expectations summarized in Section 1.4, the spectra of time series derived from other time series via linear filters can be readily calculated. In this way a large and flexible class of models can be generated which has a number of important applications and for which the parameters are easily computed. This will be illustrated by means of applications many of which have been taken from the time series literature.

5.2 THE SPECTRUM OF A MULTIVARIATE TIME SERIES—WIENER THEORY

Let

$$\mathbf{x}(t) = \begin{pmatrix} x_1(t) \\ x_2(t) \\ \vdots \\ x_p(t) \end{pmatrix}, \qquad -\infty < t < \infty,$$

be a vector of real-valued functions for which all limits of the form

$$C_{j,k}(\tau) = \lim_{T\to\infty} \frac{1}{2T} \int_{-T}^{T} x_j(t+\tau)x_k(t)\,dt \tag{5.1}$$

exist for $1 \leq j, k \leq p$, and $-\infty < \tau < \infty$. Although we treat the continuous time parameter case in this definition, all of the theory applies to discrete time series as well with the modifications spelled out in Chapter 3.

The functions $C_{j,j}(\tau)$ are the autocovariances of the time series $x_j(t)$ defined in Section 2.2. For $j \neq k$, $C_{j,k}(\tau)$ is called the *cross-covariance function* of $x_j(t)$ and $x_k(t)$. An application of the Schwarz inequality yields

$$|C_{j,k}(\tau)| \leq (C_{j,j}(0)C_{k,k}(0))^{1/2}, \qquad -\infty < \tau < \infty. \tag{5.2}$$

Consequently, all covariance functions are bounded. Moreover, by a change of variables it is easily seen that

$$C_{j,k}(-\tau) = C_{k,j}(\tau). \tag{5.3}$$

Thus, although the autocovariances are even functions, as was previously noted in Section 2.2, the cross-covariance functions are not. This has an important effect on the spectral representation of the cross-covariances.

For each $j \neq k$ there exists a (unique) measure $F_{j,k}(A)$, called the *cross-spectral distribution* of $x_j(t)$ and $x_k(t)$, such that

$$C_{j,k}(\tau) = \int e^{i\lambda\tau} F_{j,k}(d\lambda). \tag{5.4}$$

In order to deal with the discrete and continuous time parameter cases together, we will not explicitly specify the frequency range in expressions involving the frequency variable. Thus, the limits of integration are omitted in (5.4) with the understanding that they are $-\infty$ and ∞ in the continuous-time case and $-\pi$ and π for discrete time.

As indicated in Section 2.2, in order for the Fourier transform of a function to be real-valued, it is necessary that the function be even. Since the cross-covariance function fails in this respect, *the cross-spectral distribution is generally complex-valued* (*with nonvanishing imaginary part*). In this it differs from the spectral distributions $F_{j,j}(A)$.

A partial symmetry results from (5.3); namely,

$$F_{j,k}(-A) = F_{k,j}(A). \tag{5.5}$$

(Recall that $-A = \{-\lambda : \lambda \in A\}$.) To see this, note that

$$C_{j,k}(-\tau) = \int e^{i\lambda(-\tau)} F_{j,k}(d\lambda) = \int e^{i\mu\tau} F_{j,k}(-d\mu)$$

by the change of variables $\mu = -\lambda$. However, by (5.3) this is the spectral representation of $C_{k,j}(\tau)$. The uniqueness of the spectral distribution then yields (5.5).

Similarly, it can be shown that

$$F_{j,k}(A) = \overline{F_{k,j}(A)}. \tag{5.6}$$

This and other properties of the spectral distribution are easily demonstrated using the stochastic model of the next section. Consequently, we will defer the statement of these results until later.

The matrix representation of the auto- and cross-spectral distributions is quite useful. It is

$$\mathbf{F}(A) = \begin{bmatrix} F_{1,1}(A) & F_{1,2}(A) & \cdots & F_{1,p}(A) \\ F_{2,1}(A) & F_{2,2}(A) & \cdots & F_{2,p}(A) \\ \vdots & \vdots & & \vdots \\ F_{p,1}(A) & F_{p,2}(A) & \cdots & F_{p,p}(A) \end{bmatrix}. \tag{5.7}$$

Hereafter, we will denote such arrays more briefly by using the symbol $[a_{j,k}]$ to represent the matrix with element $a_{j,k}$ in the jth row and kth column. Relation (5.6) indicates that $\mathbf{F}(A)$ *is a Hermitian matrix*.

In the models we will consider, the spectral distribution matrix (5.7) or its equivalent, the matrix of auto- and cross-covariances,

$$\mathbf{C}(\tau) = [C_{j,k}(\tau)],$$

constitutes the complete set of parameters for the multivariate time series $\mathbf{x}(t)$. Consequently, the only new elements presented by the multivariate theory are the relationships between each pair of time series measured by the cross-spectra or the cross-covariances.

As we will see, this restriction to what are called *second-order parameters* still allows a rich and useful theory of interrelationships to be developed which very closely parallels the multivariate correlation theory of statistics. This will be discussed in Section 5.6.

The decomposition of $\mathbf{F}(A)$ into discrete and continuous components is accomplished as in the univariate case. The *spectral functions* are

$$p_{j,k}(\lambda) = F_{j,k}(\{\lambda\}) \tag{5.8}$$

and the *discrete spectral distributions* are then

$$F_{j,k}^{(d)}(A) = \sum_{\lambda_l \in A} p_{j,k}(\lambda_l) \tag{5.9}$$

where $\lambda_1, \lambda_2, \ldots$ are the frequencies for which $p_{j,j}(\lambda)\, p_{k,k}(\lambda) > 0$ for at least one pair of indices j, k. In matrix notation,

$$\mathbf{p}(\lambda) = [p_{j,k}(\lambda)]$$

and

$$\mathbf{F}^{(d)}(A) = [F_{j,k}^{(d)}(A)]. \tag{5.10}$$

As in Section 2.2, when the spectrum is discrete, the cross-spectral functions can be obtained from the cross-covariances by the expression

$$p_{j,k}(\lambda) = \lim_{T\to\infty} \frac{1}{2T} \int_{-T}^{T} e^{-i\lambda\tau} C_{j,k}(\tau)\, d\tau$$

in the continuous-time case and by the corresponding limit

$$p_{j,k}(\lambda) = \lim_{L\to\infty} \frac{1}{2L+1} \sum_{\tau=-L}^{L} e^{-i\lambda\tau} C_{j,k}(\tau)$$

for discrete time. The *spectral distribution functions* are

$$F_{j,k}(\lambda) = F_{j,k}((-\infty, \lambda]).$$

The *spectral density functions* are then the derivatives of these distribution functions;

$$f_{j,k}(\lambda) = \frac{dF_{j,k}(\lambda)}{d\lambda} \tag{5.11}$$

and the continuous component of $\mathbf{F}(A)$ is

$$\mathbf{F}^{(c)}(A) = [F_{j,k}^{(c)}(A)], \tag{5.12}$$

where

$$F_{j,k}^{(c)}(A) = \int_A f_{j,k}(\lambda)\, d\lambda. \tag{5.13}$$

When $\int_{-\infty}^{\infty} |C_{j,k}(\tau)|\, d\tau < \infty$ or $\int_{-\infty}^{\infty} |C_{j,k}(\tau)|^2 d\tau < \infty$ in the case of continuous time, the spectral densities can be obtained by Fourier transformation (Example 1.4);

$$f_{j,k}(\lambda) = \frac{1}{2\pi} \int_{-\infty}^{\infty} e^{-i\lambda\tau} C_{j,k}(\tau)\, d\tau. \tag{5.14}$$

In discrete time, if $\sum_{-\infty}^{\infty} |C_{j,k}(\tau)|^2 < \infty$, then

$$f_{j,k}(\lambda) = \frac{1}{2\pi} \sum_{\tau=-\infty}^{\infty} e^{-i\lambda\tau} C_{j,k}(\tau). \tag{5.15}$$

(See Example 1.2.)

An important (multidimensional) parameter of the process is the spectral density matrix

$$\mathbf{f}(\lambda) = [f_{j,k}(\lambda)].$$

The standard measures of association will be defined only when the time series has a continuous spectrum and, thus, will be functions of the elements of $\mathbf{f}(\lambda)$. For this reason, the elements of this matrix will be the objects of interest for the statistical procedures to be derived in Chapter 8.

As in the univariate case, it should be stressed that the multivariate spectrum is defined for a large class of time series models both stochastic and nonstochastic and the weakly stationary processes form a rather small subclass of these models. Consequently, the idea of a spectrum and the various measures of association among time series to be derived in this chapter are much more widely applicable than one would surmise from our rather disproportionate coverage of weakly stationary processes. However, from the viewpoint of constructing probability models which have the general spectral structure of the Wiener theory, the weakly stationary processes play a central role. We consider them next.

5.3 MULTIVARIATE WEAKLY STATIONARY STOCHASTIC PROCESSES

Let $\{X_j(t);\ j = 1, 2, \ldots, p;\ -\infty < t < \infty\}$ be a family of real-valued random variables on the same probability space. In order for the vector of stochastic processes

$$\mathbf{X}(t) = \begin{pmatrix} X_1(t) \\ X_2(t) \\ \vdots \\ X_p(t) \end{pmatrix}, \qquad -\infty < t < \infty,$$

to be a multivariate weakly stationary stochastic process, each univariate process $X_j(t)$ must be weakly stationary in the sense defined in Section 2.3 and, in addition, *the correlation between processes must be stationary.* These hypotheses can be summarized as follows.

(i) $EX_j(t) = m_j, \quad j = 1, 2, \ldots, p.$

As before, the constants m_j will all be set equal to zero by the expedient of replacing each process by the residual $X_j(t) - m_j$. Then we assume that the covariances are all finite and satisfy the conditions

(ii) $EX_j(t + \tau)X_k(t) = C_{j,k}(\tau), \quad -\infty < t, \tau < \infty, 1 \leq j, k \leq p.$

That is, the covariances depend on τ, the lag between the time arguments, but not on t. When $j = k$, this is the condition of covariance stationarity (2.20) which was basic to the theory of Chapter 2. When $j \neq k$, it is the condition of *stationary correlation* between processes.

With these assumptions, the covariance functions $C_{j,k}(\tau)$ enjoy all of the properties of the Wiener covariances defined by (5.1). Consequently, the spectral representation (5.4) and the subsequent properties of the spectrum, (5.5)–(5.13), are valid. A direct and useful method for deriving properties of the spectrum is obtained from the *multivariate version of the spectral representation theorem.* This method will be demonstrated in Section 5.4. As before, since each univariate series $X_j(t)$ is weakly stationary a spectral representation exists,

$$X_j(t) = \int e^{i\lambda t} Z_j(d\lambda).$$

The random spectral measures $Z_j(A)$ have as values complex-valued random variables for each j and each set A. They are related to the spectral distributions by the basic expression

$$EZ_j(A)\overline{Z_k(B)} = F_{j,k}(A \cap B). \tag{5.16}$$

As before, $EZ_j(A) = 0$ for all j and A.

By the Schwarz inequality, applied to the inner product

$$\langle Z_j(A), Z_k(A) \rangle = EZ_j(A)\overline{Z_k(A)},$$

we obtain the inequality

$$|F_{j,k}(A)| \leq (F_{j,j}(A)F_{k,k}(A))^{1/2}. \tag{5.17}$$

It follows from this that the cross-spectral measure $F_{j,k}(A)$ can be nonzero only if both component processes $X_j(t)$ and $X_k(t)$ have positive power on A. Auxiliary inequalities which follow from this are

$$|p_{j,k}(\lambda)| \leq (p_{j,j}(\lambda)p_{k,k}(\lambda))^{1/2} \tag{5.18}$$

and

$$|f_{j,k}(\lambda)| \leq (f_{j,j}(\lambda) f_{k,k}(\lambda))^{1/2} \tag{5.19}$$

for the spectral function and spectral densities.

In terms of the symbolic intervals discussed in Section 2.5, expressions (5.16) and (5.17) can be combined to yield

$$EZ_j(d\lambda)\overline{Z_k(d\mu)} = \begin{cases} F_{j,k}(d\lambda), & \text{if } \mu = \lambda, \\ 0, & \text{if } \mu \neq \lambda. \end{cases} \tag{5.20}$$

This expression can be used operationally as indicated in the following statement:

Extension of the Basic Property *If $g(\lambda)$ and $h(\lambda)$ are complex-valued functions for which*

$$\int |g(\lambda)|^2 F_{j,j}(d\lambda) < \infty \qquad \textit{and} \qquad \int |h(\lambda)|^2 F_{k,k}(d\lambda) < \infty,$$

then

$$\int g(\lambda) Z_j(d\lambda) \qquad \textit{and} \qquad \int h(\lambda) Z_k(d\lambda)$$

are well-defined random variables with zero means and finite variances. Moreover,

$$\begin{aligned} E\left(\int g(\lambda) Z_j(d\lambda)\right)\left(\overline{\int h(\mu) Z_k(d\mu)}\right) &= \iint g(\lambda)\overline{h(\mu)} EZ_j(d\lambda)\overline{Z_k(d\mu)} \\ &= \int g(\lambda)\overline{h(\lambda)} F_{j,k}(d\lambda). \end{aligned} \tag{5.21}$$

The equality of the first and last expressions in (5.21) is guaranteed by the theory outlined in the Appendix. Convention (5.20) leads to the correct result in this and other calculations in which it will be used later in the chapter. This will prove to be a most useful method for calculating spectra in applications of the theory. In particular, with $g(\lambda) = e^{i\lambda(t+\tau)}$ and $h(\lambda) = e^{i\lambda t}$, (5.21) yields the spectral representation (5.4) of the auto- and cross-covariances.

Properties (5.6) of the cross spectra follow easily from (5.16). Expression (5.5) requires the additional relation

$$Z_j(-A) = \overline{Z_j(A)}, \tag{5.22}$$

also noted in Section 2.4.

If we write the spectral measures in vector form

$$\mathbf{Z}(A) = \begin{pmatrix} Z_1(A) \\ Z_2(A) \\ \vdots \\ Z_p(A) \end{pmatrix},$$

then condition (5.16) implies that $\mathbf{Z}(A)$ is uncorrelated with $\mathbf{Z}(B)$ if A and B are disjoint sets. If the multivariate process $\mathbf{X}(t)$ is Gaussian, then with the additional condition $A \cap -B = \varnothing$, $\mathbf{Z}(A)$ and $\mathbf{Z}(B)$ are independent.

The spectral representation of the process can be written in vector form as

$$\mathbf{X}(t) = \int e^{i\lambda t}\mathbf{Z}(d\lambda), \tag{5.23}$$

with the understanding that the integral is applied to each coordinate of the vector $e^{i\lambda t}\mathbf{Z}(d\lambda)$.

The correlational relationship (5.16) can be written succinctly in matrix form as

$$E\mathbf{Z}(A)\mathbf{Z}(B)^* = \mathbf{F}(A \cap B), \tag{5.24}$$

where $*$ denotes the combined operations of taking the transpose and forming the complex conjugates of the vector elements. In terms of this operation, the Hermetian relation (5.6) can be written

$$\mathbf{F}(A) = \mathbf{F}(A)^*.$$

The decomposition of the multivariate spectral distribution of $\mathbf{X}(t)$ into discrete and continuous components can be given in terms of the symbolic interval $d\lambda$ as

$$\mathbf{F}(d\lambda) = \mathbf{p}(\lambda) + \mathbf{f}(\lambda)\, d\lambda,$$

where, on the right-hand side, $d\lambda$ represents the (scalar) length of this interval.

The linearity property of expectation for random matrices given in Section 1.4 permits us to establish another important property of $\mathbf{F}(A)$: If $\mathbf{a}$ denotes any $p \times 1$ vector of complex numbers, then

$$E|\mathbf{a}^*\mathbf{Z}(A)|^2 \geq 0 \qquad \text{for} \quad \text{every } A.$$

However,

$$\begin{aligned} |\mathbf{a}^*\mathbf{Z}(A)|^2 &= \mathbf{a}^*\mathbf{Z}(A)\left(\overline{\mathbf{a}^*\mathbf{Z}(A)}\right) \\ &= \mathbf{a}^*\mathbf{Z}(A)\left(\mathbf{Z}(A)^*\mathbf{a}\right) = \mathbf{a}^*\mathbf{Z}(A)\mathbf{Z}(A)^*\mathbf{a}. \end{aligned}$$

Here, we have used the fact that $\overline{a^*\mathbf{Z}(A)} = (a^*\mathbf{Z}(A))^*$ and that $*$ operates on matrix products according to the rule

$$(\mathbf{UV})^* = \mathbf{V}^*\mathbf{U}^*.$$

Then taking expectations and using the fact that the expectation of a nonnegative random variable is nonnegative, we obtain

$$\mathbf{a}^*\mathbf{F}(A)\mathbf{a} \geq 0.$$

This establishes that the matrix of spectral distributions is *nonnegative definite* (Tucker, 1962, p. 144). (An alternate terminology is *positive semidefinite*). This is an important property which carries over to the matrices of spectral functions and spectral density functions $\mathbf{p}(\lambda)$ and $\mathbf{f}(\lambda)$ as well. Among the many implications of this property are inequalities (5.17)–(5.19). Various features of the multivariate spectral parameters to be defined later also depend on this property.

If two component processes $X_j(t)$, $X_k(t)$ are uncorrelated (Section 2.6), then

$$C_{j,k}(\tau) = 0, \qquad -\infty < \tau < \infty.$$

By the uniqueness of the spectral representation of the covariances (5.4), this implies that the cross-spectral distribution is everywhere zero,

$$F_{j,k}(A) = 0 \qquad \text{all} \quad A.$$

More Than One Vector Process

Let $\mathbf{X}(t)$ and $\mathbf{Y}(t)$ be $p \times 1$ and $q \times 1$ multivariate weakly stationary processes, respectively. When we consider more than one vector process it will be tacitly assumed that all components of all processes are stationarily correlated. In fact, all properties of such processes can be derived from the theory for a single weakly stationary vector process by forming the "stacked" process

$$\mathbf{W}(t) = \begin{pmatrix} \mathbf{X}(t) \\ \mathbf{Y}(t) \\ \vdots \end{pmatrix}.$$

In particular, the $p \times q$ matrix of covariances of $\mathbf{X}(t)$ and $\mathbf{Y}(t)$ has elements

$$C_{j,k}^{X,Y}(\tau) = EX_j(t+\tau)Y_k(t).$$

In vector notation,

$$\mathbf{C}^{X,Y}(\tau) = E\mathbf{X}(t+\tau)\mathbf{Y}(t)^*. \tag{5.25}$$

If the spectral distribution of these processes is defined by

$$\mathbf{F}^{X,Y}(A \cap B) = E\mathbf{Z}^X(A)\mathbf{Z}^Y(B)^*,$$

then the matrix version of the calculation which yielded (5.21) can be used to derive the spectral representation of $\mathbf{C}^{X,Y}(\tau)$ as follows:

$$\begin{aligned}\mathbf{C}^{X,Y}(\tau) &= E\mathbf{X}(t+\tau)\mathbf{Y}(t)^* \\ &= E\left(\int e^{i\lambda(t+\tau)}\mathbf{Z}^X(d\lambda)\right)\left(\int e^{i\mu t}\mathbf{Z}^Y(d\mu)\right)^* \\ &= \iint e^{i\lambda\tau}e^{i(\lambda-\mu)t}E\mathbf{Z}^X(d\lambda)\mathbf{Z}^Y(d\mu)^* \\ &= \int e^{i\lambda\tau}\mathbf{F}^{X,Y}(d\lambda).\end{aligned}$$

The vector processes $\mathbf{X}(t)$ and $\mathbf{Y}(t)$ are said to be *uncorrelated* if every component of $\mathbf{X}(t)$ is uncorrelated with every component of $\mathbf{Y}(t)$. Equivalent conditions are $\mathbf{C}^{X,Y}(\tau) = \mathbf{0}$ for all τ, thus $\mathbf{F}^{X,Y}(A) = \mathbf{0}$ for every A, where $\mathbf{0}$ is the $p \times q$ matrix with all zero elements.

The decomposition

$$\mathbf{F}^{X,Y}(d\lambda) = \mathbf{p}^{X,Y}(\lambda) + \mathbf{f}^{X,Y}(\lambda)\,d\lambda$$

holds in the case of two vector processes, where $\mathbf{p}^{X,Y}(\lambda)$ and $\mathbf{f}^{X,Y}(\lambda)$ are obtained from $\mathbf{F}^{X,Y}(A)$ as in (5.8) and (5.11).

5.4 LINEAR FILTERS FOR MULTIVARIATE TIME SERIES

We again restrict our attention to the weakly stationary model, although the results we will derive for spectra are valid for the more inclusive model of Section 5.2. Let $\mathbf{X}(t)$ and $\mathbf{Y}(t)$ be weakly stationary processes with p and q components, respectively. The new feature of a *multivariate linear filter* with input $\mathbf{X}(t)$ and output $\mathbf{Y}(t)$ is that each component of $\mathbf{Y}(t)$ will, in general, be influenced by every component of $\mathbf{X}(t)$. The "black box" representation of such a filter is pictured in Fig. 5.1. The nature of the influence of $X_k(t)$

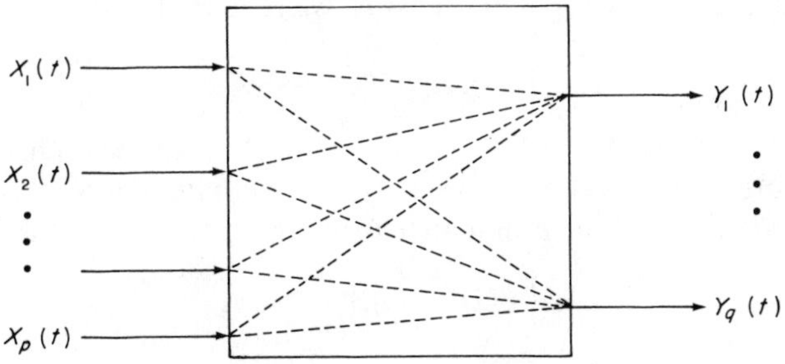

Fig. 5.1 *Representation of a multivariate linear filter as a black box with p input leads and q output leads. Each output is coupled with each input.*

on $Y_j(t)$ is that of an ordinary linear filter in the sense discussed in Chapter 4. Then, the total influence of all components of $\mathbf{X}(t)$ on $Y_j(t)$ is simply the sum of the contributions of each component. This can be stated more precisely by using the spectral representations of the input and output processes: Let $B_{j,k}(\lambda)$ denote the transfer function of the linear filter from $X_k(t)$ to $Y_j(t)$. This would be represented by the "black box" with all inputs except $X_k(t)$ turned off and measurements being made only at the $Y_j(t)$ terminal. With all inputs running, what would be observed at this terminal is

$$\begin{aligned} Y_j(t) &= \sum_{k=1}^{p} \int e^{i\lambda t} B_{j,k}(\lambda) Z_k{}^X(d\lambda) \\ &= \int e^{i\lambda t} \sum_{k=1}^{p} B_{j,k}(\lambda) Z_k{}^X(d\lambda), \end{aligned} \tag{5.26}$$

where $Z_k{}^X(d\lambda)$ is the random spectral measure of $X_k(t)$. That is, the spectral measure of $Y_j(t)$ is

$$Z_j{}^Y(d\lambda) = \sum_{k=1}^{p} B_{j,k}(\lambda) Z_k{}^X(d\lambda).$$

This can be written conveniently in matrix notation as

$$\mathbf{Z}^Y(d\lambda) = \mathbf{B}(\lambda)\mathbf{Z}^X(d\lambda), \tag{5.27}$$

where $\mathbf{B}(\lambda)$ is the $(q \times p)$-dimensional matrix,

$$\mathbf{B}(\lambda) = [B_{j,k}(\lambda)].$$

If we let $\mathbf{L}$ denote the multivariate linear filter represented by the "black box," then $\mathbf{B}(\lambda)$ is its *transfer function* and (5.26) can be written in matrix form as

$$\mathbf{Y}(t) = \mathbf{L}(\mathbf{X}(t)) = \int e^{i\lambda t}\mathbf{B}(\lambda)\mathbf{Z}^X(d\lambda). \tag{5.28}$$

The multivariate filter will *match* the input if each term of the output in (5.26) has finite variance (power). That is, we require that

$$\int |B_{j,k}(\lambda)|^2 F_{k,k}^X(d\lambda) < \infty, \tag{5.29}$$

for $j = 1, 2, \ldots, q$ and $k = 1, 2, \ldots, p$. As we show in the Appendix, along with a more theoretical discussion of multivariate filters, this *matching condition* can be represented in matrix form as

$$\operatorname{tr} \int \mathbf{B}(\lambda)\mathbf{F}^X(d\lambda)\mathbf{B}(\lambda)^* < \infty, \tag{5.30}$$

where tr denotes the trace of the matrix. Moreover, the *spectral distribution of the output* can be calculated using matrix algebra and the linearity property of expectation for random matrices (Section 1.4): From (5.27),

$$\begin{aligned}\mathbf{F}^Y(d\lambda) &= E(\mathbf{B}(\lambda)\mathbf{Z}^X(d\lambda))(\mathbf{B}(\lambda)\mathbf{Z}^X(d\lambda))^* \\ &= E\mathbf{B}(\lambda)\mathbf{Z}^X(d\lambda)\mathbf{Z}^X(d\lambda)^*\mathbf{B}(\lambda)^* \\ &= \mathbf{B}(\lambda)(E\mathbf{Z}^X(d\lambda)\mathbf{Z}^X(d\lambda)^*)\mathbf{B}(\lambda).^*\end{aligned}$$

Thus,

$$\mathbf{F}^Y(d\lambda) = \mathbf{B}(\lambda)\mathbf{F}^X(d\lambda)\mathbf{B}(\lambda)^*. \tag{5.31}$$

This implies that the *spectral functions* and *spectral density functions* of input and output are

$$\mathbf{p}^Y(\lambda) = \mathbf{B}(\lambda)\mathbf{p}^X(\lambda)\mathbf{B}(\lambda)^* \tag{5.32}$$

and

$$\mathbf{f}^Y(\lambda) = \mathbf{B}(\lambda)\mathbf{f}^X(\lambda)\mathbf{B}(\lambda)^*. \tag{5.33}$$

Example 5.1 *Filters Which Transform Component Time Series Independently*

Suppose $\mathbf{X}(t)$ and $\mathbf{Y}(t)$ are both p-dimensional processes and let L_j be a univariate filter with transfer function $B_j(\lambda)$ such that

$$Y_j(t) = L_j(X_j(t))$$

for $j = 1, 2, \ldots, p$. This situation occurs, for example, when each component has its own linear transmission channel which is completely separate from the channels of the other components. A prime example of this would be a telephone trunk line which carries thousands of individual, separate conversations within a single cable. Moreover, it is often the case that when multivariate time series are being prepared for spectrum analysis, different preprocessing filters are used on each component. The input and output spectra are then related by the expressions of this example. This model in some instances also applies to the same linear mechanism within which the component signals are polarized into noninterferring modes of transmission. Thus, two beams of light polarized at right angles to one another would be transformed simultaneously but independently by a piece of colored glass. To a great extent, the horizontal and vertical components of most types of seismic waves travel through the earth and are recorded independently. Coding techniques allow many signals to be mixed, transmitted, then unscrambled into their original form.

This type of filter has a simple matrix representation. The transfer function is

$$\mathbf{B}(\lambda) = \begin{bmatrix} B_1(\lambda) & 0 & \cdots & 0 \\ 0 & B_2(\lambda) & \cdots & 0 \\ \vdots & \vdots & \cdots & \vdots \\ 0 & 0 & \cdots & B_p(\lambda) \end{bmatrix}.$$

That is, the one-dimensional transfer functions appear on the diagonal of the matrix and the off-diagonal elements are all zero. From the component representation

$$Z_j^Y(d\lambda) = B_j(\lambda)Z_j^X(d\lambda),$$

the output spectral distributions are seen to be

$$\begin{aligned} F_{j,k}^Y(d\lambda) &= EZ_j^Y(d\lambda)\overline{Z_k^Y(d\lambda)} \\ &= B_j(\lambda)\overline{B_k(\lambda)}F_{j,k}^X(d\lambda). \end{aligned} \tag{5.34}$$

This is also an immediate algebraic consequence of the form of $\mathbf{B}(\lambda)$ and expression (5.31). If $\mathbf{L}$ is the multivariate linear filter with this transfer function, then $\mathbf{L}$ matches any input $\mathbf{X}(t)$ for which L_j matches the component $X_j(t)$ for each $j = 1, 2, \ldots, p$.

A Method for Calculating Spectral Parameters

An important method for computing spectral parameters will be introduced in the next example.

Example 5.2 *The Multivariate Distributed Lag Model and a Method for Computing Spectra*

A discrete-time, weakly stationary, time series model that has been used in various fields, notably geophysics (Hamon and Hannan, 1963) and economics [see Fishman (1969) for a bibliography] can be written in the form

$$\mathbf{Y}(t) = \sum_{s=-r_1}^{r_2} \mathbf{a}(s)\mathbf{X}(t-s) + \boldsymbol{\eta}(t), \tag{5.35}$$

where $\mathbf{Y}(t)$ and $\boldsymbol{\eta}(t)$ are $(q \times 1)$-dimensional weakly stationary processes and $\mathbf{X}(t)$ is a $(p \times 1)$-dimensional process. This is the *multivariate distributed lag model.* In economics the coordinates of $\mathbf{X}(t)$ are called the *exogenous variables*, those of $\mathbf{Y}(t)$ the *endogeneous variables*, and $\boldsymbol{\eta}(t)$ is a *vector process of residuals.* For simplicity we assume all processes have zero means. A key assumption concerning this model is that the processes $\mathbf{X}(t)$ and $\boldsymbol{\eta}(t)$ are uncorrelated.

The objective of a statistical analysis of this model is to estimate, for given r_1 and r_2, the coefficients of the $r_1 + r_2 + 1$, $(q \times p)$-dimensional matrices $\mathbf{a}(s)$. Excellent treatments of this estimation problem are given by Fishman (1969) and Hannan (1970). We will be content to demonstrate a direct method for calculating the spectrum of $\mathbf{Y}(t)$ and the transfer function of the linear filter operating on $\mathbf{X}(t)$ which uses the spectral representations of the processes. This method of calculation is of fundamental importance for the determination of the spectral parameters of a large class of time series models.

The method is based on the following observations: Let

$$\mathbf{W}(t) = \mathbf{Y}(t) - \sum_{s=-r_1}^{r_2} \mathbf{a}(s)\mathbf{X}(t-s) - \boldsymbol{\eta}(t).$$

Then, $\mathbf{W}(t) \equiv 0$. This implies that the covariance matrix of $\mathbf{W}(t)$, $\mathbf{C}^W(\tau)$, is identically zero for all τ, thus

$$\mathbf{F}^W(A) = 0 \qquad \text{for} \quad \text{all } A.$$

However, then,

$$\mathbf{Z}^W(A) = 0 \qquad \text{for} \quad \text{all } A. \tag{5.36}$$

Now, applying the spectral representation (5.23) to both sides of (5.35), we obtain

$$\int e^{i\lambda t}\mathbf{Z}^Y(d\lambda) = \sum_{s=-r_1}^{r_2} \mathbf{a}(s) \int e^{i\lambda(t-s)}\mathbf{Z}^X(d\lambda) + \int e^{i\lambda t}\mathbf{Z}^\eta(d\lambda),$$

or

$$\mathbf{W}(t) = \int e^{i\lambda t}[\mathbf{Z}^Y(d\lambda) - \mathbf{B}(\lambda)\mathbf{Z}^X(d\lambda) - \mathbf{Z}^\eta(d\lambda)], \tag{5.37}$$

where

$$\mathbf{B}(\lambda) = \sum_{s=-r_1}^{r_2} \mathbf{a}(s)e^{-i\lambda s}. \tag{5.38}$$

Now, the bracketed expression in (5.37) is $\mathbf{Z}^W(d\lambda)$. Thus, if we apply (5.36) to $\mathbf{Z}^W(d\lambda)$, *the original equation* (5.35) *is equivalent to the symbolic equation*

$$\mathbf{Z}^Y(d\lambda) = \mathbf{B}(\lambda)\mathbf{Z}^X(d\lambda) + \mathbf{Z}^\eta(d\lambda) \qquad \textit{for} \quad \textit{all } \lambda. \tag{5.39}$$

This is the key equation of the method. All spectral parameters will be obtained from this expression by the application of standard rules of matrix algebra and expectation to $\mathbf{Z}^Y(d\lambda)$, $\mathbf{Z}^X(d\lambda)$ and $\mathbf{Z}^\eta(d\lambda)$ *as though they were vectors of valid, complex-valued random variables with finite variances.*

Note that this equation could have been written from (5.35) "by inspection." The form of the transfer function of the filter is a consequence of an easy multivariate extension of the rule for calculating transfer functions given in Section 4.2. Thus, it is generally a simple matter to obtain expressions such as (5.39) directly from the time domain representations of the processes.

We first calculate the spectral distribution $\mathbf{F}^{Y,X}(d\lambda)$ by multiplying both sides of (5.39) on the right by $\mathbf{Z}^X(d\lambda)^*$ and taking expectations:

$$\begin{aligned}\mathbf{F}^{Y,X}(d\lambda) &= E\mathbf{Z}^Y(d\lambda)\mathbf{Z}^X(d\lambda)^* \\ &= E[\mathbf{B}(\lambda)\mathbf{Z}^X(d\lambda)\mathbf{Z}^X(d\lambda)^* + \mathbf{Z}^\eta(d\lambda)\mathbf{Z}^X(d\lambda)^*] \\ &= \mathbf{B}(\lambda)\mathbf{F}^X(d\lambda).\end{aligned}$$

We have used the fact that $E\mathbf{Z}^\eta(d\lambda)\mathbf{Z}^X(d\lambda)^* = 0$ because $\mathbf{X}(t)$ and $\boldsymbol{\eta}(t)$ are uncorrelated processes.

Recall that

$$\mathbf{F}^X(d\lambda) = \mathbf{p}^X(\lambda) + \mathbf{f}^X(\lambda)\, d\lambda \qquad \text{and} \qquad \mathbf{F}^{X,Y}(d\lambda) = \mathbf{p}^{Y,X}(\lambda) + \mathbf{f}^{Y,X}(\lambda)\, d\lambda.$$

Thus, in the pure discrete case, the expression

$$\mathbf{B}(\lambda) = \mathbf{p}^{Y,X}(\lambda)\mathbf{p}^X(\lambda)^{-1}$$

determines the transfer function at those values of λ for which $\mathbf{p}^X(\lambda)$ is nonsingular. However, we will be primarily interested in the case of pure continuous spectra, since it is common practice to estimate and remove the discrete spectral component before analysis. Then,

$$\mathbf{B}(\lambda) = \mathbf{f}^{Y,X}(\lambda)\mathbf{f}^X(\lambda)^{-1} \tag{5.40}$$

whenever $\mathbf{f}^X(\lambda)$ is nonsingular. In this expression and all comparable expressions hereafter, when $d\lambda$ represents the length of the infinitesimal interval it is treated as though it were a positive number. Thus, it is simply cancelled when it appears to the same power in numerator and denominator of a ratio. Now, estimates of the spectral densities will provide an estimate $\hat{\mathbf{B}}(\lambda)$ of $\mathbf{B}(\lambda)$ at a finite set of frequencies and, since (5.38) represents this function as a Fourier series, estimates of the coefficient matrices can be obtained from a discrete version of the inversion formula,

$$\hat{\mathbf{a}}(s) = \frac{1}{2\pi}\int_{-\pi}^{\pi} e^{i\lambda s}\hat{\mathbf{B}}(\lambda)\, d\lambda.$$

This is the estimation procedure detailed by Hannan (1970, p. 475).

Returning to (5.39), the spectral distribution of $\mathbf{Y}(t)$ can be calculated as

$$\begin{aligned}\mathbf{F}^Y(d\lambda) &= E[\mathbf{B}(\lambda)\mathbf{Z}^X(d\lambda) + \mathbf{Z}^\eta(d\lambda)][\mathbf{B}(\lambda)\mathbf{Z}^X(d\lambda) + \mathbf{Z}^\eta(d\lambda)]^* \\ &= E\{\mathbf{B}(\lambda)\mathbf{Z}^X(d\lambda)\mathbf{Z}^X(d\lambda)^*\mathbf{B}(\lambda)^* + \mathbf{Z}^\eta(d\lambda)\mathbf{Z}^X(d\lambda)^*\mathbf{B}(\lambda)^* \\ &\quad + \mathbf{B}(\lambda)\mathbf{Z}^X(d\lambda)\mathbf{Z}^\eta(d\lambda)^* + \mathbf{Z}^\eta(d\lambda)\mathbf{Z}^\eta(d\lambda)^*\}.\end{aligned}$$

The expectation of the second and third terms are zero, since $\mathbf{X}(t)$ and $\boldsymbol{\eta}(t)$ are uncorrelated. Consequently, moving the expectation inside the braces, we obtain

$$\mathbf{F}^Y(d\lambda) = \mathbf{B}(\lambda)\mathbf{F}^X(d\lambda)\mathbf{B}(\lambda)^* + \mathbf{F}^\eta(d\lambda).$$

In the case of continuous spectra, this yields the following expression for the spectral densities,

$$\mathbf{f}^Y(\lambda) = \mathbf{B}(\lambda)\mathbf{f}^X(\lambda)\mathbf{B}(\lambda)^* + \mathbf{f}^\eta(\lambda).$$

Now, substituting (5.40) for $\mathbf{B}(\lambda)$ into this equation, we obtain

$$\mathbf{f}^Y(\lambda) = \mathbf{f}^{Y,X}(\lambda)\mathbf{f}^X(\lambda)^{-1}\mathbf{f}^{X,Y}(\lambda) + \mathbf{f}^\eta(\lambda),$$

where we have used the easily established result

$$\mathbf{f}^{Y,X}(\lambda)^* = \mathbf{f}^{X,Y}(\lambda).$$

This provides the following expression for the residual spectral density in terms of the spectral densities of the observed processes,

$$\mathbf{f}^\eta(\lambda) = \mathbf{f}^Y(\lambda) - \mathbf{f}^{Y,X}(\lambda)\mathbf{f}^X(\lambda)^{-1}\mathbf{f}^{X,Y}(\lambda). \tag{5.41}$$

Since $\boldsymbol{\eta}(t)$ measures the difference between $\mathbf{Y}(t)$ and the best "explanation" of $\mathbf{Y}(t)$ in terms of a linear function of $\mathbf{X}(t-s)$, $-r_1 \leq s \leq r_2$, the elements of $\mathbf{f}^\eta(\lambda)$ provide a measure of how good this "explanation" is as a function of frequency. A perfect fit, for example, would yield $\mathbf{f}^\eta(\lambda) = \mathbf{0}$ for all λ. An estimate of the residual spectral density can be obtained from estimates of the spectra of the observable processes by means of (5.41).

5.5 THE BIVARIATE SPECTRAL PARAMETERS, THEIR INTERPRETATIONS AND USES

Hereafter, we will restrict our attention to multivariate stochastic processes $\mathbf{X}(t)$ with continuous spectra, unless otherwise specified. Consequently, the basic (second-order) parameter is the spectral density matrix

$$\mathbf{f}(\lambda) = [f_{j,k}(\lambda)].$$

In this section we will be interested in the interrelationships between *pairs* of component series $X_j(t)$ and $X_k(t)$. These relationships are necessarily determined by the three *bivariate parameters* $f_{j,j}(\lambda)$, $f_{k,k}(\lambda)$, and $f_{j,k}(\lambda)$ for each $j \neq k$. We will take j and k to be fixed but arbitrary indices in this section. These parameters provide the information available in the correlation between $X_j(t)$ and $X_k(t)$ in the most useable form. Two sets of real-valued parameters will be defined which are equivalent to the three bivariate spectral parameters but which display the correlational information in different ways.

The two sets of parameters depend on the two standard representations of a complex number in terms of real numbers. The cartesian form leads to the representation

$$f_{j,k}(\lambda) = c_{j,k}(\lambda) - iq_{j,k}(\lambda), \tag{5.42}$$

where the real-valued functions $c_{j,k}(\lambda)$ and $q_{j,k}(\lambda)$ are called the *cospectral density* (*cospectrum*) and *quadrature spectral density* (*quadspectrum*), respectively. As we will see, the minus sign in (5.42) is a consequence of the choice of the sign in the exponent of the spectral representation

$$C_{j,k}(\tau) = \int e^{i\lambda\tau} f_{j,k}(\lambda)\, d\lambda.$$

In all cases of practical interest, this expression can be inverted to yield

$$f_{j,k}(\lambda) = \frac{1}{2\pi}\int_{-\infty}^{\infty} e^{-i\lambda\tau} C_{j,k}(\tau)\, d\tau \tag{5.43}$$

in the continuous-time case. Now, we expand the exponential to obtain $e^{-i\lambda\tau} = \cos \lambda\tau - i \sin \lambda\tau$ and write

$$C_{j,k}(\tau) = \frac{C_{j,k}(\tau) + C_{j,k}(-\tau)}{2} + \frac{C_{j,k}(\tau) - C_{j,k}(-\tau)}{2}.$$

The first term on the right-hand side of this expression is an even function and the second an odd function. Substituting these expressions into (5.43) we obtain

$$f_{j,k}(\lambda) = \frac{1}{2\pi}\int_{-\infty}^{\infty} \cos \lambda\tau \left\{\frac{C_{j,k}(\tau) + C_{j,k}(-\tau)}{2}\right\} d\tau$$
$$- i\frac{1}{2\pi}\int_{-\infty}^{\infty} \sin \lambda\tau \left\{\frac{C_{j,k}(\tau) - C_{j,k}(-\tau)}{2}\right\} d\tau.$$

The products yielding odd functions integrate to zero in this result since the integrals are over symmetric intervals. Thus, we obtain the *continuous-time inversion formulas*

$$c_{j,k}(\lambda) = \frac{1}{2\pi}\int_{-\infty}^{\infty} \cos \lambda\tau \left\{\frac{C_{j,k}(\tau) + C_{j,k}(-\tau)}{2}\right\} d\tau \tag{5.44}$$

and

$$q_{j,k}(\lambda) = \frac{1}{2\pi}\int_{-\infty}^{\infty} \sin \lambda\tau \left\{\frac{C_{j,k}(\tau) - C_{j,k}(-\tau)}{2}\right\} d\tau. \tag{5.45}$$

Note that this computation and the definition of $q_{j,k}(\lambda)$ given by (5.45) account for the minus sign in (5.41).

The *discrete-time versions* of these formulas are

$$c_{j,k}(\lambda) = \frac{1}{2\pi}\sum_{\tau=-\infty}^{\infty} \cos \lambda\tau \left\{\frac{C_{j,k}(\tau) + C_{j,k}(-\tau)}{2}\right\}, \tag{5.46}$$

$$q_{j,k}(\lambda) = \frac{1}{2\pi}\sum_{\tau=-\infty}^{\infty} \sin \lambda\tau \left\{\frac{C_{j,k}(\tau) - C_{j,k}(-\tau)}{2}\right\}. \tag{5.47}$$

These expressions play an important role in the computation of cross-spectral density estimators for one of the methods of spectral estimation to be discussed in Chapter 8. Note that the real-valued parameters $c_{j,k}(\lambda)$, $q_{j,k}(\lambda)$, $f_{j,j}(\lambda)$, $f_{k,k}(\lambda)$ are equivalent to the original set of bivariate spectral parameters.

The polar representation of $f_{j,k}(\lambda)$ yields another set of spectral parameters. Write

$$f_{j,k}(\lambda) = |f_{j,k}(\lambda)| e^{i\vartheta_{j,k}(\lambda)}, \tag{5.48}$$

and let

$$\rho_{j,k}(\lambda) = |f_{j,k}(\lambda)| / (f_{j,j}(\lambda) f_{k,k}(\lambda))^{1/2}. \tag{5.49}$$

The functions $\vartheta_{j,k}(\lambda)$ and $\rho_{j,k}(\lambda)$ are called the *phase* [*phase lead of* $X_j(t)$ *over* $X_k(t)$] and *coherence* (*coefficient of coherence*), respectively. The function

$$\gamma_{j,k}(\lambda) = f_{j,k}(\lambda)/(f_{j,j}(\lambda) f_{k,k}(\lambda))^{1/2}$$

is called the *complex coherence*. It will play a role in the computation of higher order coherences in the next section. By property (5.19), the coherence satisfies the inequality

$$0 \le \rho_{j,k}(\lambda) \le 1 \tag{5.50}$$

wherever the ratio (5.49) is well defined. At frequencies for which $f_{j,j}(\lambda) f_{k,k}(\lambda) = 0$, we will define $\rho_{j,k}(\lambda) = 0$.

It is easily seen that $\rho_{j,k}(\lambda)$, $\vartheta_{j,k}(\lambda)$, $f_{j,j}(\lambda)$, and $f_{k,k}(\lambda)$ constitute a set of spectral parameters equivalent to the original set. These parameters are obtainable from the spectral densities, cospectrum, and quadrature spectrum by means of the relations

$$\rho_{j,k}(\lambda) = \left(\frac{c_{j,k}^2(\lambda) + q_{j,k}^2(\lambda)}{f_{j,j}(\lambda) f_{k,k}(\lambda)} \right)^{1/2} \tag{5.51}$$

and

$$\vartheta_{j,k}(\lambda) = -\operatorname{Arctan}(q_{j,k}(\lambda)/c_{j,k}(\lambda)). \tag{5.52}$$

[The minus sign again results from the fact that $q_{j,k}(\lambda) = -\operatorname{Im} f_{j,k}(\lambda)$.] By expanding (5.48) in cartesian form, the cospectrum and quadspectrum can be obtained from the spectral densities, phase and coherence by means of the expressions

$$c_{j,k}(\lambda) = \rho_{j,k}(\lambda) \cos \vartheta_{j,k}(\lambda) (f_{j,j}(\lambda) f_{k,k}(\lambda))^{1/2} \tag{5.53}$$

$$q_{j,k}(\lambda) = \rho_{j,k}(\lambda) \sin \vartheta_{j,k}(\lambda) (f_{j,j}(\lambda) f_{k,k}(\lambda))^{1/2}. \tag{5.54}$$

The phase and coherence are, perhaps, the most useful parameters for measuring the relationship between $X_j(t)$ and $X_k(t)$ because the values of these functions can be interpreted quantitatively. We consider this next.

Interpretation of Phase and Coherence

Suppose that the random spectral measures of $X_j(t)$ and $X_k(t)$ could be represented in the form

$$Z_j(d\lambda) = e^{i\varphi_j(\lambda)}|Z_j(d\lambda)|, \qquad Z_k(d\lambda) = e^{i\varphi_k(\lambda)}|Z_k(d\lambda)|,$$

where $\varphi_j(\lambda)$ and $\varphi_k(\lambda)$ are nonrandom phase functions. That is, the phase functions are the same for all sample functions of the process. We would then have

$$f_{j,k}(\lambda)\, d\lambda = EZ_j(\lambda)\overline{Z_k(\lambda)} = e^{i(\varphi_j(\lambda)-\varphi_k(\lambda))}E|Z_j(d\lambda)Z_k(d\lambda)|,$$

from which it would follow that

$$\vartheta_{j,k}(\lambda) = \varphi_j(\lambda) - \varphi_k(\lambda).$$

That is, in this special case, the phase function is simply the difference in phase of the two time series at frequency λ or, more precisely, the phase lead of $X_j(t)$ over $X_k(t)$ at λ. In the usual situation wherein $\varphi_j(\lambda)$ and $\varphi_k(\lambda)$ are random quantities, this interpretation can be largely retained if now $\vartheta_{j,k}(\lambda)$ is thought of as an (ensemble) average of the random differences $\varphi_j(\lambda) - \varphi_k(\lambda)$. *Thus, $\vartheta_{j,k}(\lambda)$ is commonly interpreted as being the average phase lead of $X_j(t)$ over $X_k(t)$ at frequency λ.*

Recall from Section 4.2 that the phase shift of a linear filter at frequency λ can be given in time units by dividing by λ. In the same sense, the parameter

$$t_{j,k}(\lambda) = \vartheta_{j,k}(\lambda)/\lambda$$

is the time lead of the harmonic of $X_j(t)$ at frequency λ over that of $X_k(t)$.

An important property of the phase parameter is the following: If the univariate processes $X_j(t)$ and $X_k(t)$ are independently passed through linear filters with transfer functions $B_j(\lambda)$ and $B_k(\lambda)$ and if $Y_j(t)$ and $Y_k(t)$ are the outputs, then

$$f_{j,k}^Y(\lambda) = B_j(\lambda)\overline{B_k(\lambda)}f_{j,k}^X(\lambda)$$

by (5.34). Thus, if $\vartheta_j(\lambda)$ and $\vartheta_k(\lambda)$ are the phase shifts of the two filters, the phase of the output is related to that of the input by the expression

$$\vartheta_{j,k}^Y(\lambda) = \vartheta_{j,k}^X(\lambda) + \vartheta_j(\lambda) - \vartheta_k(\lambda).$$

Consequently, if it is important that the phase relationships remain undisturbed when the time series are processed by linear filters, filters with the same phase shift must be used on both series. A class of symmetric filters, called *nonnegative definite filters*, are important from this viewpoint in that they have zero phase shift for all frequencies. This is nearly true of all properly designed symmetric filters as well. See the discussion of this topic in Section A6.2.

The interpretation of coherence is somewhat more involved, but correspondingly more important than that of phase. As might be guessed, the term "coherence" is borrowed from the study of light. The term was first applied to time series by Wiener (1930). Because of the symbolic representation

$$\rho_{j,k}(\lambda) = |EZ_j(d\lambda)\overline{Z_k(d\lambda)}|/(E|Z_j(d\lambda)|^2 E|Z_k(d\lambda)|^2)^{1/2}$$

[see (5.20) and (5.49)], it is reasonable to expect that the coefficient of coherence will have the properties of the absolute value of a correlation coefficient at each frequency λ. In particular, the extreme values, zero and one, should correspond to complete lack of correlation and the maximum degree of correlation possible for a definition of correlation that makes sense in the time series context. Moreover, values of coherence other than zero and one should have some reasonable quantitative interpretation. We investigate the basis for such an interpretation in what follows.

The correlation coefficient measures the degree of *linear* association between two random variables. That is, it represents the degree to which one random variable can be represented as a linear function of the other. The term "association" rather than "dependence" is used because a large correlation need not indicate a causal relationship between the random variables. Similarly, the term "regression" is used to describe a *directed* relationship of one random variable with another. That is, whereas we speak of the linear association between X_1 and X_2, treating X_1 and X_2 on equal footing, when we speak of the linear regression of X_1 on X_2 we think of X_1 as the dependent and X_2 the independent variable in some sort of functional relationship. This distinction marks the difference between correlation theory and regression theory. The form of regression considered here differs from the theory more commonly taught in statistics wherein the independent variable is assumed to be nonrandom. The terms and interpretations of statistical correlation and regression theory will carry over to the time series context with little change. We do not distinguish between the two topics in this section in order to concentrate on the interpretations of the parameters. In the multivariate theory of the next section, these topics will be treated separately.

It is reasonable to expect that the coefficient of coherence will, in some sense, measure the degree of *linear* association between the time series $X_j(t)$ and $X_k(t)$. The only concept of linear function we have available for time series is that of a linear filter. Consequently, we can ask the question: To what degree can $X_k(t)$ be represented as the output of a linear filter with input $X_j(t)$? If mean-squared error is taken as the measure of the difference between $X_k(t)$ and filtered versions of $X_j(t)$, this question can be rephrased to read: How small can we make

$$\sigma_L^2 = E(X_k(t) - L(X_j(t)))^2$$

by proper choice of the linear filter L? [Note that this expression does not depend upon t, since $X_k(t) - L(X_j(t))$ is weakly stationary.] If the linear filter $\tilde{L}$ which minimizes $\sigma_L{}^2$ can be found, then the degree of linear association between $X_j(t)$ and $X_k(t)$ can be assessed by comparing $\sigma_{\tilde{L}}{}^2$, the power in the residual time series $X_k(t) - \tilde{L}(X_j(t))$, with the power $E(X_k(t))^2$ of $X_k(t)$ itself. The coefficient of coherence actually makes this comparison possible frequency by frequency.

To see this, it is first necessary to construct the linear filter $\tilde{L}$. Since filters are uniquely determined by their transfer functions, it suffices to compute the transfer function $\tilde{B}_{j,k}(\lambda)$ of $\tilde{L}$. This can easily be done by virtue of the following observation. Let L be any linear filter which matches $X_j(t)$ and let $B(\lambda)$ be its transfer function. Then by the spectral representations of $X_k(t)$ and $L(X_j(t))$,

$$\sigma_L{}^2 = E\left\{\int e^{i\lambda t}[Z_k(d\lambda) - B(\lambda)Z_j(d\lambda)]\right\}\overline{\left\{\int e^{i\mu t}[Z_k(d\mu) - B(\mu)Z_k(d\mu)]\right\}}$$

$$= \int E|Z_k(d\lambda) - B(\lambda)Z_j(d\lambda)|^2.$$

Now, the minimum value of $\sigma_L{}^2$ *will be achieved for the linear filter whose transfer function minimizes the integrand of this expression for all frequencies* λ. That is, the problem of minimizing $\sigma_L{}^2$ is reduced to an infinite number of identical minimization problems, each involving the determination of a complex number $\tilde{B}_{j,k}(\lambda)$ which minimizes the quantity

$$E|Z_k(d\lambda) - B(\lambda)Z_j(d\lambda)|^2$$

among all possible complex numbers $B(\lambda)$. However, viewing $Z_k(d\lambda)$ and $Z_j(d\lambda)$ as random variables in $L_2(P)$ as before, this quantity is the squared distance between $Z_k(d\lambda)$ and an element in the linear subspace generated by $Z_j(d\lambda)$, since all such elements are simply complex multiples of $Z_j(d\lambda)$. The minimum distance is achieved by the orthogonal projection of $Z_k(d\lambda)$ onto this subspace. By the criterion for determining orthogonal projections given in Section 1.3, this means that the projection $\tilde{B}_{j,k}(\lambda)Z_j(d\lambda)$ has the property

$$(Z_k(d\lambda) - \tilde{B}_{j,k}(\lambda)Z_j(d\lambda)) \perp Z_j(d\lambda),$$

since $Z_j(d\lambda)$ is the only generator of the subspace. That is,

$$E(Z_k(d\lambda) - \tilde{B}_{j,k}(\lambda)Z_j(d\lambda))\overline{Z_j(d\lambda)} = 0.$$

From this it follows that

$$\tilde{B}_{j,k}(\lambda) = EZ_k(d\lambda)\overline{Z_j(d\lambda)}/E|Z_j(\lambda)|^2.$$

Because of (5.20) and the restriction to continuous spectra, this becomes

$$\tilde{B}_{j,k}(\lambda) = f_{k,j}(\lambda)/f_{j,j}(\lambda). \tag{5.55}$$

This is the transfer function of the filter $\tilde{L}$ which minimizes $\sigma_L{}^2$. Note that $\tilde{L}$ matches $X_j(t)$, since the condition $\rho_{j,k}(\lambda) \le 1$ implies

$$\int |\tilde{B}_{j,k}(\lambda)|^2 f_{j,j}(\lambda)\, d\lambda = \int (|f_{j,k}(\lambda)|^2 / f_{j,j}(\lambda))\, d\lambda$$
$$= \int \rho_{j,k}^2(\lambda) f_{k,k}(\lambda)\, d\lambda < \infty.$$

The gain and phase shift functions of $\tilde{L}$ are

$$|\tilde{B}_{j,k}(\lambda)| = |f_{j,k}(\lambda)| / f_{j,j}(\lambda) = \rho_{j,k}(\lambda)(f_{k,k}(\lambda)/f_{j,j}(\lambda))^{1/2} \tag{5.56}$$

and

$$\tilde{\vartheta}_{j,k}(\lambda) = \arg(f_{j,k}(\lambda)) = -\vartheta_{j,k}(\lambda). \tag{5.57}$$

Thus, another interpretation of the phase angle $\vartheta_{j,k}(\lambda)$ is that it is the negative of the phase shift of the filter which transforms $X_j(t)$ into the best linear approximation to $X_k(t)$. In this sense, $X_j(t)$ leads $X_k(t)$ at frequency λ by the angle $\vartheta_{j,k}(\lambda)$.

In actual fact, this is the angle by which $X_j(t)$ leads $\tilde{X}_k(t) = \tilde{L}(X_j(t))$ at frequency λ and the degree to which it is a reasonable measure of the angular separation of $X_j(t)$ and $X_k(t)$ will depend on how close $\tilde{X}_k(t)$ and $X_k(t)$ are at this frequency. To determine this, let

$$U_k(t) = X_k(t) - \tilde{X}_k(t).$$

This is a weakly stationary time series with spectral measure

$$Z_k{}^U(d\lambda) = Z_k{}^X(d\lambda) - \tilde{B}_{j,k}(\lambda) Z_j{}^X(d\lambda).$$

Applying the Pythagorean theorem to the orthogonal elements $Z_k{}^U(d\lambda)$ and $\tilde{B}_{j,k}(\lambda) Z_j{}^X(d\lambda) = Z_k{}^{\tilde{X}}(d\lambda)$, we obtain

$$E|Z_k{}^X(d\lambda)|^2 = E|Z_j{}^U(d\lambda) + \tilde{B}_{j,k}(\lambda) Z_j{}^X(d\lambda)|^2$$
$$= E|Z_k{}^U(d\lambda)|^2 + |\tilde{B}_{j,k}(\lambda)|^2 E|Z_j{}^X(d\lambda)|^2.$$

Consequently, the spectral densities of the processes $X_k(t)$, $\tilde{X}_k(t)$, and $U_k(t)$ are related by the expression

$$f_{k,k}^X(\lambda) = f_{k,k}^U(\lambda) + f_{k,k}^{\tilde{X}}(\lambda), \tag{5.58}$$

where

$$f_{k,k}^{\tilde{X}}(\lambda) = |\tilde{B}_{j,k}(\lambda)|^2 f_{j,j}^X(\lambda). \tag{5.59}$$

Now, by the definition of $\sigma_{\tilde{L}}{}^2$, the minimum value of the error is

$$\sigma_{\tilde{L}}{}^2 = \int f_{k,k}^U(\lambda)\, d\lambda.$$

Therefore, a frequency by frequency assessment of the magnitude of this error relative to the power in $X_k(t)$ can be obtained by means of the ratio of spectral densities $f_{k,k}^U(\lambda)/f_{k,k}^X(\lambda)$. With only minor abuse of language, this ratio can be described as the proportion of the power in $X_k(t)$ at frequency λ which *cannot* be explained by the linear regression of $X_k(t)$ on $X_j(t)$ [i.e., by $\tilde{L}(X_j(t))$]. Because of (5.58), the proportion of the power at frequency λ which *can* be so explained is $f_{k,k}^{\tilde{X}}(\lambda)/f_{k,k}^X(\lambda)$. However, from (5.56) and (5.59) we easily obtain

$$f_{k,k}^{\tilde{X}}(\lambda) = \rho_{j,k}^2(\lambda) f_{k,k}^X(\lambda).$$

Thus, this proportion is $\rho_{j,k}^2(\lambda)$. Moreover, repeating the above computations with j and k interchanged, the symmetry of the coefficient of coherence in j and k implies that $\rho_{j,k}^2(\lambda)$ is also the proportion of the power in $X_j(t)$ at frequency λ which can be explained by the linear regression of $X_j(t)$ on $X_k(t)$. *Thus, the squared coefficient of coherence $\rho_{j,k}^2(\lambda)$ can be interpreted as the proportion of the power at frequency λ in either time series $X_j(t)$, $X_k(t)$ which can be explained by its linear regression on the other.*

This means that linear filters can be constructed by which 100 $\rho_{j,k}^2(\lambda)$% of the power at frequency λ can be removed from either series by subtracting off the appropriate linearly filtered version of the other. This is the most useful interpretation of coherence.

In addition, by (5.58) the proportion of the power in the residual series at frequency λ will be $1 - \rho_{j,k}^2(\lambda)$. This yields the useful expression

$$f_{k,k}^U(\lambda) = (1 - \rho_{j,k}^2(\lambda)) f_{k,k}^X(\lambda) \qquad (5.60)$$

for the residual spectral density.

Note that if $\rho_{j,k}(\lambda) = 0$, the two series are completely unrelated at frequency λ in the sense that no linearly filtered version of one can be used to reduce the power in the other at that frequency. On the other hand, if $\rho_{j,k}(\lambda) = 1$, one series is exactly a linearly filtered version of the other at frequency λ. In fact, if $\rho_{j,k}(\lambda) = 1$ for all λ, both time series can be represented as the output of a linear filter with the other as input. For then, $f_{k,k}^U(\lambda) \equiv 0$ which implies that $\sigma_{\tilde{L}}^2 = E(X_k(t) - \tilde{L}(X_j(t)))^2 = 0$. It follows that

$$X_k(t) = \tilde{L}(X_j(t)), \qquad -\infty < t < \infty.$$

The same argument applies with j and k interchanged. More generally, the coherence will vary with frequency indicating a changing pattern of linear association. Regions of high coherence usually have special significance in applications. We will now look at some examples in which the phase and coherence play a special role.

Example 5.3 *An Experiment in Optics*

An experiment traditionally performed in courses in optics is the demonstration of the nature of light coherence. This phenomenon, which makes the modern laser possible, has an analog in the mathematical model we have developed and, in fact, the mathematical coherence parameter is a useful quantitative description of light coherence. For simplicity we will consider two time series $X_1(t)$, $X_2(t)$ which represent the amplitudes of corresponding components of the electromagnetic vectors for two monochromatic light beams. The two beams emanate from a common source and are constrained to travel different paths to a photometer as pictured in Fig. 5.2. The principle of

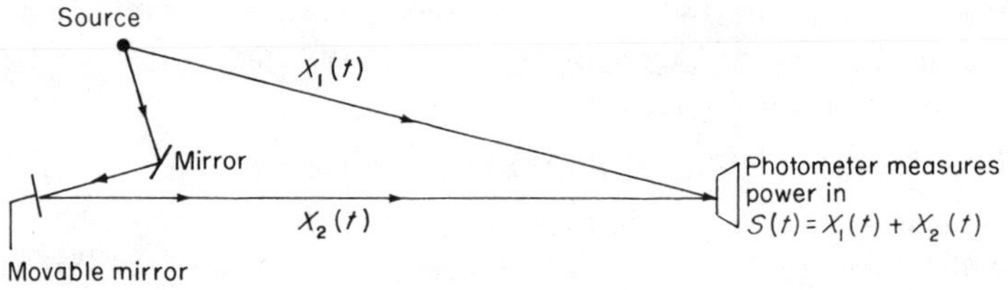

Fig. 5.2 *Schematic model of optical experiment to demonstrate coherence.*

this simple mechanism is identical to that of the Michelson interferometer. The actual operation of the Michelson interferometer is more complex, however [see, e.g., Jenkins and White (1950, p. 239)].

At the point where the two light beams are picked up by the photometer, the amplitudes are added and the power in the resulting sum $S(t) = X_1(t) + X_2(t)$ is measured. If λ denotes the frequency (velocity $\times$ wave number) of the light rays, then the photometer (effectively) records the spectral density $f_S(\lambda)$ of $S(t)$. In terms of our model, the random spectral measure of $S(t)$ is the sum of the spectral measures of $X_1(t)$ and $X_2(t)$;

$$Z_S(d\lambda) = Z_1(d\lambda) + Z_2(d\lambda).$$

Thus, by the properties of expectation,

$$\begin{aligned} f_S(\lambda)\, d\lambda &= EZ_S(d\lambda)\overline{Z_S(d\lambda)} \\ &= EZ_1(d\lambda)\overline{Z_1(d\lambda)} + EZ_1(d\lambda)\overline{Z_2(d\lambda)} + EZ_2(d\lambda)\overline{Z_1(d\lambda)} + EZ_2(d\lambda)\overline{Z_2(d\lambda)} \\ &= [f_{1,1}(\lambda +) f_{1,2}(\lambda) + f_{2,1}(\lambda) + f_{2,2}(\lambda)]\, d\lambda. \end{aligned}$$

However, since $f_{2,1}(\lambda) = \overline{f_{1,2}(\lambda)}$ and $\bar{u} + u = 2\,\mathrm{Re}\, u$, we obtain

$$f_S(\lambda) = f_{1,1}(\lambda) + f_{2,2}(\lambda) + 2c_{1,2}(\lambda).$$

If we assume that the two light beams are adjusted to have the same power, it will follow that $f_{1,1}(\lambda) = f_{2,2}(\lambda)$. Then because of relation (5.53), we have

$$f_S(\lambda) = 2f_{1,1}(\lambda)[1 + \rho_{1,2}(\lambda) \cos \vartheta_{1,2}(\lambda)].$$

Now, if the source consists of two separate light producers, for example, two different light bulbs, then the two light rays will be completely incoherent (in theory) which implies that $\rho_{1,2}(\lambda) = 0$. The light power recorded at the photometer would then be

$$f_S(\lambda) = 2f_{1,1}(\lambda).$$

That is, the power would simply be twice that of each component ray. On the other hand, if the two light rays were perfectly coherent, a property which could be obtained by splitting a single ray into two beams, say, we would have $\rho_{1,2}(\lambda) = 1$ and

$$f_S(\lambda) = 2f_{1,1}(\lambda)[1 + \cos \vartheta_{1,2}(\lambda)].$$

Now, by varying the length of the path of $X_2(t)$ by operating the moveable mirror, the phase angle $\vartheta_{1,2}(\lambda)$, which measures the lead of $X_1(t)$ over $X_2(t)$, will vary. When the two amplitudes are perfectly in phase, which occurs when $\vartheta_{1,2}(\lambda) = \pi k$ for an even integer k, we will have

$$f_S(\lambda) = 4f_{1,1}(\lambda).$$

Thus, the power will be twice that for incoherent light. On the other hand, if the amplitudes are 180° out of phase [$\vartheta_{1,2}(\lambda) = \pi k$ for an odd integer k], the amplitudes will "cancel" and we will observe

$$f_S(\lambda) = 0.$$

Thus, as the mirror is moved, the observed light power will vary from 0 to $4f_{1,1}(\lambda)$.

In actual experiments it is difficult to achieve perfectly coherent light rays, so, in fact, we will have $0 < \rho_{1,2}(\lambda) < 1$ and the light power will vary over the less extreme range $2f_{1,1}(\lambda)(1 - \rho_{1,2}(\lambda))$ to $2f_{1,1}(\lambda)(1 + \rho_{1,2}(\lambda))$ as the path length of $X_2(t)$ is changed.

Example 5.4 *Computing the Transfer Function of an In-Service Linear Filter. Degree of Linearity, Signal-to-Noise Ratio*

The standard laboratory method for determining the transfer function $B(\lambda)$ of a linear filter L parallels our procedure for calculating transfer functions. For several values of λ, sinusoids $e^{i\lambda t}$ are fed into the filter and the amplitudes and phases of the outputs $B(\lambda)e^{i\lambda t}$ are measured. When the filter cannot be taken out of service or if the filter is a complicated physical system, such as the earth between two observation points, it is necessary to be able to determine the transfer function by observing the normal input $X_1(t)$ and output $X_2(t)$ of the filter while it is in operation. This can be done if the input is sufficiently "broad-band"; that is if $f_{1,1}(\lambda) > 0$ over a wide range of frequencies λ. To see this, note that

$$Z_2(d\lambda) = B(\lambda)Z_1(d\lambda).$$

Thus,

$$\begin{aligned} f_{2,1}(\lambda)\,d\lambda &= EZ_2(d\lambda)\overline{Z_1(d\lambda)} \\ &= B(\lambda)E|Z_1(d\lambda)|^2 = B(\lambda)f_{1,1}(\lambda)\,d\lambda. \end{aligned}$$

It follows that

$$B(\lambda) = f_{2,1}(\lambda)/f_{1,1}(\lambda) \qquad \text{wherever} \quad f_{1,1}(\lambda) > 0.$$

Consequently, if the spectra are estimated from observed lengths of the input and output series, the gain and phase function of L can be estimated by means of the expressions

$$\begin{aligned} |B(\lambda)| &= |f_{1,2}(\lambda)|/f_{1,1}(\lambda) = (c_{1,2}^2(\lambda) + q_{1,2}^2(\lambda))^{1/2}/f_{1,1}(\lambda), \\ \vartheta(\lambda) &= -\vartheta_{1,2}(\lambda) = -\operatorname{Arctan}(q_{1,2}(\lambda)/c_{1,2}(\lambda)), \end{aligned}$$

where $f_{1,1}(\lambda)$, $c_{1,2}(\lambda)$, and $q_{1,2}(\lambda)$ are replaced by the corresponding estimates (see Chapter 8). For an example of the determination of the transfer function of a linear system with servo control by cross-spectral methods; see Goodman and Katz (1958).

By the theory of coherence given above, if $X_2(t) = L(X_1(t))$, where L is a linear filter, then $\rho_{1,2}(\lambda) = 1$ for all λ. It follows that deviations of $\rho_{1,2}(\lambda)$ from unity indicate the presence of nonlinearity in the system. (We use the term *nonlinear* to mean that it is not true that $X_2(t) = L(X_1(t))$ for a linear filter L. This is a mild abuse of standard terminology.) There are an infinite variety of types of nonlinearity and coherence does little to indicate which type is operating in the system. However, if the type of nonlinearity is known, its strength can often be related to the value of the coefficient of coherence.

An important nonlinear model is that of a linear system L contaminated additively by random noise. This model can be represented by the expression

$$X_2(t) = Y(t) + N(t),$$

where $Y(t) = L(X_1(t))$ and the input $X_1(t)$, and noise $N(t)$ are uncorrelated time series. A schematic representation of this system is given in Fig. 5.3.

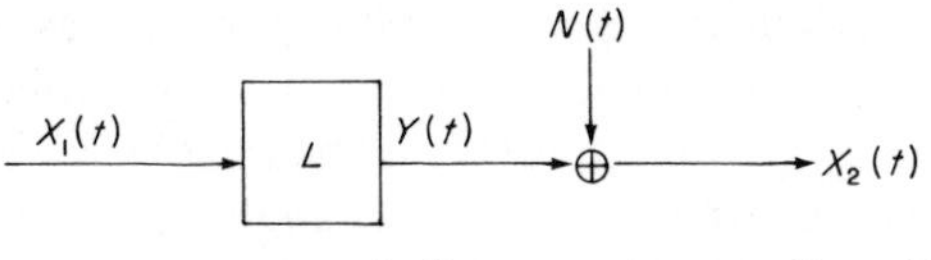

Fig. 5.3 *Schematic representation of a linear system contaminated by additive noise.*

We will assume that both $X_1(t)$ and $N(t)$ are weakly stationary. An important measure of the output signal fidelity is the signal-to-noise density ratio

$$\alpha(\lambda) = f_Y(\lambda)/f_N(\lambda),$$

which provides a frequency-by-frequency comparison of the power in the output signal relative to that of the noise. We will now show that the coefficient of coherence between the "observable" time series $X_1(t)$ and $X_2(t)$ is related to $\alpha(\lambda)$ in a simple way.

The spectral measures of the process $X_2(t)$ can be expressed as

$$Z_2(d\lambda) = B(\lambda)Z_1(d\lambda) + Z_N(d\lambda),$$

where $Z_1(d\lambda)$ and $Z_N(d\lambda)$ are the spectral measures of $X_1(t)$ and $N(t)$. Thus,

$$EZ_2(d\lambda)\overline{Z_1(d\lambda)} = B(\lambda)E|Z_1(d\lambda)|^2 + EZ_N(d\lambda)\overline{Z_1(d\lambda)}$$

and

$$\begin{aligned} E|Z_2(d\lambda)|^2 = |B(\lambda)|^2E|Z_1(d\lambda)|^2 &+ B(\lambda)EZ_1(d\lambda)\overline{Z_N(d\lambda)} \\ &+ \overline{B(\lambda)}EZ_N(d\lambda)\overline{Z_1(d\lambda)} + E|Z_N(d\lambda)|^2. \end{aligned}$$

Since $X_1(t)$ and $N(t)$ are assumed to be uncorrelated,

$$EZ_N(d\lambda)\overline{Z_1(d\lambda)} = EZ_1(d\lambda)\overline{Z_N(d\lambda)} = 0.$$

Thus, the cross-spectral density of $X_2(t)$ and $X_1(t)$ and spectral density of $X_2(t)$ are

$$f_{2,1}(\lambda) = B(\lambda)f_{1,1}(\lambda) \tag{5.61}$$

and

$$f_{2,2}(\lambda) = |B(\lambda)|^2 f_{1,1}(\lambda) + f_N(\lambda).$$

It follows that the squared coefficient of coherence of $X_1(t)$ and $X_2(t)$ is

$$\begin{aligned} \rho_{1,2}^2(\lambda) &= |f_{2,1}(\lambda)|^2/f_{2,2}(\lambda)f_{1,1}(\lambda) \\ &= |B(\lambda)|^2 f_{1,1}(\lambda)/(|B(\lambda)|^2 f_{1,1}(\lambda) + f_N(\lambda)) = \alpha(\lambda)/(1+\alpha(\lambda)). \end{aligned}$$

We have used the relation $f_Y(\lambda) = |B(\lambda)|^2 f_{1,1}(\lambda)$ and the definition of $\alpha(\lambda)$ to obtain the last expression. Thus,

$$\alpha(\lambda) = \rho_{1,2}^2(\lambda)/(1 - \rho_{1,2}^2(\lambda)).$$

Since $\rho_{1,2}^2(\lambda)$ can be estimated from samples of the processes $X_1(t)$ and $X_2(t)$ by techniques to be given in Chapter 8, the signal-to-noise ratio can be estimated. Note that by using (5.61), the characteristics of the linear filter can also be estimated in terms of the observable processes as before. For regions of the spectrum for which $\rho_{1,2}(\lambda)$ is close to one, thus $\alpha(\lambda)$ is large, the output of the linear filter is the dominant component of $X_2(t)$. Thus, a signal with improved *overall* signal-to-noise ratio $\int f_Y(\lambda)\, d\lambda / \int f_N(\lambda)\, d\lambda$ can be achieved by band-pass filtering $X_2(t)$ to eliminate bands of the spectrum for which $\alpha(\lambda)$ is relatively small. An application of these ideas to a problem in seismology was given by Koopmans (1961). We next look at a classical time series problem which is easily solved by the methods developed above.

Example 5.5 *The Historical Filtering Problem*

Let $X(t)$ and $Y(t)$ be stationarily correlated univariate time series where $Y(t)$ is thought of as being an observable process which is a distorted version of the unobservable process $X(t)$. The problem is to construct the linear filter $\tilde{L}$ which, when applied to $Y(t)$, best reproduces $X(t)$. By "best" we will again mean that $\tilde{L}$ minimizes

$$E(X(t) - L(Y(t)))^2$$

among all linear filters L which match $Y(t)$. Since $\tilde{L}$ will be allowed to operate on $Y(t)$ for all times t, $-\infty < t < \infty$, in theory, a complete sample function must be available before the process $\tilde{X}(t) = \tilde{L}(Y(t))$ can be constructed. Consequently, this problem is called the *historical filtering problem* in contrast to the more difficult, *real-time filtering problem* in which the filter is restricted to operate on the past and present of the process. We will consider this problem in Chapter 7.

By the argument given earlier in this section, if $C(\lambda)$ is the transfer function of L, then

$$E(X(t) - L(Y(t)))^2 = \int E|Z_X(d\lambda) - C(\lambda)Z_Y(d\lambda)|^2,$$

and $\tilde{C}(\lambda)$, the transfer function of $\tilde{L}$, minimizes

$$E|Z_X(d\lambda) - C(\lambda)Z_Y(d\lambda)|^2$$

for each λ. The derivation leading to (5.55) produces

$$\tilde{C}(\lambda) = f_{X,Y}(\lambda)/f_Y(\lambda)$$

in this case, where $f_{X,Y}(\lambda)$ and $f_Y(\lambda)$ are the cross-spectral densities of $X(t)$ and $Y(t)$ and the spectral density of $Y(t)$, respectively. Moreover, if $\eta(t) = X(t) - \tilde{X}(t)$, then the degree to which the optimal filter fails to reproduce $X(t)$ is

$$E(\eta(t))^2 = \int f_\eta(\lambda)\, d\lambda,$$

where $f_\eta(\lambda)$ is the spectral density of $\eta(t)$. However, this is the analog of the residual spectral density (5.60), thus can be expressed in the form

$$f_\eta(\lambda) = (1 - \rho_{X,Y}^2(\lambda))f_X(\lambda),$$

where $\rho_{X,Y}(\lambda)$ is the coefficient of coherence between $X(t)$ and $Y(t)$ and $f_X(\lambda)$ is the spectral density of $X(t)$. Thus, as would be expected, the reproducibility of $X(t)$ by a linear filtered version of $Y(t)$ depends on the degree of linear association between the two series as measured by the coherence.

If

$$Y(t) = X(t) + N(t), \tag{5.62}$$

where $N(t)$ is a noise process with spectral density $f_N(\lambda)$, uncorrelated with $X(t)$, then it is easy to check that

$$f_{X,Y}(\lambda) = f_X(\lambda) \qquad \text{and} \qquad f_Y(\lambda) = f_X(\lambda) + f_N(\lambda).$$

Thus,

$$\tilde{C}(\lambda) = f_X(\lambda)/(f_X(\lambda) + f_N(\lambda)).$$

Note that this transfer function is real-valued. Hence, the filter $\tilde{L}$ is symmetric. Thus it is clear in this special case that this filter cannot be used in real-time filtering. When $Y(t)$ has form (5.62), alternate forms for the transfer function are easily seen to be

$$\tilde{C}(\lambda) = \alpha(\lambda)/(1 + \alpha(\lambda)),$$

where $\alpha(\lambda)$ is the signal-to-noise density ratio $\alpha(\lambda) = f_X(\lambda)/f_N(\lambda)$ and

$$\tilde{C}(\lambda) = \rho^2_{X,Y}(\lambda).$$

Both of these expressions make sense, intuitively, since where the signal-to-noise ratio is high, and thus $X(t)$ and $Y(t)$ are nearly equal, the filter $\tilde{L}$ passes $Y(t)$ virtually unchanged. However, where the signal-to-noise ratio is low, which means that $N(t)$ is the dominant component of $Y(t)$, the filter suppresses the noise by passing almost none of the power of $Y(t)$.

Since $X(t)$ and $N(t)$ are generally not observable separately, it is not always possible to estimate the transfer function $\tilde{C}(\lambda)$ from actual data. An exception is provided by the model of the last example in which the signal $X(t)$ is the output of a linear system with observable input $W(t)$. Then, even if the transfer characteristics of the linear system are unknown, the linear filter which, when applied to $Y(t)$, best reproduces $X(t)$ has the transfer function

$$\tilde{C}(\lambda) = \rho^2_{W,Y}(\lambda).$$

This is an immediate consequence of a very useful property of coherence (to be considered next) which implies that $\rho^2_{X,Y}(\lambda) = \rho^2_{W,Y}(\lambda)$. Thus, $\tilde{C}(\lambda)$ can be estimated from observations on $W(t)$ and $Y(t)$ and the impulse response function of the corresponding convolution filter can be obtained by Fourier transformation. Consequently, the best noise suppression filter can be constructed to a good degree of approximation in this case.

An Invariance Property of Coherence

We saw in Section 1.4 that the correlation coefficient is a generalization of the cosine of the angle between vectors to the "angle" between random variables. Since the coefficient of coherence $\rho_{j,k}^X(\lambda)$ is essentially the correlation coefficient of the random spectral measures $Z_j^X(d\lambda)$ and $Z_k^X(d\lambda)$ of two time series, it will also have the properties of the cosine. Cosines are invariant under scale changes of the component vectors, which corresponds to multiplying $Z_j^X(d\lambda)$ and $Z_k^X(d\lambda)$ by complex quantities $B_j(\lambda)$ and $B_k(\lambda)$, respectively. However, this is equivalent to the modification of the spectral measures which results from the application of independent linear filters to $X_j(t)$ and $X_k(t)$. *That is, if* $Y_j(t) = L_j(X_j(t))$ *and* $Y_k(t) = L_k(X_k(t))$, *where* L_j *and* L_k *are any linear filters matching the corresponding inputs, then*

$$\rho_{j,k}^Y(\lambda) = \rho_{j,k}^X(\lambda).$$

This equation is valid for all λ *for which the transfer functions of both filters are nonzero.* This property of coherence, which we will call *invariance under linear filtering*, can be easily established by appealing directly to the definition of coherence and to expression (5.34). It has important practical consequences: Often, it is necessary to compute the degree of association between two time series which have been passed through a series of linear filters with possibly unknown characteristics. Because of this invariance property, the coefficient of coherence of the outputs is the same as that of the inputs. The next example illustrates the usefulness of this property.

Example 5.6 *The Detection of a Coherent Source of Ocean Waves*

Munk *et al.* (1959) considered the possibility of detecting and locating sources of distant underwater disturbances or storm centers by computing the spectral parameters of low-frequency ocean wave amplitudes at a number of stations. For convenience, we will deal with only two stations and will let $X_1(t)$ and $X_2(t)$ denote the recorded amplitudes at these stations. A simple but reasonable model of the mechanisms generating these time series is the following. At the source of the disturbance the energy impressed on the sea surface can be represented by a time series $Y(t)$. The resulting waves travel different paths to the two recording stations and if the sea is assumed to transmit energy linearly, the contributions to the variations in sea height due to the disturbance at the two stations will be $U_1(t) = L_1(Y(t))$ and $U_2(t) = L_2(Y(t))$, respectively. Because both filters have the same input, the invariance of coherence under linear filtering implies that

$$\rho_{1,2}^U(\lambda) \equiv 1.$$

Now it is reasonable to assume that

$$X_j(t) = U_j(t) + N_j(t), \qquad j = 1, 2,$$

where $N_j(t)$ is the contribution to sea height due to causes other than the distant source modeled by $Y(t)$. Moreover, by the definition of the $N_j(t)$'s it is reasonable to assume that $U_j(t)$ and $N_k(t)$ are uncorrelated for $j, k = 1, 2$. Then, any correlation between $X_1(t)$ and $X_2(t)$ can be attributed either to the correlation between $U_1(t)$ and $U_2(t)$ or to that between $N_1(t)$ and $N_2(t)$. The simplest situation is that of rather widely separated recording stations for which $N_1(t)$ and $N_2(t)$ are locally generated, thus, uncorrelated. This precludes the existence of a second distant coherent source, for example. Then the presence of the original source would make itself known by a coefficient of coherence $\rho^X_{1,2}(\lambda)$ near unity over regions of the spectrum for which the power in $U_1(t)$ and $U_2(t)$ dominates that of $N_1(t)$ and $N_2(t)$. Since the power in locally generated sea waves will generally occupy a higher frequency range than that of waves which have traveled a very long distance, the computed coefficient of coherence provides an effective indicator for detecting distant coherent sources.

Under the given simplifying assumptions, the coefficient of coherence between $X_1(t)$ and $X_2(t)$ can be shown to be

$$\rho^X_{1,2}(\lambda) = \alpha_1(\lambda)\alpha_2(\lambda)/((1 + \alpha_1{}^2(\lambda))(1 + \alpha_2{}^2(\lambda)))^{1/2},$$

where $\alpha_j(\lambda)$ is the signal-to-noise density ratio for $U_j(t)$ and $N_j(t)$,

$$\alpha_j(\lambda) = f^U_{j,j}(\lambda)/f^N_{j,j}(\lambda), \qquad j = 1, 2.$$

Consequently, it is possible to interpret the values of the coefficient of coherence directly in terms of the relative signal and noise strengths.

Example 5.7 *An Economics Model and Its Spectral Parameters*

As another illustration of the use of the method for calculating spectral parameters given in the last section we will calculate the spectra, coherence, and phase for discrete time series $X(t)$ and $Y(t)$, which represent the deviations from their mean values of the price and available quantity of a given commodity, respectively. These series are assumed to be related by the equations

$$Y(t) = \alpha X(t-1) + U(t), \qquad X(t) = -\beta Y(t) + V(t), \tag{5.63}$$

where $U(t)$ and $V(t)$ are taken to be zero-mean white noise processes, uncorrelated with one another, with variances $\sigma_U{}^2$, $\sigma_V{}^2$. To obtain a reasonably realistic model, the "loading parameters" α and β are both assumed to be positive. A further restriction will be indicated presently.

The equations for the spectral measures of the various processes can be read off directly from (5.63),

$$Z_Y(d\lambda) = \alpha e^{-i\lambda} Z_X(d\lambda) + Z_U(d\lambda), \qquad Z_X(d\lambda) = -\beta Z_Y(d\lambda) + Z_V(d\lambda).$$

Here, we have used the fact that $e^{-i\lambda}$ is the transfer function of the linear filter which takes $X(t)$ into $X(t-1)$. This is immediate from the usual rule for calculating transfer functions. Rewriting these equations as a pair of linear equations in the unknowns $Z_X(d\lambda)$ and $Z_Y(d\lambda)$, it is easy to obtain the solutions

$$Z_X(d\lambda) = \frac{\beta Z_U(d\lambda) + Z_V(d\lambda)}{1 + \alpha\beta e^{-i\lambda}}, \qquad Z_Y(d\lambda) = \frac{\alpha e^{-i\lambda} Z_V(d\lambda) + Z_U(d\lambda)}{1 + \alpha\beta e^{-i\lambda}}.$$

Thus, the spectral density of $X(t)$ is

$$f_X(\lambda)\,d\lambda = E|Z_X(d\lambda)|^2 = \frac{E[\beta Z_U(d\lambda) + Z_V(d\lambda)]\overline{[\beta Z_U(d\lambda) + Z_V(d\lambda)]}}{|1 + \alpha\beta e^{-i\lambda}|^2}$$

$$= \frac{1}{2\pi} \frac{\beta^2 \sigma_U{}^2 + \sigma_V{}^2}{|1 + \alpha\beta e^{-i\lambda}|^2}\,d\lambda.$$

This yields,

$$f_X(\lambda) = \frac{1}{2\pi} \frac{\beta^2 \sigma_U{}^2 + \sigma_V{}^2}{1 + \alpha^2\beta^2 + 2\alpha\beta\cos\lambda}.$$

Similarly,

$$f_Y(\lambda) = \frac{1}{2\pi} \frac{\alpha^2 \sigma_V{}^2 + \sigma_U{}^2}{1 + \alpha^2\beta^2 + 2\alpha\beta\cos\lambda}$$

and

$$f_{X,Y}(\lambda) = \frac{1}{2\pi} \frac{\beta\sigma_U{}^2 + \alpha e^{i\lambda}\sigma_V{}^2}{|1 + \alpha\beta e^{-i\lambda}|^2}$$

$$= \frac{1}{2\pi} \frac{\beta\sigma_U{}^2 + \alpha\sigma_V{}^2\cos\lambda + i\alpha\sigma_V{}^2\sin\lambda}{1 + \alpha^2\beta^2 + 2\alpha\beta\cos\lambda}.$$

In these calculations we have used the relation

$$EZ_U(d\lambda)Z_V(d\lambda) = 0$$

and the expressions

$$f_U(\lambda) = \sigma_U{}^2/2\pi, \qquad f_V(\lambda) = \sigma_V{}^2/2\pi, \qquad -\pi < \lambda \le \pi.$$

for the spectral densities of the white noise processes.

Now to guarantee that these spectral densities do not "blow up" for some value of λ, that is, the denominators do not go to zero, it is necessary to take $\alpha\beta \ne 1$. As we will show in our discussion of finite parameter models in Chapter 7, it is actually necessary to take $|\alpha\beta| < 1$ in order for $X(t)$ and $Y(t)$ to be realizable processes, certainly a desirable attribute in a realistic model.

Consequently, combining this with the earlier restriction on the parameters, we assume that $0 < \alpha\beta < 1$.

Finally, the squared coefficient of coherence of $X(t)$ and $Y(t)$ is

$$\rho_{X,Y}^2(\lambda) = \frac{|f_{X,Y}(\lambda)|^2}{f_X(\lambda) f_Y(\lambda)} = \frac{(\beta\sigma_U{}^2 + \alpha\sigma_V{}^2 \cos\lambda)^2 + (\alpha\sigma_V{}^2 \sin\lambda)^2}{(\beta^2\sigma_U{}^2 + \sigma_V{}^2)(\alpha^2\sigma_V{}^2 + \sigma_U{}^2)}$$

$$= \frac{\beta^2\sigma_U{}^4 + \alpha^2\sigma_V{}^4 + 2\alpha\beta\sigma_U{}^2\sigma_V{}^2 \cos\lambda}{\beta^2\sigma_U{}^4 + \alpha^2\sigma_V{}^4 + (1 + \alpha^2\beta^2)\sigma_U{}^2\sigma_V{}^2},$$

and the phase angle is

$$\vartheta_{X,Y}(\lambda) = \text{Arctan}\,\frac{\text{Im}\, f_{X,Y}(\lambda)}{\text{Re}\, f_{X,Y}(\lambda)} = \text{Arctan}\left(\frac{\alpha\sigma_V{}^2 \sin\lambda}{\beta\sigma_U{}^2 + \alpha\sigma_V{}^2 \cos\lambda}\right).$$

This model predicts the lowest coherence between price and quantity in the high frequencies where the power in both series is greatest. This is, in general, in accord with real systems wherein short-term variations in price have less influence on the quantity in stock than do longer term variations. Moreover, as long as $\text{Re}\, f_{X,Y}(\lambda)$ is positive, $\vartheta_{X,Y}(\lambda) > 0$ and it follows that variations in price lead those in quantity as would be expected. Thus, this simple model displays a number of features one would expect of a real economic system. To determine the adequacy of the model for a real system, it would be possible to compare the predicted spectral parameters computed above with estimates of the coherence, phase, and spectral densities based on actual observations of the two processes. The estimation of these parameters will be taken up in Chapter 8.

5.6 THE MULTIVARIATE SPECTRAL PARAMETERS, THEIR INTERPRETATIONS AND USES

There is a close relationship between the spectral parameters of time series analysis and the correlational parameters of multivariate statistics due to the virtually identical representations of inner products and linear transformations for the vectors of spectral measures $Z_1(d\lambda), \ldots, Z_p(d\lambda)$ of a multivariate time series on one hand and vectors of zero-mean random variables with finite variances on the other. Thus, as we saw, the coefficient of coherence is essentially the modulus of a correlation coefficient with the properties and interpretation one would expect of this parameter. The remaining multivariate spectral parameters are defined to retain this parallelism and, in fact, the relationship is so close that most of the spectral theory can be taken, with only minor modifications, from texts on multivariate analysis such as the one by Anderson (1958). See also Cramér (1951a). We will cover the essential features of the theory without going into great depth. Additional details are covered by Hannan (1970) and Koopmans (1964b). Due to an important difference in emphasis, we will treat the correlation theory and regression theory separately.

Multivariate Correlation Analysis

If $\mathbf{X}(t)$ is a multivariate weakly stationary process with $p > 2$ components, then it is often important to account for the interaction among several of the components when the power in one of them is to be determined or when the association between two of the components is to be assessed. It may be the case, for example, that most of the power in a given series can be removed by subtracting off a function of various other of the components. This would indicate a relationship among the components which might arise, for example, as the result of a common "driving mechanism."

Again the type of association to be considered is *linear association*. Thus, we will account for the influence of components $X_{m_1}(t), X_{m_2}(t), \ldots, X_{m_q}(t)$ on $X_j(t)$ by constructing the linear function of these series which best approximates $X_j(t)$. That is, if

$$Y_j(t) = \mathbf{L}\big(X_{m_1}(t), \ldots, X_{m_q}(t)\big) \tag{5.64}$$

denotes a multivariate linear filter with q inputs and one output, as described in Section 5.4, then the filter which best approximates $X_j(t)$ is the one which minimizes

$$E(X_j(t) - Y_j(t))^2. \tag{5.65}$$

Let $\tilde{X}_{j\cdot\mathbf{m}}(t)$ denote the output of the minimizing filter, where $\mathbf{m} = (m_1, m_2, \ldots, m_q)'$. Then, to measure the strength of the linear regression of $X_j(t)$ on $X_{m_1}(t), \ldots, X_{m_q}(t)$ we can compare the power of $\tilde{X}_{j\cdot\mathbf{m}}(t)$ with the power of $X_j(t)$. This comparison can be made frequency by frequency by using the coefficient of coherence of the two time series. This parameter, called the *multiple coherence*, is defined in terms of the spectral densities and cross-spectral densities of $X_j(t)$ and $\tilde{X}_{j\cdot\mathbf{m}}(t)$. Consequently, it is important to be able to calculate these densities.

The problem of describing the unconditional linear relationship between two component time series $X_j(t)$ and $X_k(t)$ was solved in the last section by the introduction of the coefficient of coherence. When the vector time series has other components, it is likely that part of the coherence between $X_j(t)$ and $X_k(t)$ can be attributed to the linear relationship each has with some or all of the others. To determine the coherence between $X_j(t)$ and $X_k(t)$ which is *not* attributable to the regression of these processes on $X_{m_1}(t), \ldots, X_{m_q}(t)$, say, it is reasonable to adjust each of the two processes by subtracting off its best "explanation" as a linear function of the $X_{m_j}(t)$'s. Thus, the *partial coherence* is the coefficient of coherence of the residual processes $X_j(t) - \tilde{X}_{j\cdot\mathbf{m}}(t)$ and $X_k(t) - \tilde{X}_{k\cdot\mathbf{m}}(t)$. Again, it is necessary to compute the spectral densities from which this parameter is calculated. To do this, we will use the spectral representation of $\mathbf{X}(t)$ and that of a multidimensional linear filter to obtain the random spectral measure of $\tilde{X}_{j\cdot\mathbf{m}}(t)$.

In spectral form, Eq. (5.64) can be written

$$Z_j{}^Y(d\lambda) = \sum_{r=1}^{q} B_{j,r}(\lambda) Z_{m_r}^X(d\lambda), \tag{5.66}$$

where the $1 \times q$ vector $\mathbf{B}_j(\lambda) = [B_{j,r}(\lambda)]$ is the transfer function of $\mathbf{L}$. The argument of the last section can again be applied to show that

$$E(X_j(t) - Y_j(t))^2 = \int E\left| Z_j{}^X(d\lambda) - \sum_{r=1}^{q} B_{j,r}(\lambda) Z_{m_r}^X(d\lambda)\right|^2.$$

Thus, the minimum of this expression occurs at the transfer function $\tilde{\mathbf{B}}_j(\lambda)$ which determines the projection of $Z_j{}^X(d\lambda)$ on the linear subspace generated by $Z_{m_1}^X(d\lambda), \ldots, Z_{m_q}^X(d\lambda)$ for each λ. By the criterion for determining projections given in Section 1.3, the elements of the transfer function must satisfy the conditions

$$E\left[Z_j{}^X(d\lambda) - \sum_{r=1}^{q} \tilde{B}_{j,r}(\lambda) Z_{m_r}^X(d\lambda)\right]\overline{[Z_{m_s}^X(d\lambda)]} = 0, \qquad s = 1, 2, \ldots, q.$$

In terms of the spectral density functions, these equations are

$$\sum_{r=1}^{q} \tilde{B}_{j,r}(\lambda) f_{m_r, m_s}^X(\lambda) = f_{j, m_s}^X(\lambda), \qquad s = 1, 2, \ldots, q \tag{5.67}$$

or, in vector form,

$$\tilde{\mathbf{B}}_j(\lambda)\mathbf{f}_{\mathbf{m}}{}^X(\lambda) = \mathbf{f}_{j,\mathbf{m}}^X(\lambda),$$

where $\mathbf{f}_{\mathbf{m}}{}^X(\lambda)$ and $\mathbf{f}_{j,\mathbf{m}}^X(\lambda)$ denote the $q \times q$ matrix and $1 \times q$ vector of spectral densities indicated in (5.67). Thus, when the inverse exists,

$$\tilde{\mathbf{B}}_j(\lambda) = \mathbf{f}_{j,\mathbf{m}}^X(\lambda)\mathbf{f}_{\mathbf{m}}{}^X(\lambda)^{-1}. \tag{5.68}$$

When the inverse does not exist, it is still possible to define the transfer function by a formula of this type with the inverse replaced by what is known as a *pseudo inverse* [see, e.g., Koopmans (1964b)]. We will not be concerned with this, since the case in which the inverse exists is by far the more important one in practice.

Now, by the definition of $\tilde{X}_{j\cdot\mathbf{m}}(t)$ and by (5.66), the spectral measure of this process is

$$Z_{j\cdot\mathbf{m}}^{\tilde{X}}(d\lambda) = \mathbf{f}_{j,\mathbf{m}}^X(\lambda)\mathbf{f}_{\mathbf{m}}{}^X(\lambda)^{-1}\mathbf{Z}_{\mathbf{m}}{}^X(d\lambda), \tag{5.69}$$

where

$$\mathbf{Z_m}^X(d\lambda) = (Z_{m_1}^X(d\lambda), \ldots, Z_{m_q}^X(d\lambda))'.$$

Moreover,

$$\begin{aligned} E|Z_{j\cdot\mathbf{m}}^{\hat{X}}(d\lambda)|^2 &= E(\mathbf{f}_{j,\mathbf{m}}^X(\lambda)\mathbf{f_m}^X(\lambda)^{-1}\mathbf{Z_m}^X(d\lambda))(\mathbf{f}_{j,\mathbf{m}}^X(\lambda)\mathbf{f_m}^X(\lambda)^{-1}\mathbf{Z_m}^X(d\lambda))^* \\ &= f_{j,\mathbf{m}}^X(\lambda)\mathbf{f_m}^X(\lambda)^{-1}\mathbf{f}_{j,\mathbf{m}}^X(\lambda)^*\, d\lambda, \end{aligned}$$

since

$$E\mathbf{Z_m}^X(d\lambda)\mathbf{Z_m}^X(d\lambda)^* = \mathbf{f_m}^X(\lambda)\, d\lambda.$$

It follows, that the spectral density function of $\tilde{X}_{j\cdot\mathbf{m}}(t)$ is

$$f_{j,j\cdot\mathbf{m}}^{\tilde{X}}(\lambda) = \mathbf{f}_{j,\mathbf{m}}^X(\lambda)\mathbf{f_m}^X(\lambda)^{-1}\mathbf{f}_{j,\mathbf{m}}(\lambda)^*. \tag{5.70}$$

Thus, the multiple coherence of $X_j(t)$ *on* $X_{m_1}(t), \ldots, X_{m_q}(t)$, *which is the proportion of the power (density) at frequency* λ *attributable to the linear regression of* $X_j(t)$ *on* $X_{m_1}(t), \ldots, X_{m_q}(t)$, *is given by*

$$\begin{aligned} R_{j\cdot\mathbf{m}}^2(\lambda) &= f_{j,j\cdot\mathbf{m}}^{\tilde{X}}(\lambda)/f_{j,j}^X(\lambda) \\ &= \mathbf{f}_{j,\mathbf{m}}^X(\lambda)\mathbf{f_m}^X(\lambda)^{-1}\mathbf{f}_{j,\mathbf{m}}^X(\lambda)^*/f_{j,j}^X(\lambda). \end{aligned} \tag{5.71}$$

By the Pythagorean theorem, since $(Z_j^X(d\lambda) - Z_{j\cdot\mathbf{m}}^{\tilde{X}}(d\lambda)) \perp Z_{j\cdot\mathbf{m}}^{\tilde{X}}(d\lambda)$, we have

$$E|Z_j^X(d\lambda)|^2 = E|Z_{j\cdot\mathbf{m}}^{\tilde{X}}(d\lambda)|^2 + E|Z_j^X(d\lambda) - Z_{j\cdot\mathbf{m}}^{\tilde{X}}(d\lambda)|^2.$$

Thus,

$$f_{j,j}^X(\lambda) = f_{j,j\cdot\mathbf{m}}^{\tilde{X}}(\lambda) + f_{j,j\cdot\mathbf{m}}^U(\lambda), \tag{5.72}$$

where $f_{j,j\cdot\mathbf{m}}^U(\lambda)$ is the spectral density of the residual process

$$U_{j\cdot\mathbf{m}}(t) = X_j(t) - \tilde{X}_{j\cdot\mathbf{m}}(t).$$

It follows from (5.71) and (5.72) that $0 \le R_{j\cdot\mathbf{m}}^2(\lambda) \le 1$ as one would expect, and, moreover, *the proportion of the power (density) in* $X_j(t)$ *at frequency* λ *not attributable to its linear regression on* $X_{m_1}(t), \ldots, X_{m_q}(t)$ *is*

$$f_{j,j\cdot\mathbf{m}}^U(\lambda)/f_{j,j}^X(\lambda) = 1 - R_{j\cdot\mathbf{m}}^2(\lambda).$$

This also yields the useful expression

$$f_{j,j\cdot\mathbf{m}}^U(\lambda) = (1 - R_{j\cdot\mathbf{m}}^2(\lambda))f_{j,j}^X(\lambda) \tag{5.73}$$

for the residual spectral density.

The *partial coherence* of $X_j(t)$ and $X_k(t)$ with the regression on $X_{m_1}(t), \ldots, X_{m_q}(t)$ removed is the coefficient of coherence of $U_{j\cdot\mathbf{m}}(t)$ and $U_{k\cdot\mathbf{m}}(t)$. This parameter can now be computed. Moreover, the computation can be simplified by the observation that $Z_{j\cdot\mathbf{m}}^U(d\lambda) \perp Z_{k\cdot\mathbf{m}}^{\tilde{X}}(d\lambda)$, since $Z_{k\cdot\mathbf{m}}^{\tilde{X}}(d\lambda)$

is in the linear subspace generated by $Z_{m_1}^X(d\lambda), \ldots, Z_{m_q}^X(d\lambda)$. We can use this fact to write

$$EZ_{j\cdot\mathbf{m}}^U(d\lambda)\overline{Z_{k\cdot\mathbf{m}}^U(d\lambda)} = E(Z_j^X(d\lambda) - Z_{j\cdot\mathbf{m}}^{\tilde{X}}(d\lambda))\overline{(Z_k^X(d\lambda))}$$
$$= (f_{j,k}^X(\lambda) - \mathbf{f}_{j,\mathbf{m}}^X(\lambda)\mathbf{f}_\mathbf{m}^X(\lambda)^{-1}\mathbf{f}_{k,\mathbf{m}}^X(\lambda)^*)\, d\lambda.$$

Thus, we obtain

$$f_{j,k\cdot\mathbf{m}}^U(\lambda) = f_{j,k}^X(\lambda) - \mathbf{f}_{j,\mathbf{m}}^X(\lambda)\mathbf{f}_\mathbf{m}^X(\lambda)^{-1}\mathbf{f}_{k,\mathbf{m}}^X(\lambda)^*. \tag{5.74}$$

The $(p-q)\times(p-q)$ matrix $\mathbf{f}_\mathbf{m}^U(\lambda)$ with elements given by (5.74) as j and k range over the indices $1, 2, \ldots, p$ excluding $m_1, m_2, \ldots, m_q$, is called the *residual spectral matrix*. The *complex partial coherence* is

$$\gamma_{j,k\cdot\mathbf{m}}(\lambda) = f_{j,k\cdot\mathbf{m}}^U(\lambda)/(f_{j,j\cdot\mathbf{m}}^U(\lambda)f_{k,k\cdot\mathbf{m}}^U(\lambda))^{1/2}$$
$$= (1 - R_{j,k\cdot\mathbf{m}}^2(\lambda))\gamma_{j,k}(\lambda)/((1 - R_{j\cdot\mathbf{m}}^2(\lambda))(1 - R_{k\cdot\mathbf{m}}^2(\lambda)))^{1/2}, \tag{5.75}$$

where we have let

$$R_{j,k\cdot\mathbf{m}}^2(\lambda) = f_{j,k\cdot\mathbf{m}}^{\tilde{X}}(\lambda)/f_{j,k}^X(\lambda) = \mathbf{f}_{j,\mathbf{m}}^X(\lambda)\mathbf{f}_\mathbf{m}^X(\lambda)^{-1}\mathbf{f}_{k,\mathbf{m}}^X(\lambda)^*/f_{j,k}^X(\lambda), \tag{5.76}$$

and $\gamma_{j,k}(\lambda)$ is the complex coherence of $X_j(t)$ and $X_k(t)$. The partial coherence is then

$$\rho_{j,k\cdot\mathbf{m}}(\lambda) = |1 - R_{j,k\cdot\mathbf{m}}^2(\lambda)|\rho_{j,k}(\lambda)/((1 - R_{j\cdot\mathbf{m}}^2(\lambda))(1 - R_{k\cdot\mathbf{m}}^2(\lambda)))^{1/2}. \tag{5.77}$$

The phase angle between the residual processes $U_j(t)$ and $U_k(t)$ is

$$\vartheta_{j,k\cdot\mathbf{m}}(\lambda) = \psi_{j,k\cdot\mathbf{m}}(\lambda) + \vartheta_{j,k}(\lambda), \tag{5.78}$$

where

$$\psi_{j,k\cdot\mathbf{m}}(\lambda) = \arg(1 - R_{j,k\cdot\mathbf{m}}^2(\lambda)).$$

It can be seen from expressions (5.71), (5.76), and (5.77) that the multiple and partial coherence are elementary functions of the elements of the spectral density matrix $\mathbf{f}^X(\lambda)$. The standard procedure for obtaining statistical estimates of these parameters is to first obtain estimates of the elements of $\mathbf{f}^X(\lambda)$, then enter these estimates in the above formulas in place of the corresponding parameters. Since this must be done for a variety of values of λ and, often, for several selections of the indices j, k, and elements in $\mathbf{m}$, the number of computations required can be quite large. Fortunately, convenient digital computer programs exist for carrying out these calculations.

For example, matrices of estimated spectra and cross spectra can be calculated by means of the program BMDX92 (Dixon, 1969). This is used as input to the multiple time series spectral analysis program BMDX68 which computes the estimates of the residual spectral matrix, coherences, multiple coherences, and matrix transfer function (5.68), among other things.

Example 5.8 *Calculation of the Spectral Parameters in a Simple Case*

When **m** consists of a single index, the expressions for the spectral parameters are especially simple. Suppose $R_{1\cdot 3}^2(\lambda)$ and $\rho_{1,2\cdot 3}(\lambda)$ are to be calculated. Now,

$$\mathbf{f_m}^X(\lambda) = [f_{3,3}^X(\lambda)]$$

and it follows that

$$\begin{aligned} R_{1\cdot 3}^2(\lambda) &= f_{1,\mathbf{m}}^X(\lambda)\mathbf{f_m}^X(\lambda)^{-1}\overline{f_{1,\mathbf{m}}^X(\lambda)}/f_{1,1}^X(\lambda) \\ &= |f_{1,3}^X(\lambda)|^2/f_{1,1}^X(\lambda)f_{3,3}^X(\lambda) = \rho_{1,3}^2(\lambda). \end{aligned}$$

The equality of the multiple and ordinary coherence agrees with the fact that the interpretations of the two parameters are the same in this case. From (5.76) and the definition of complex coherence,

$$\begin{aligned} &[1 - R_{1,2\cdot 3}^2(\lambda)]\gamma_{1,2}(\lambda) \\ &\quad = \big(f_{3,3}^X(\lambda)f_{1,2}^X(\lambda) - f_{1,3}^X(\lambda)f_{3,2}^X(\lambda)\big)/f_{3,3}^X(\lambda)\big(f_{1,1}^X(\lambda)f_{2,2}^X(\lambda)\big)^{1/2}. \end{aligned}$$

Thus, from the second expression on the right-hand side of (5.75), we obtain

$$\gamma_{1,2\cdot 3}(\lambda) = \big(\gamma_{1,2}(\lambda) - \gamma_{1,3}(\lambda)\overline{\gamma_{3,2}(\lambda)}\big)/((1 - \rho_{1,3}^2(\lambda))(1 - \rho_{2,3}^2(\lambda))^{1/2}.$$

This is the complex analog of a well-known expression for correlation coefficients. The partial coefficient of coherence is, then,

$$\rho_{1,2\cdot 3}(\lambda) = |\gamma_{1,2\cdot 3}(\lambda)|.$$

Example 5.9 *An Application of Partial Coherence to a Biomedical Problem*

Gersh and Goddard (1970) considered the problem of determining the location of an epileptic focus in the brain of a cat based on EEG records from electrodes implanted in six deep sites in the brain. The goal of the study was to determine whether any one of the six sites could be interpreted as "driving" the others, based on recordings made during an induced epileptic seizure. The data, then, consisted of a six-dimensional time series recorded over the period of the seizure—some 8 seconds in the given record.

If, for example, site 1 is "driving" the remaining sites 2–6, one might postulate a model of the type pictured in Fig. 5.4. The noise processes $N_j(t)$, $j = 2, \ldots, 6$ are taken to be uncorrelated with each other and with $X_1(t)$. This model could be "identified" over regions of the spectrum in which the noise power is relatively low by the fact that the pairwise coherences $\rho_{j,k}(\lambda)$ would all be reasonably large as would all partial coherences $\rho_{j,k\cdot l}(\lambda)$ regressed on single series other than $X_1(t)$, i.e., for $l \neq 1$. This is true, since the dominant variation in each pair of series would be due to $X_1(t)$. On the other hand, $\rho_{j,k\cdot 1}(\lambda)$ would be relatively small, since upon

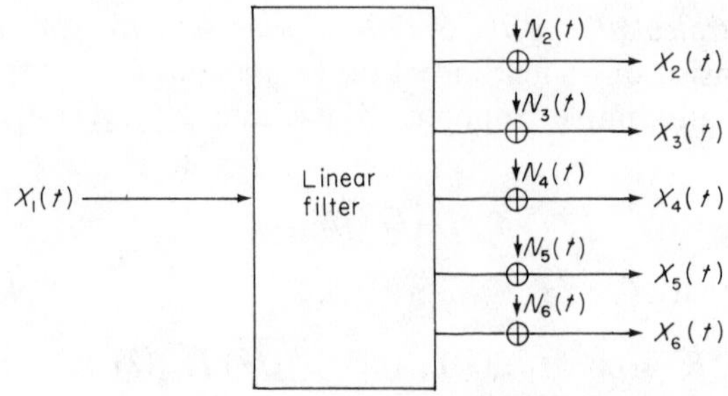

Fig. 5.4 *Schematic representation of a six-dimensional time series in which* $X_1(t)$ *"drives" the remaining components.*

removing the effects of $X_1(t)$, the comparison would be, principally, between the incoherent residual noises.

By computing the coherences for each pair of series and for all pairs regressed on each single series, Gersh and Goddard were able to identify one of the series as having the characteristics of the "driving" series in this model.

Multivariate Spectral Regression Analysis

In the regression context, the component processes of the multivariate time series are partitioned into two categories—the dependent processes and the independent processes. To emphasize this distinction, the dependent components will be assembled into a $(q \times 1)$-dimensional vector process $\mathbf{Y}(t)$ and the independent processes will constitute a $(p \times 1)$-vector process $\mathbf{X}(t)$. It is now assumed that

$$\mathbf{Y}(t) = \mathbf{L}(\mathbf{X}(t)) + \boldsymbol{\eta}(t), \tag{5.79}$$

where $\mathbf{L}$ is a multivariate linear filter with unknown $(q \times p)$-dimensional transfer function $\mathbf{B}(\lambda)$ and $\boldsymbol{\eta}(t)$ is an unobservable $(q \times 1)$-dimensional process uncorrelated with $\mathbf{X}(t)$. Thus $\mathbf{Y}(t)$ is assumed to arise from a fixed but unknown linear transformation of $\mathbf{X}(t)$ which is disturbed by an error process $\boldsymbol{\eta}(t)$. The extent of the deviation of $\mathbf{Y}(t)$ from a linear function of $\mathbf{X}(t)$ is measured by the unknown spectral density matrix $\mathbf{f}^{\eta}(\lambda)$. This matrix and the transfer function $\mathbf{B}(\lambda)$, which indicates how the linear dependence is parcelled out to the various input and output time series, are the principal parameters of interest.

Except for the more general form of the linear transformation, model (5.79) is identical to the distributed lag model of Example 5.2. Precisely the same argument can be used to calculate the transfer function of $\mathbf{L}$ and the spectral density function of $\boldsymbol{\eta}(t)$. Recall from Eqs. (5.40) and (5.41) that

$$\mathbf{B}(\lambda) = \mathbf{f}^{Y,X}(\lambda)\mathbf{f}^{X}(\lambda)^{-1} \tag{5.80}$$

and

$$\mathbf{f}^{\eta}(\lambda) = \mathbf{f}^{Y}(\lambda) - \mathbf{f}^{Y,X}(\lambda)\mathbf{f}^{X}(\lambda)^{-1}\mathbf{f}^{X,Y}(\lambda), \tag{5.81}$$

where

$$\mathbf{f}^{X}(\lambda)\,d\lambda = E\mathbf{Z}^{X}(d\lambda)\mathbf{Z}^{X}(d\lambda)^{*}, \qquad \mathbf{f}^{Y}(\lambda)\,d\lambda = E\mathbf{Z}^{Y}(d\lambda)\mathbf{Z}^{Y}(d\lambda)^{*},$$

and

$$\mathbf{f}^{X,Y}(\lambda)\,d\lambda = E\mathbf{Z}^{X}(d\lambda)\mathbf{Z}^{Y}(d\lambda)^{*}.$$

In the regression context, processes $\mathbf{X}(t)$ and $\mathbf{Y}(t)$ are observable. Consequently, estimates of $\mathbf{f}^{X}(\lambda)$, $\mathbf{f}^{Y}(\lambda)$, and $\mathbf{f}^{X,Y}(\lambda)$ can be obtained by the technique to be discussed in Chapter 8. By replacing the matrices of parameters in (5.80) and (5.81) by their estimators, estimates of $\mathbf{B}(\lambda)$ and $\mathbf{f}^{\eta}(\lambda)$ can be computed. We will discuss some of the properties of these estimators in Chapter 8.

A matrix parameter comparable to the coefficient of coherence can be defined which provides a more convenient measure of the extent of the linear regression of $\mathbf{Y}(t)$ on $\mathbf{X}(t)$ than does $\mathbf{f}^{\eta}(\lambda)$. If these processes were one-dimensional, the complex coherence would be

$$\gamma(\lambda) = f^{X,Y}(\lambda)/(f^{X}(\lambda)f^{Y}(\lambda))^{1/2}.$$

The fact that $\mathbf{f}^{X}(\lambda)$ and $\mathbf{f}^{Y}(\lambda)$ are nonnegative definite matrices (see Section 5.3) makes it possible to define a comparable expression in the multidimensional case. We need a few facts about nonnegative definite (Hermetian) matrices which can be found, for example, in the book by Graybill (1969): The inverse of a nonnegative definite matrix, when it exists, is again nonnegative definite. The eigenvalues of such a matrix are all real and nonnegative. Thus, the inverse will fail to exist only when one or more of the eigenvalues are zero. Nonnegative definite matrices have well-defined square roots. That is, if $\mathbf{A}$ is a nonnegative definite matrix, there is a nonnegative definite matrix $\mathbf{C}$ such that $\mathbf{CC} = \mathbf{A}$.

Combining these results, we can define the *matrix complex coherence* by

$$\boldsymbol{\gamma}(\lambda) = \mathbf{f}^{X}(\lambda)^{-1/2}\mathbf{f}^{X,Y}(\lambda)\mathbf{f}^{Y}(\lambda)^{-1/2}, \tag{5.82}$$

where $\mathbf{A}^{-1/2}$ denotes the square root of the inverse of the matrix $\mathbf{A}$. Now the matrix parameter

$$\boldsymbol{\rho}^{2}(\lambda) = \boldsymbol{\gamma}(\lambda)^{*}\boldsymbol{\gamma}(\lambda) = \mathbf{f}^{Y}(\lambda)^{-1/2}\mathbf{f}^{Y,X}(\lambda)\mathbf{f}^{X}(\lambda)^{-1}\mathbf{f}^{X,Y}(\lambda)\mathbf{f}^{Y}(\lambda)^{-1/2} \tag{5.83}$$

is comparable to the squared coherence. It can be shown to be a nonnegative definite matrix, consequently, all its eigenvalues are nonnegative. Moreover, from (5.81) we have

$$\mathbf{f}^{\eta}(\lambda) = \mathbf{f}^{Y}(\lambda)^{1/2}(\mathbf{I} - \boldsymbol{\rho}^{2}(\lambda))\mathbf{f}^{Y}(\lambda)^{1/2}, \tag{5.84}$$

where $\mathbf{I}$ is the $q \times q$ identity matrix. When $\boldsymbol{\eta}(t)$ has no power in any component at frequency λ, i.e., $\mathbf{f}^{\eta}(\lambda) = \mathbf{0}$, we have

$$\boldsymbol{\rho}^2(\lambda) = \mathbf{I}$$

[provided $\mathbf{f}^{Y}(\lambda)$ is nonsingular]. In this case all eigenvalues of $\boldsymbol{\rho}^2(\lambda)$ are equal to 1. Now, it is easily shown from (5.84) that $\mathbf{I} - \boldsymbol{\rho}^2(\lambda)$ is also nonnegative definite. Since the eigenvalues of this matrix are one minus the eigenvalues of $\boldsymbol{\rho}^2(\lambda)$, it follows that the eigenvalues of $\boldsymbol{\rho}^2(\lambda)$ are all between zero and one. Thus the complete linear dependence of $\mathbf{Y}(t)$ on $\mathbf{X}(t)$ at frequency λ corresponds to the extreme case in which all eigenvalues are one. On the other hand, if all eigenvalues of $\boldsymbol{\rho}^2(\lambda)$ are zero, then $\boldsymbol{\rho}^2(\lambda)$ is the zero matrix and (5.84) implies that $\mathbf{f}^{\eta}(\lambda) = \mathbf{f}^{Y}(\lambda)$. In this case $\mathbf{Y}(t)$ is completely "explained" by the error term $\boldsymbol{\eta}(t)$ at frequency λ and no linear relationship between $\mathbf{Y}(t)$ and $\mathbf{X}(t)$ exists.

More generally, (5.84) is seen to be the matrix version of (5.60). The eigenvalues of $\boldsymbol{\rho}^2(\lambda)$ can be used to assess the relative degree of linear regression of $\mathbf{Y}(t)$ on $\mathbf{X}(t)$. Statistical estimates of these eigenvalues can be used to test various hypotheses about this regression. For example, the hypothesis of no linear regression at frequency λ can be tested by comparing the estimated largest eigenvalue or the sum of the eigenvalues of $\boldsymbol{\rho}^2(\lambda)$ with zero. The second possibility is equivalent to testing the trace of $\boldsymbol{\rho}^2(\lambda)$ to be zero. This hypothesis can also be framed as a hypothesis concerning $\mathbf{B}(\lambda)$, since it is easily seen that

$$\boldsymbol{\rho}^2(\lambda) = \mathbf{B}(\lambda)\mathbf{f}^{X}(\lambda)\mathbf{B}(\lambda)^*.$$

If $\mathbf{f}^{X}(\lambda)$ is nonsingular, the hypothesis of no regression at frequency λ is equivalent to the hypothesis $\mathbf{B}(\lambda) = \mathbf{0}$.

When the $\mathbf{Y}(t)$ series is one-dimensional, i.e., $q = 1$, expression (5.83) yields

$$\boldsymbol{\rho}^2(\lambda) = \mathbf{f}^{Y,X}(\lambda)\mathbf{f}^{X}(\lambda)^{-1}\mathbf{f}^{Y,X}(\lambda)^*/f^{Y}(\lambda).$$

Comparing this with (5.71), it is seen that $\boldsymbol{\rho}^2(\lambda)$ is simply the multiple coherence of $Y(t)$ with $\mathbf{X}(t)$ as we would expect. Thus, the multiple coherence is an important regression parameter in this special case. When $p = 1$ it can be shown that

$$\boldsymbol{\rho}^2(\lambda) = \mathbf{f}^{X,Y}(\lambda)\mathbf{f}^{Y}(\lambda)^{-1}\mathbf{f}^{Y,X}(\lambda)/f^{X}(\lambda), \tag{5.85}$$

the multiple coherence of $X(t)$ with $\mathbf{Y}(t)$. This indicates that the matrix parameter $\boldsymbol{\rho}^2(\lambda)$ possesses a symmetry in the time series $\mathbf{X}(t)$ and $\mathbf{Y}(t)$ roughly analogous to that displayed by the coefficient of coherence.

Finally, although we have derived the above regression theory model based on the assumption that all time series are weakly stationary processes, it is important to note that it is also possible to derive a statistical regression theory based on a model in which the $\mathbf{X}(t)$ time series is nonrandom. In this theory, which has been developed by Brillinger (1970), the randomness is assumed to be due entirely to the error process $\boldsymbol{\eta}(t)$ appearing in (5.79). The resulting model is closer to the standard regression theory of statistics than is the model we have discussed. In particular, although the spectral density function of the error process $\mathbf{f}^{\eta}(\lambda)$ is well defined, none of the other spectral densities need be. However, expressions (5.80) and (5.81) are still used to construct estimates of $\mathbf{B}(\lambda)$ and $\mathbf{f}^{\eta}(\lambda)$ based on the same functions of the observed time series as would be used if the $\mathbf{X}(t)$ series were stochastic. Although the estimation procedures are essentially identical, the statistical distributors of the estimates are different [see Brillinger (1970)].

Some Spectral Regression Examples

The following three examples will give some idea of the scope of applicability of spectral regression methods. Complete descriptions of the studies and their conclusions are not feasible here. Further details can be found in the indicated references.

Example 5.10 *An Application to Metallurgy*

The earliest time series study in which spectral regression techniques were used is attributed to Tick (1955) in which the variability of the hot metal output of a blast furnace was evaluated as a function of (i.e., was regressed on) such variables as hot blast temperature, wind rate, and amounts of ore, coke, and limestone. These variables unavoidably vary with time and are correlated with one another to different degrees making the time series model a natural choice. The dependence of the variability (power) of hot metal output on the regression time series was evaluated by integrating the appropriate estimated residual spectral densities over frequency.

Example 5.11 *A Study of the Relationship between Sun Spots and Meteorological Data*

A contribution to the long-standing controversy concerning the influence of solar energy indicators, such as sun-spot numbers, on terrestrial time series, such as temperature and rainfall data, was made by Brillinger (1969). In this study the independent time series $\mathbf{X}(t)$ was taken to be the one-dimensional series of monthly relative sun-spot numbers. Three different sets of dependent series $\mathbf{Y}(t)$ were used; (i) Santa Fe, New Mexico rainfall ($p = 1$), (ii) English rainfall ($p = 1$), and (iii) temperatures at 14 European stations ($p = 14$). The degree of linear regression of each dependent series on sun-spot numbers was evaluated by estimating the generalized coherence in form (5.85). Although the estimates showed a considerable variability with frequency, they were smaller than 0.5 at all frequencies in all three cases. The sun-spot coherence was larger with temperature at almost all frequencies than with rainfall which supports the conclusion of other investigators that little relationship between rainfall and sun-spot numbers exists. A comprehensive summary of evidence for and against the existence of relationships between solar energy parameters and various terrestrial series is given by Monin and Vulis (1971).

Example 5.12 *A Study of Sea Level Data*

Groves and Hannan (1968) studied records of sea level, surface atmospheric pressure, and wind velocity components at Kwajelein and Eniwetok in the Marshall Islands for the purpose of determining oceanic influences on sea level records free from local weather noise. To do this, the two sea level series were taken as the dependent series and were regressed on the six series of pressure and wind velocities. The residual series of sea levels, with weather effects accounted for, were then compared. It was found that the partial coherence of sea level records at the two islands with weather effects removed was smaller than the ordinary coherence of these records. This indicates that most of the coherence in sea level records is probably due to coherent weather patterns rather than to coherent patterns of water movement from non-weather-induced phenomena.

As with all studies of this type, the interpretation of results is subject to the statistical uncertainty of the estimates of the parameters. We will consider methods for evaluating and controlling this uncertainty in Chapters 8 and 9.

APPENDIX TO CHAPTER 5

A5.1 The Multidimensional Spectral Representation

The mathematical setting for the spectral representation of a multivariate weakly stationary process is somewhat more involved than that for the representation of one-dimensional processes, but many of the formulas carry over with little change. We will touch on only a few of the details here. More extensive analyses are given by Koopmans (1964a,b) and Wiener and Masani (1957, 1958).

A vector analog of the space $L_2(P)$ is required. Take the process $\mathbf{X}(t)$ to be p-dimensional. For each t the vector $\mathbf{X}(t)$ is an element of the product space $\mathbf{L}_2(P) = [L_2(P)]^p$ of column vectors $\mathbf{X} = (X_1, \ldots, X_p)'$ with $EX_j = 0$ and $E|X_j|^2 < \infty$. This vector space is endowed with the norm

$$\|\mathbf{X}\| = \left[\sum_{j=1}^{p} E|X_j|^2\right]^{1/2}.$$

Starting with finite linear combinations of the form

$$\mathbf{Y} = \sum_k \mathbf{A}_k \mathbf{X}(t_k), \tag{A5.1}$$

where the $\mathbf{A}_k$'s are $p \times p$ matrices of complex numbers, and forming all possible limits of Cauchy sequences of such elements in the vector norm, we obtain the *space generated by the process* $\mathscr{M}^{\mathbf{X}}$. It is easily seen that

$$\mathscr{M}^{\mathbf{X}} = [\mathscr{M}^{\mathbf{X}}]^p,$$

where $\mathscr{M}^{\mathbf{X}}$ is the linear subspace of $L_2(P)$ generated by the elements of the component processes of $\mathbf{X}(t)$. On the other hand, if $\mathbf{Z}(A)$ is the vector of spectral measures of the component processes, (A5.1) becomes

$$\begin{aligned}\mathbf{Y} &= \sum_k \mathbf{A}_k \int e^{i\lambda t_k} \mathbf{Z}(d\lambda)\\ &= \int \mathbf{B}(\lambda)\mathbf{Z}(d\lambda),\end{aligned}$$

where

$$\mathbf{B}(\lambda) = \sum_k e^{i\lambda t_k} \mathbf{A}_k.$$

In the limit, every element of $\mathscr{M}^{\mathbf{X}}$ can be represented in the form

$$\mathbf{Y} = \int \mathbf{B}(\lambda)\mathbf{Z}(d\lambda) \tag{A5.2}$$

for some $p \times p$ matrix of complex-valued functions $\mathbf{B}(\lambda)$.

Next we define a generalized inner product, the *Grammian matrix*, by

$$\langle\!\langle \mathbf{X}, \mathbf{Y} \rangle\!\rangle_P = E\mathbf{X}\mathbf{Y}^*.$$

Then, by (A5.2), if $\mathbf{X} = \int \mathbf{B}(\lambda)\mathbf{Z}(d\lambda)$ and $\mathbf{Y} = \int \mathbf{C}(\lambda)\mathbf{Z}(d\lambda)$, we have

$$\begin{aligned}\langle\!\langle \mathbf{X}, \mathbf{Y} \rangle\!\rangle_P &= E\left(\int \mathbf{B}(\lambda)\mathbf{Z}(d\lambda)\right)\left(\int \mathbf{C}(\lambda)\mathbf{Z}(d\lambda)\right)^*\\ &= \int \mathbf{B}(\lambda)\mathbf{F}(d\lambda)\mathbf{C}(\lambda)^*.\end{aligned}$$

Note that $\|\mathbf{X}\|^2 = \operatorname{tr} \langle\!\langle \mathbf{X}, \mathbf{X} \rangle\!\rangle_P = \operatorname{tr} \int \mathbf{B}(\lambda)\mathbf{F}(d\lambda)\mathbf{B}(\lambda)^*$.

Now, if $\mathbf{L}_2(\mathbf{F})$ is the class of all $p \times p$ matrix-valued functions $\mathbf{B}(\lambda)$ with complex entries such that

$$\operatorname{tr} \int \mathbf{B}(\lambda)\mathbf{F}(d\lambda)\mathbf{B}(\lambda)^* < \infty,$$

and if the Grammian of $\mathbf{L}_2(\mathbf{F})$ is defined by

$$\langle\!\langle \mathbf{B}(\lambda), \mathbf{C}(\lambda) \rangle\!\rangle_{\mathbf{F}} = \int \mathbf{B}(\lambda)\mathbf{F}(d\lambda)\mathbf{C}(\lambda)^*,$$

then the mapping $\mathbf{Y} \leftrightarrow \mathbf{B}(\lambda)$ defined by (A5.2) is a one-to-one correspondence between $\mathscr{M}^{\mathbf{X}}$ and $\mathbf{L}_2(\mathbf{F})$ which preserves the Grammian. This is the desired multivariate generalization of the spectral representation defined in the appendix to Chapter 2.

A5.2 Multivariate Linear Filters

To simplify the discussion of linear filters we will only consider filters with the same number of inputs and outputs. The generalized shift operator $\mathbf{U}_t$ is taken to be the $p \times p$ matrix operator with the univariate operator U_t, which was defined in Section A4.1, repeated down the main diagonal and 0's in the off-diagonal positions. Then a multivariate linear filter $\mathbf{L}$ is a linear operator on a domain $\mathscr{D}(\mathbf{L}) \subset \mathscr{M}^{\mathbf{X}}$ to $\mathscr{M}^{\mathbf{X}}$ which satisfies the condition

$$\mathbf{L}\mathbf{U}_t = \mathbf{U}_t \mathbf{L}, \qquad -\infty < t < \infty.$$

By the same argument as the one given in Section A4.1, $\mathbf{L}$ is completely determined by its value at $\mathbf{X}(0)$. Since $\mathbf{L}(\mathbf{X}(0)) \in \mathscr{M}^{\mathbf{X}}$ it follows that

$$\mathbf{L}(\mathbf{X}(0)) = \int \mathbf{B}(\lambda)\mathbf{Z}(d\lambda).$$

The matrix function $\mathbf{B}(\lambda)$ is, as before, the transfer function of the filter.

Now, $\mathbf{U} \in \mathscr{M}^{\mathbf{X}}$ is in $\mathscr{D}(\mathbf{L})$ provided

$$\operatorname{tr} \int \mathbf{C}(\lambda)\mathbf{B}(\lambda)\mathbf{F}(d\lambda)\mathbf{B}(\lambda)^*\mathbf{C}(\lambda)^* < \infty,$$

where $\mathbf{U} = \int \mathbf{C}(\lambda)\mathbf{Z}(d\lambda)$. Then, as in the one-dimensional case, it can be shown that

$$\mathbf{L}(\mathbf{U}) = \int \mathbf{C}(\lambda)\mathbf{B}(\lambda)\mathbf{Z}(d\lambda).$$

In particular, since it is easily seen that $\mathbf{X}(t) \leftrightarrow e^{i\lambda t}\mathbf{I}$, where $\mathbf{I}$ is the $p \times p$ identity matrix, we have

$$\mathbf{L}(\mathbf{X}(t)) = \int e^{i\lambda t}\mathbf{B}(\lambda)\mathbf{Z}(d\lambda).$$

This is the spectral representation of the filter output. The *matching condition*

$$\operatorname{tr} \int \mathbf{B}(\lambda)\mathbf{F}(d\lambda)\mathbf{B}(\lambda)^* < \infty$$

is now seen to be simply the requirement that $\mathbf{X}(t) \in \mathscr{D}(\mathbf{L})$.

Again, there is a unique correspondence between the class of all linear filters on $\mathscr{M}^{\mathbf{X}}$ and $\mathbf{L}_2(\mathbf{F})$. Every element of this collection is the transfer function of some multidimensional linear filter. The various operations on filters detailed in Chapter 4 have natural extensions to the multivariate case but we will not pursue this topic further.

CHAPTER

6

Digital Filters

6.1 INTRODUCTION

We will use the term "digital filter" to mean linear filter in discrete time. Thus, with minor modifications, the theory of Chapter 4 will apply in this chapter as well. The reason for treating filters in continuous and discrete time separately stems not from the difference in theory but rather from the difference in application. Whereas continuous-time filters are used primarily as models for physical filters, digital filters are used for the purposeful modification of discrete-time data. Consequently, it is not only important to understand the operation of these filters but also to know how to select the "parameters" of the filters in order to achieve specific objectives of data modification. This selection of parameters is called *filter design*.

Because of the many kinds of filtering operations investigators have found necessary or useful, the literature on digital filtering is extensive and diverse. In economics, for example, much attention is given to isolating or removing seasonal trends in order to detect weaker features of the spectrum. Geophysicists are concerned with removing tidal effects and other low-frequency power to improve the characteristics of spectral estimates. (How this improves the estimates will be discussed in Chapter 9.) In many fields the suppression of extraneous noise to better define a weak signal is a problem of importance. (This is the "classical" filtering problem and the construction of digital filters to accomplish this goal is discussed in Chapters 5 and 7.) The removal or resolution of polynomial trends is also of importance in many fields of application.

Often the techniques used in different areas for essentially the same purpose have been developed independently and are somewhat different. Because of this, a complete coverage of digital filtering in a single chapter is not feasible. Consequently, we will attempt to describe the basic features all digital filters share and to provide the rationale for a variety of "standard" filter construction techniques. Some of the more interesting special purpose filters will be given as examples. A general reference for this material is Blackman (1965).

6.2 GENERAL PROPERTIES OF DIGITAL FILTERS

Most of the digital filters encountered in practice are of *convolution* type,

$$L(x(t)) = \sum_{j=-\infty}^{\infty} c_j x(t-j), \qquad t = 0, \pm 1, \ldots, \tag{6.1}$$

where the c_j's are real numbers, called the *filter weights*, satisfying the condition

$$\sum_{j=-\infty}^{\infty} c_j^2 < \infty. \tag{6.2}$$

An example of a digital filter which is not of convolution type will be given in conjunction with our discussion of prediction theory in Chapter 7. However, as is indicated in the Appendix, for time series with continuous spectra the class of convolution filters is quite adequate for all practical purposes. Again, although the theory is more generally applicable, we will restrict our attention to time series which are zero-mean, weakly stationary stochastic processes except where otherwise indicated.

The *principle for evaluating transfer functions* in discrete time is the following straightforward analog of the continuous-time version: *Apply the digital filter to the input* $x(t) = e^{i\lambda t}$, $t = 0 \pm 1, \ldots$. *Then the output will be* $y(t) = B(\lambda)e^{i\lambda t}$, *where* $B(\lambda)$ *is the transfer function of the filter*. It follows, upon applying this criterion to (6.1), that the *transfer function of L* is

$$B(\lambda) = \sum_{j=-\infty}^{\infty} c_j e^{-i\lambda j}, \qquad -\pi < \lambda \le \pi. \tag{6.3}$$

From the discussion of Fourier series in Example 1.2, this is seen to be a periodic function of period 2π, thus is completely defined by its values in the interval $-\pi < \lambda \le \pi$, Moreover, the filter weights are uniqely determined from the transfer function by the expression

$$c_k = \frac{1}{2\pi} \int_{-\pi}^{\pi} e^{i\lambda k} B(\lambda)\, d\lambda, \qquad k = 0, \pm 1, \ldots. \tag{6.4}$$

Parseval's relation implies that

$$\sum_{k=-\infty}^{\infty} c_k^{\,2} = \frac{1}{2\pi}\int_{-\pi}^{\pi} |B(\lambda)|^2 \, d\lambda.$$

Now, from Example 1.2, any function $B(\lambda)$ for which the condition $\int_{-\pi}^{\pi} |B(\lambda)|^2 \, d\lambda < \infty$ holds has a Fourier series expansion, thus is the transfer function of a convolution filter with weights determined by (6.4). Consequently, in theory, a large variety of filter characteristics can be achieved by using convolution filters. These filters will match any weakly stationary process with continuous spectrum and bounded spectral density; for if

$$f_X(\lambda) \le M,$$

then

$$\int_{-\pi}^{\pi} |B(\lambda)|^2 f_X(\lambda) \, d\lambda \le M \int_{-\pi}^{\pi} |B(\lambda)|^2 \, d\lambda.$$

If the filter weights satisfy the somewhat stronger condition

$$\sum_{j=-\infty}^{\infty} |c_j| < \infty, \tag{6.5}$$

then the filter has a bounded, continuous, transfer function and, consequently, matches any weakly stationary input with finite power. To see this, note that

$$\int_{-\pi}^{\pi} |B(\lambda)|^2 F_X(d\lambda) \le K^2 \int_{-\pi}^{\pi} F_X(d\lambda) < \infty,$$

where we have taken $|B(\lambda)| \le K$. Since our principal interest will be focused on filters with only finitely many nonzero filter weights, the distinction between (6.2) and (6.5) and the concern for matching filters to inputs are rather academic and we will ignore them henceforth. The more theoretical relevance of these conditions is discussed briefly in the Appendix.

Translation of Notation to an Arbitrary Sampling Interval Δt

Here and throughout the rest of the book (with the exception of Chapter 9), discrete time series are treated as though they were sampled versions of a continuous time series with $\Delta t = 1$. This permits us to use the most convenient and economical notation for all expressions. Nothing is lost by this convention, since it is a simple matter to convert the spectral densities, transfer functions, etc. to the correct frequency units when $\Delta t \ne 1$. This is done automatically by most spectral analysis computer programs.

For example, treating the sampled autocovariance sequence $C(k) = C_X(k\,\Delta t)$ as though $\Delta t = 1$, the spectral density is computed by the expression

$$f(\lambda) = \frac{1}{2\pi} \sum_{k=-\infty}^{\infty} C(k) e^{-i\lambda k}, \qquad -\pi < \lambda \leq \pi. \tag{6.6}$$

[See expression (3.23).] However, the correct frequency variable is

$$\mu = \lambda/\Delta t, \qquad -\pi/\Delta t < \mu \leq \pi/\Delta t,$$

and the correct spectral density $f_{\Delta t}(\mu)$ is

$$f_{\Delta t}(\mu) = \frac{\Delta t}{2\pi} \sum_{k=-\infty}^{\infty} C_X(k\,\Delta t) e^{-i\mu k\,\Delta t}.$$

With the change of variable $\lambda = \mu\,\Delta t$ in (6.6) we see, upon comparing the resulting expression with the last one, that

$$f_{\Delta t}(\mu) = \Delta t\, f(\mu\,\Delta t), \qquad -\pi/\Delta t < \mu \leq \pi/\Delta t. \tag{6.7}$$

In general, all spectral densities, including the cross-spectral densities introduced in Chapter 5, are rescaled according to this equation. Spectral functions are rescaled according to the expression

$$p_{\Delta t}(\mu) = p(\mu\,\Delta t). \tag{6.8}$$

Viewed slightly differently, if A represents any set of frequencies in the correct frequency units [in the interval $(-\pi/\Delta t, \pi/\Delta t)$] and if we define the set $A\,\Delta t$ by the expression

$$A\,\Delta t = \{\gamma\,\Delta t : \gamma \in A\},$$

then

$$F_{\Delta t}(A) = F(A\,\Delta t), \tag{6.9}$$

where $F_{\Delta t}$ is the spectral distribution of the discrete-time process in the correct frequency units and F is the spectral distribution with Δt assumed to be one. Then, the spectral distribution functions, obtained by setting $A = (-\pi/\Delta t, \mu]$, satisfy the relation

$$F_{\Delta t}(\mu) = F(\mu\,\Delta t).$$

Now, (6.7) is obtained by differentiating both sides of this expression. Equation (6.8) is an immediate consequence of (6.9).

The transfer functions are rescaled according to the equation

$$B_{\Delta t}(\mu) = B(\mu\,\Delta t), \tag{6.10}$$

since, by (6.3),

$$B_{\Delta t}(\mu) = \sum_{k=-\infty}^{\infty} c_k e^{-i\mu k\,\Delta t} = B(\mu\,\Delta t).$$

Expression (6.4) for the filter weights can be put in the appropriate form by the change of variables $\mu = \lambda/\Delta t$;

$$c_k = \frac{2\pi}{\Delta t}\int_{-\pi/\Delta t}^{\pi/\Delta t} e^{i\mu k\,\Delta t} B_{\Delta t}(\mu)\,d\mu, \qquad k = 0, \pm 1, \ldots .$$

The notational convenience of assuming $\Delta t = 1$ can be appreciated by comparing these two expressions.

The Gain and Phase Functions for Digital Filters and Two Important Examples

The properties of linear filters given in Chapter 4 carry over to digital filters without change. Recall that if L is a linear filter with transfer function $B(\lambda)$ and $Y(t) = L(X(t))$, then the spectral measure and spectral distribution of the output are, respectively,

$$Z_Y(d\lambda) = B(\lambda)Z_X(d\lambda) \qquad \text{and} \qquad F_Y(d\lambda) = |B(\lambda)|^2 F_X(d\lambda).$$

In particular, the spectral function and spectral density functions are related by the expressions

$$p_Y(\lambda) = |B(\lambda)|^2 p_X(\lambda), \qquad f_Y(\lambda) = |B(\lambda)|^2 f_X(\lambda).$$

Now, however, these functions need only be given for the Nyquist range $-\pi < \lambda \leq \pi$, since they are periodically repeated outside of this interval. The transfer function satisfies the property

$$B(-\lambda) = \overline{B(\lambda)}$$

in order that real-valued inputs result in real-valued outputs. This leads to the properties

$$|B(-\lambda)| = |B(\lambda)|,$$
$$\vartheta(-\lambda) = -\vartheta(\lambda), \qquad \vartheta(\lambda) = \arg B(\lambda),$$

for the gain and phase shift functions.

It is easily seen from (6.3) that if we take

$$c_{-j} = c_j$$

for all j, then $B(\lambda)$ is real-valued and thus the phase shift can assume only the values 0 or $\pm\pi$. As in Section 4.2, digital filters with this property are called symmetric filters. A special subclass of symmetric filters, called *nonnegative definite filters* introduce no phase shift into the data, i.e., have $\vartheta(\lambda) = 0$ for all λ, and for this reason are of importance in data processing. Since the topic of nonnegative definite filters is somewhat specialized, it is discussed in the Appendix (Section A6.2).

Example 6.1 *The Difference Operator*

The difference operator Δ is defined by

$$\Delta(X(t)) = X(t) - X(t-1).$$

This is the digital filter with weights $c_0 = 1$, $c_1 = -1$, and $c_j = 0$ otherwise. The transfer function of Δ is

$$\begin{aligned} B(\lambda) = 1 - e^{-i\lambda} &= e^{-i\lambda/2}(e^{i\lambda/2} - e^{-i\lambda/2}) \\ &= 2ie^{-i\lambda/2}\sin(\lambda/2) \\ &= 2e^{i((\pi-\lambda)/2)}\sin(\lambda/2) \\ &= \begin{cases} 2|\sin(\lambda/2)|\,e^{i((\pi-\lambda)/2)}, & \lambda > 0, \\ 2|\sin(\lambda/2)|\,e^{-i((\pi-\lambda)/2)}, & \lambda < 0. \end{cases} \end{aligned}$$

Thus, the gain and phase shift functions are

$$|B(\lambda)| = 2|\sin(\lambda/2)|$$

and

$$\vartheta(\lambda) = \begin{cases} (\pi - \lambda)/2, & \lambda > 0, \\ -(\pi - \lambda)/2, & \lambda < 0. \end{cases}$$

A graph of the squared gain function is given in Fig. 6.1.

As would be expected, the difference operator is a high-pass filter with properties similar to those of the derivative, since it is the discrete analog of the derivative. In particular, the squared gain function is of order λ^2 near $\lambda = 0$ and rises well above 1 at the high frequencies. Thus, it deviates substantially from the squared gain function of an ideal high-pass filter.

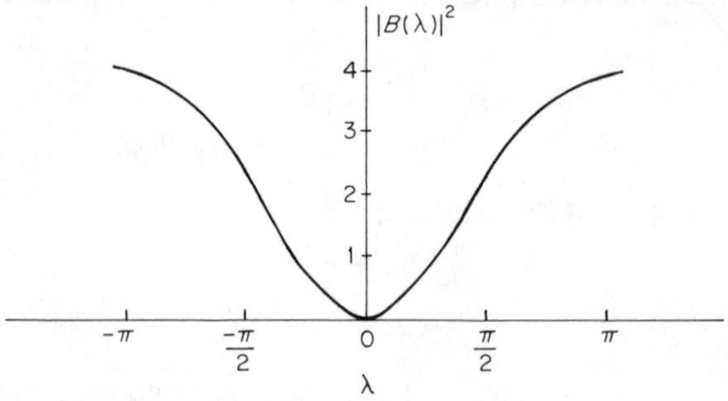

Fig. 6.1 *The squared gain function for the difference operator.*

Example 6.2 *The Simple Averaging Filter*

For a positive integer R, let

$$Y(k) = \frac{1}{R} \sum_{j=0}^{R-1} X(k-j), \qquad k = 0, \pm 1, \ldots .$$

This is a discrete convolution filter with $c_0 = c_1 = \cdots = c_{R-1} = 1/R$ and $c_j = 0$ otherwise. The transfer function is

$$B(\lambda) = \frac{1}{R} \sum_{j=0}^{R-1} e^{-i\lambda j}$$
$$= e^{i\lambda(R-1)/2} \sin(\lambda R/2)/R \sin(\lambda/2), \qquad -\pi < \lambda \leq \pi.$$

[We have used (1.18) to obtain this expression.] Thus,

$$|B(\lambda)| = \left| \frac{\sin(\lambda R/2)}{R \sin(\lambda/2)} \right| .$$

For values of λ for which $\sin(\lambda R/2) \geq 0$, the phase shift is $\vartheta(\lambda) = -((R-1)/2)\lambda$. These frequency components are displaced in time by an amount $\tau(\lambda) = -(R-1)/2$ which depends on the number of nonzero filter weights—a mildly unpleasant property. When $\sin(\lambda R/2) < 0$, $\vartheta(\lambda) = -((R-1)/2)\lambda + \pi$ if $\lambda > 0$ and the negative of this if $\lambda < 0$.

For $R = 2$, the trigonometric identity $\sin \lambda = 2 \sin(\lambda/2)\cos(\lambda/2)$ can be used to reduce the gain function to the form

$$|B(\lambda)| = |\cos(\lambda/2)| .$$

The phase shift is of simple form in this case,

$$\vartheta(\lambda) = \begin{cases} -\lambda/2, & \text{if } \lambda > 0, \\ \lambda/2, & \text{if } \lambda < 0. \end{cases}$$

A filter with virtually no phase shift for any value of R can be obtained by using a *symmetric average*: Take R to be an odd integer and let

$$Y(k) = \frac{1}{R} \sum_{j=-(R+1)/2}^{(R-1)/2} X(k-j).$$

The transfer function of this filter is

$$B(\lambda) = \frac{\sin(\lambda R/2)}{R \sin(\lambda/2)}, \qquad -\pi < \lambda \leq \pi.$$

Thus, the symmetric average has the same gain function as the unsymmetric average, but the phase function is now $\vartheta(\lambda) = 0$ when $\sin(\lambda R/2) \geq 0$ and $\vartheta(\lambda) = \pm\pi$ otherwise. As is seen from the graph of $|B(\lambda)|^2$ given in Fig. 6.2,

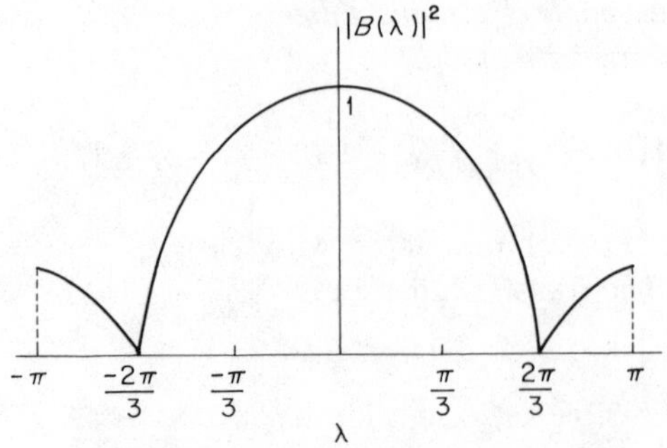

Fig. 6.2 *Graph of the squared gain function of a simple averaging filter of R = 3 terms.*

the averaging filters are low-pass filters. Moreover, the values of this function are relatively small over the frequency range for which $\vartheta(\lambda) \neq 0$. Since this means that relatively little power will be transmitted by the filter in this range, the symmetric average will act as though $\vartheta(\lambda) = 0$ for all λ. This property is characteristic of "well-designed" symmetric filters. (See Section A6.2 for additional details.)

The z-Transform and Some Useful Notation for Linear Filters

If

$$Y(t) = \sum_{j=-\infty}^{\infty} a_j X(t-j) \tag{6.11}$$

is a digital filter, the *z-transform of the filter* is the complex-valued function

$$\mathscr{A}(z) = \sum_{j=-\infty}^{\infty} a_j z^j.$$

Important properties of the filter, such as stability, can be related to properties of the z-transform. This transform will also play an important role in prediction theory, to be discussed in Chapter 7.

The transfer function of the filter, to be denoted by $A(\lambda)$, is related to the z-transform by the expression

$$A(\lambda) = \mathscr{A}(e^{-i\lambda}).$$

Thus, the transfer function is the "value" of the z-transform on the unit circle.

The *backward shift operator* is the linear filter B such that

$$B(X(t)) = X(t-1).$$

This filter is easily seen to have transfer function $e^{-i\lambda}$. Consequently, if we define the *linear combination filter*

$$\mathscr{A}(B) = \sum_{j=-\infty}^{\infty} a_j B^j,$$

where $B^{-j} = (B^{-1})^j$ and B^{-1} inverts B, then $\mathscr{A}(B)$ is the linear filter with transfer function $A(\lambda) = \mathscr{A}(e^{-i\lambda})$ given above, i.e., (6.11) can be written as

$$Y(t) = \mathscr{A}(B)(X(t)).$$

This compact notation has some interesting operational properties which are used to advantage, for example, by Whittle (1963) and Box and Jenkins (1970). We will not use this notation extensively in this book. However, an illustration of its convenience is given in the following discussion of recursive filters.

Recursive Filters

A linear filter is said to be *recursive* if the output at time t depends (linearly) on a fixed number of previous input and output values. For example,

$$Y(t) = \sum_{j=1}^{q} (-d_j)\, Y(t-j) + \sum_{k=0}^{p} c_k\, X(t-k)$$

is a recursive filter with input $X(t)$ and output $Y(t)$. This can be represented compactly by the equation

$$\mathscr{D}(B)(Y(t)) = \mathscr{C}(B)(X(t)), \tag{6.12}$$

where

$$\mathscr{D}(B) = \sum_{j=0}^{q} d_j B^j, \qquad d_0 = 1,$$

and

$$\mathscr{C}(B) = \sum_{k=0}^{p} c_k B^k.$$

Now, by the rules for dividing polynomials, we can obtain, formally,

$$\mathscr{A}(B) = \mathscr{C}(B)/\mathscr{D}(B) = \sum_{k=0}^{\infty} a_k B^k. \tag{6.13}$$

Then, under a condition to be specified presently, the recursive filter is equivalent to the filter

$$Y(t) = \mathscr{A}(B)(X(t)). \tag{6.14}$$

Thus, these operators, viewed as linear combination filters, can be manipulated algebraically using the standard techniques for adding, multiplying, and dividing polynomials and infinite series.

A sufficient condition for $\mathscr{A}(B)$ *to satisfy* (6.2) *is that the polynomial* $\mathscr{D}(z)$ *have all of its zeros outside of the unit circle.* For then, $\mathscr{A}(z) = \mathscr{C}(z)/\mathscr{D}(z)$ is analytic for $|z| < \rho$, where ρ is the minimum of the absolute values of the zeros of $\mathscr{D}(z)$. Then for every ρ', $1 < \rho' < \rho$,

$$\mathscr{A}(\rho') = \sum_{k=0}^{\infty} a_k(\rho')^k < \infty, \tag{6.15}$$

where the series is the power series expansion of $\mathscr{A}(z)$ about $z = 0$ evaluated at ρ'. Since the terms of (6.15) tend to zero, there exists a constant $K > 1$ such that

$$|a_k| \leq K(1/\rho')^k \qquad \text{for all } k. \tag{6.16}$$

Thus, in fact, condition (6.5), which is stronger than (6.2), is satisfied and $\mathscr{A}(B)$ is a well-defined linear filter which matches any input. Moreover, it is immediate from the theory of sequential filters given in Section 4.3 that the input and output of (6.14) satisfy the recursive relation (6.12).

The filter $\mathscr{A}(B)$ is also stable, since if $|X(t)| \leq M$ for all t, then

$$|Y(t)| = |\mathscr{A}(B)(X(t))| \leq M \sum_{j=0}^{\infty} |a_j| < \infty.$$

It is a realizable filter, i.e., depends only on the past and present of the input, because of the one-sided (power series) expansion of $\mathscr{A}(z)$. This expansion depended on the zeros of $\mathscr{D}(z)$ being outside of the unit circle. Thus, this condition can also be viewed as guaranteeing the stability and realizability of the recursive filter.

Recursive filters can be designed to have exceptionally good response characteristics by proper selection of the filter weights c_k and d_j and the integers p and q. A good discussion of the design of recursive low-pass filters is given by Enochson and Otnes (1968).

As we will show in Chapter 7, the best linear predictor of a weakly stationary process can be represented as a recursive filter for an important class of processes. Moreover, the Kalman filter (Kalman, 1960), which solves a special case of the real-time filtering problem to be discussed in Chapter 7, is of the recursive type. Consequently, recursive filters play an important role in time series analysis. The following example introduces a simple but extensively used recursive filter.

Example 6.3 *The Exponential Smoothing Filter*

The exponential smoothing filter is defined by the expression

$$Y(t) = (1 - \alpha)X(t) + \alpha Y(t - 1).$$

Putting this in the form of (6.12) we have $\mathscr{D}(z) = 1 - \alpha z$ and $\mathscr{C}(z) = 1 - \alpha$. The only zero of $\mathscr{D}(z)$ is $z = 1/\alpha$. Consequently, this is a valid digital filter if $|1/\alpha| > 1$, thus $|\alpha| < 1$. In fact,

$$\mathscr{A}(B) = \mathscr{C}(B)/\mathscr{D}(B) = (1 - \alpha)/(1 - \alpha B) = (1 - \alpha) \sum_{k=0}^{\infty} \alpha^k B^k$$

from the familiar sum of a geometric series.

The transfer function of this filter is

$$A(\lambda) = (1 - \alpha)/(1 - \alpha e^{-i\lambda}),$$

with squared gain and phase shift functions

$$|A(\lambda)|^2 = (1 - \alpha)^2/(1 - 2\alpha \cos \lambda + \alpha^2),$$
$$\vartheta(\lambda) = -\text{Arctan}((\alpha \sin \lambda)/(1 - \alpha \cos \lambda)).$$

The squared gain function of this filter is plotted for several values of α in Fig. 6.3. It is seen that this is indeed a smoothing (low-pass) filter with

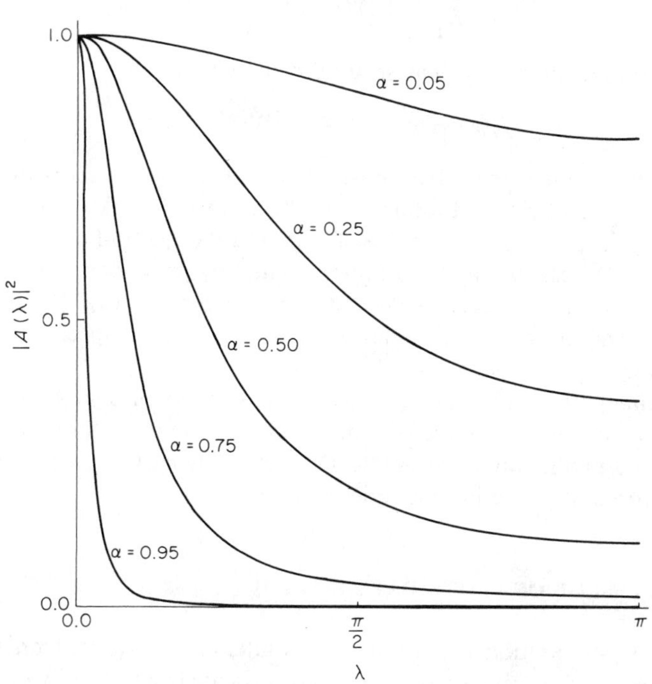

Fig. 6.3 *Graph of squared gain function of the exponential smoothing filter for several values of α.*

characteristics selectable to a degree by varying α. This filter has received considerable attention in economics and business [see Brown (1962) for further details].

Simulation

Another important use of recursive filters is to simulate time series with prescribed spectral characteristics. By means of random number generators it is easy to construct sequences of numbers on a digital computer which can be thought of as the values of independent, identically distributed random variables with virtually any distribution. In particular these numbers can be viewed as a realization of a white noise sequence $\eta(t)$. As we have seen, the spectrum of such a process is continuous with spectral density function

$$f_\eta(\lambda) = \sigma^2/2\pi, \qquad -\pi < \lambda \leq \pi,$$

where σ^2 is the variance of $\eta(t)$.

Now, if $\eta(t)$ is used as the input of the recursive filter

$$Y(t) = \sum_{j=1}^{q}(-d_j)Y(t-j) + \sum_{k=0}^{p} c_k \eta(t-k), \tag{6.17}$$

then the output time series has spectral density function

$$f_Y(\lambda) = (\sigma^2/2\pi)|\mathscr{C}(e^{-i\lambda})/\mathscr{D}(e^{-i\lambda})|^2 \tag{6.18}$$

as is easily obtained from the above theory. If all of the zeros of $\mathscr{D}(z)$ are outside of the unit circle, it follows that $Y(t)$ is a valid weakly stationary time series. Moreover, by proper choice of p, q, and the c_k's and d_j's, the rational function (6.18) can be made to approximate any given continuous spectral density function as closely as desired. Then, algorithm (6.17) will generate a sample function of a time series with the appropriate spectrum whatever white noise sequence is employed.

This method of generating time series is used for many purposes. It provides a convenient source of time series for demonstrating the effects of filtering and prediction, for example, and for studying the properties of statistical procedures used in time series analysis.

6.3 THE EFFECT OF FINITE DATA LENGTH

As we saw in Section 4.3, virtually every filter of importance from the viewpoint of processing a time series can be constructed from a family of low-pass filters with a variety of cutoff points. Consequently, it seems reasonable to construct a family of *ideal* low-pass filters and use the members of this family for all data processing purposes. In theory, this can easily be done. In fact, if we require the filters to be symmetric, the transfer function of the low-pass filter with cutoff frequency λ_0 will be

$$B(\lambda) = \begin{cases} 1, & |\lambda| \le \lambda_0, \\ 0, & |\lambda| > \lambda_0. \end{cases}$$

This function is square-integrable, thus corresponds to the convolution filter with (square-summable) weights

$$a_k = \frac{1}{2\pi} \int_{-\pi}^{\pi} B(\lambda) e^{i\lambda k}\, d\lambda = \begin{cases} \sin \lambda_0 k / \pi k, & k = \pm 1, \pm 2, \ldots, \\ \lambda_0 / \pi, & k = 0. \end{cases}$$

Unfortunately, the use of such a filter in practice is restricted by the fact that only a finite segment of the time series will be available. That is, instead of having the input $X(t)$ available for all integer values of t we will have only a finite number of observations $X(1), X(2), \ldots, X(N)$, say. If we try to use the ideal low-pass filter on this data by setting $X(t) = 0$ for $t \le 0$ and $t \ge N + 1$, for example, we will clearly not obtain the same result as if the entire input series were used. Since the ideal characteristics of the filter are based on the assumption that the entire input is used, the actual output will deviate from the ideal in some way. We will now undertake a short analysis of this effect.

A finite length of data can be viewed as the result of multiplying the original time series by a *data window* defined by the indicator function

$$I_{[1,N]}(t) = \begin{cases} 1, & \text{if } 1 \le t \le N, \\ 0, & \text{otherwise.} \end{cases}$$

Then, using the spectral representation for $X(t)$, the output of a linear filter with filter weights a_k to the input $I_{[1,N]}(t)X(t)$ is

$$\begin{aligned} U(t) &= \sum_{k=-\infty}^{\infty} a_k I_{[1,N]}(t-k) X(t-k) \\ &= \int_{-\pi}^{\pi} e^{i\lambda t} A_{N,t}(\lambda) Z_X(d\lambda), \end{aligned} \tag{6.19}$$

where

$$A_{N,t}(\lambda) = \sum_{k=-\infty}^{\infty} a_k I_{[1,N]}(t-k) e^{-i\lambda k} = \sum_{k=t-N}^{t-1} a_k e^{-i\lambda k}.$$

The time series $U(t)$ is the observed output of the filter when the data is of finite length and is to be compared to the desired output

$$Y(t) = \sum_{k=-\infty}^{\infty} a_k X(t-k) = \int_{-\pi}^{\pi} e^{i\lambda t} A(\lambda) Z_X(d\lambda),$$

where $A(\lambda) = \sum_{k=-\infty}^{\infty} a_k e^{-i\lambda k}$.

A reasonable measure of the deviation of $U(t)$ from the ideal is $E(U(t) - Y(t))^2$. This quantity depends on t and can be evaluated as follows:

$$E(U(t) - Y(t))^2 = E\left|\int_{-\pi}^{\pi} e^{i\lambda t}[A_{N,t}(\lambda) - A(\lambda)]Z_X(d\lambda)\right|^2$$

$$= \int_{-\pi}^{\pi} \left|\sum_{k<t-N,\,k>t-1} a_k e^{-i\lambda k}\right|^2 F_X(d\lambda).$$

If $X(t)$ has a continuous spectrum with spectral density function bounded from above and away from zero, say,

$$0 < m \leq f_X(\lambda) \leq M,$$

then

$$m \int_{-\pi}^{\pi} \left|\sum_{k<t-N,\,k>t-1} a_k e^{-i\lambda k}\right|^2 d\lambda \leq E(U(t) - Y(t))^2$$

$$\leq M \int_{-\pi}^{\pi} \left|\sum_{k<t-N,\,k>t-1} a_k e^{-i\lambda k}\right|^2 d\lambda.$$

By Parseval's relation, this is equivalent to

$$2\pi m \sum_{k<t-N,\,k>t-1} a_k^2 \leq E(U(t) - Y(t))^2 \leq 2\pi M \sum_{k<t-N,\,k>t-1} a_k^2. \quad (6.20)$$

Thus, in this commonly occurring case, the magnitude of the error at time t can be conveniently described by means of the quantity

$$\sum_{k<t-N,\,k>t-1} a_k^2 = \sum_{k=-\infty}^{t-N-1} a_k^2 + \sum_{k=t}^{\infty} a_k^2. \quad (6.21)$$

Figure 6.4 is a plot of the weights of a hypothetical digital filter with the weights entering into the sum (6.21) cross-hatched. Note that the magnitude of the error depends on the data length N and on how rapidly the filter weights tend to zero with increasing values of $|k|$. Typically, the larger filter weights

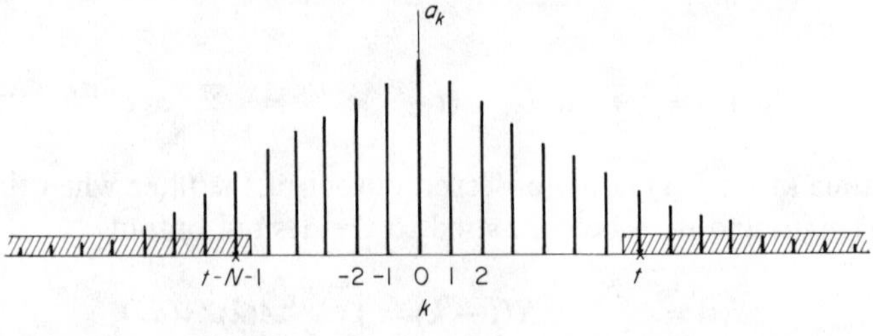

Fig. 6.4 *Graph of filter weights of a hypothetical digital filter. Cross-hatched weights contribute to the error at time t of applying the filter to data of finite length N.*

(in absolute value) have indices k near zero. Consequently, the largest errors will be encountered for values of t outside of the range $1 \le t \le N$.

The most compelling reason for not using the ideal low-pass filters discussed above is that the rate of decrease of the filter weights is very slow ($a_k^2 \approx 1/k^2$). Thus, the actual outputs of these filters will differ substantially from the ideal. Put in another way, any attempt to explain the outputs of these filters to a finite length of data by the characteristics of the ideal filter will be in error. The magnitude of the error, which can be estimated from (6.20), will be uncomfortably large for all values of t. Thus, it is desirable to work with low-pass filters which are less than ideal but for which the actual filter characteristics are more closely described over some range of t values by the transfer function of the corresponding linear filter.

The recursive filters discussed in the last section are a substantial improvement over the ideal low-pass filters from this standpoint. It is seen from (6.16) that the filter weights are zero for $k < 0$ and fall off exponentially fast as $k \to \infty$. Thus, by (6.20), we obtain

$$E(U(t) - Y(t))^2 \le 2\pi M \sum_{k=t}^{\infty} a_k^2 \le Cl^t$$

for $1 \le t \le N$, where $l = (\rho')^{-2}$ and $C = 2\pi MK(1 - l)^{-1}$. It follows that after an initialization period for times near $t = 1$, the actual output of the recursive filter is very close to the output that would have been obtained for an input extending into the infinite past.

Note *that if* α *and* β *are positive integers such that* $\alpha + \beta + 1 < N$ *and if a digital filter has filter weights* a_k *which are zero for* $k < -\alpha$ and $k > \beta$, *then* $U(t) = Y(t)$ *for* $\beta + 1 \le t \le N - \alpha$. That is, this section of the output data agrees exactly with the output of the linear filter to the entire input series and, consequently, the transfer function of the filter and the theory of Chapter 4 can be used to describe exactly its properties. In practice, the output is computed only for the range of times $\beta + 1 \le t \le N - \alpha$. The fact that all of the theory of linear filters applies without approximation outweighs the disadvantage of the loss of β points from the beginning and α points from the end of the output series. For this reason, this type of filter has been used extensively in practice. We will consider such filters further in the next section.

Example 6.4 *Effect of Smoothing a Time Series*

The effect of applying a simple symmetric averaging filter with $R = 20$ (see Example 6.6) to a 100-year segment of a time series recorded at yearly intervals is illustrated in Fig. 6.5. Here, $\alpha = \beta = 10$ and it is seen that 10 data points are lost from each end of the smoothed sequence.

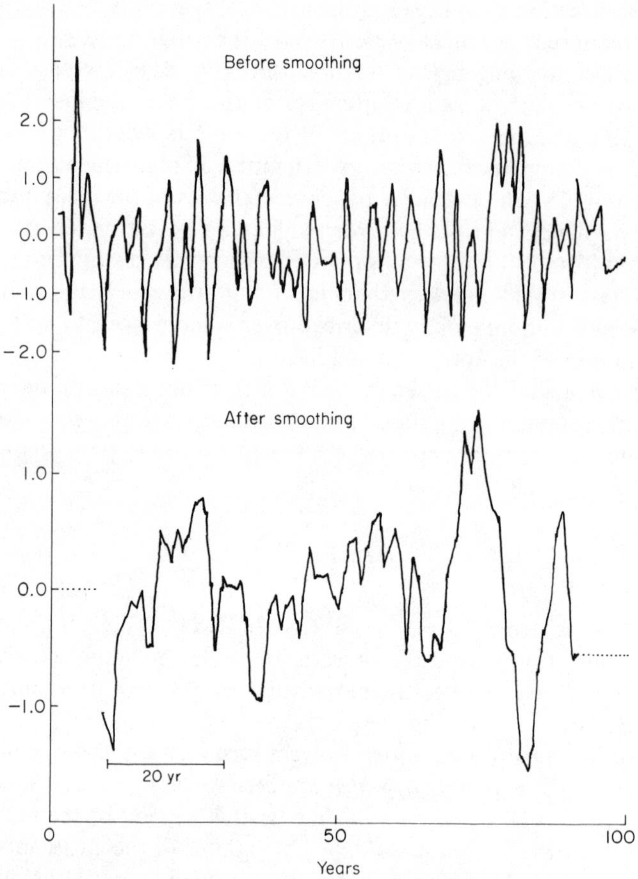

Fig. 6.5 *Time series of certain yearly recorded data before and after smoothing with a 20-year symmetric average. Note loss of 10 years of data from each end of the smoothed series.*

Linear Filtering by Fast Fourier Transform

It should be noted that the extreme efficiency of the fast Fourier transform algorithm (to be discussed in Chapter 9) has made a different but rather natural kind of filtering not only computationally feasible but, for large N, actually superior to convolution filtering. The relationship

$$Y(t) = \int_{-\pi}^{\pi} e^{i\lambda t} C(\lambda) Z_X(d\lambda) \tag{6.22}$$

for the output of a linear filter with input

$$X(t) = \int_{-\pi}^{\pi} e^{i\lambda t} Z_X(d\lambda) \tag{6.23}$$

and transfer function $C(\lambda)$ suggests the following procedure: First, the finite Fourier transform of $X(t)$ is obtained by the fast Fourier transform algorithm,

$$Z_{X,\nu} = \frac{1}{N} \sum_{t=1}^{N} e^{-i\lambda_\nu t} X(t), \tag{6.24}$$

where $\lambda_\nu = 2\pi\nu/N$, $-[(N-1)/2] \le \nu \le [N/2]$. Note that the inverse transform of (6.24) is a discrete analog of (6.23). Consequently, it makes sense to carry through this analogy to the filter output given by (6.22). To do this, $Z_{X,\nu}$ is multiplied by the transfer function $C(\lambda)$ evaluated at the points λ_ν and the product is transformed back into the time domain by the fast Fourier transform to obtain

$$W(t) = \sum_{\nu=-[(N-1)/2]}^{[N/2]} e^{i\lambda_\nu t} C(\lambda_\nu) Z_{X,\nu}. \tag{6.25}$$

It is reasonable to expect that $W(t)$ and $Y(t)$ will be "close together" because of the similarity in their generating expressions. Again, using the mean-squared error criterion, we will indicate in the Appendix *that if the weights of the filter are* b_k, $k = 0, \pm 1, \ldots,$ *and if the spectral density of the input satisfies the inequality* $0 < m \le f_X(\lambda) \le M$, *then*

$$2\pi m \mathscr{S}_{N,t} \le E(W(t) - Y(t))^2 \le 2\pi M \mathscr{S}_{N,t}, \tag{6.26}$$

where

$$\mathscr{S}_{N,t} = \sum_{k=-\infty}^{t-N-1} b_k^2 + \sum_{k=t}^{\infty} b_k^2 + \sum_{k=t-N}^{t-1} (b_k - \tilde{b}_k)^2$$

and

$$\tilde{b}_k = \sum_{p=-\infty}^{\infty} b_{k+pN}.$$

Again assuming that the weights drop off reasonably monotonically in absolute value as $|k| \to \infty$, the largest values of $\mathscr{S}_{N,t}$ occur at the beginning and end of the range $1 \le t \le N$. Thus, by dropping a number of output values for t near 1 and N, the number depending on the rate of decrease of the b_k's, this procedure will produce an output close to (6.22). It follows that the characteristics of the filtering operation will be well described by the transfer function $C(\lambda)$.

Since the various ideal filters with rectangularly shaped gain functions have weights which decrease rather slowly with $|k|$, when they are used it will be necessary to discard a rather large number of output values or suffer a substantial discrepancy between the actual and ideal outputs. *However, if we select a filter with finitely many weights—that is, if* $b_k = 0$ *for* $|k| \ge K$, *say—then it is easy to verify that*

$$\mathscr{S}_{N,t} = 0 \qquad \text{for} \quad K + 1 \le t \le N - K + 1.$$

Thus, if we discard exactly $K + 1$ *values from each end of the output sequence,* $W(t)$ *and* $Y(t)$ *agree for the remaining time points.* From our previous results, this means that for filters with finitely many weights, the fast Fourier transform method and the convolution method both lead to the correct output for the same segment of output data. Thus, aside from considerations of computational accuracy, the choice between the two methods would reasonably be based on computing speed.

We will now argue that the fast Fourier transform method is faster for moderate to large values of N and the relative time saving increases with N if K is also increased proportionately with N. The rationale for linking K and N in this fashion is discussed in the next section.

A convenient if somewhat crude measure of computing speed is the *order* of the number of elementary operations (additions and multiplications) required by the method for a given number of data points. For N data points the number of operations is said to have order $g(N)$ if the actual number of operations divided by $g(N)$ tends to a finite, positive limit as $N \to \infty$.

The convolution filtering method requires $2K + 1$ additions and multiplications for each value of t, thus $(N - 2K)(2K + 1) = 2N^2\alpha(1 - 2\alpha)$ in total, where we have taken $K = \alpha N$, $0 < \alpha < \frac{1}{2}$. Thus, the number of operations is of order N^2. However, as we will show in Chapter 9, the number of operations required for a commonly used version of the fast Fourier transform is of order $N \log N$. Thus, the two fast Fourier transformations [three, if $C(\lambda_\nu)$ is to be determined from the filter weights] and N multiplications carried out in the present method still require an order of $N \log N$ operations. This increases more slowly with N than does N^2. Thus, there will be a sample size beyond which the fast Fourier transform method requires fewer operations thus less computing time than the convolution method.

Unfortunately, an order argument does not provide much information about the sample size for which the fast Fourier transform method becomes more efficient than the convolution method in a given practical situation. There are many factors which control this cross-over point and little, short of comparing actual computing times, will determine it. Moreover, the average filter user will be restricted to "ready made" computer routines and the particular method of calculation built into the program. Thus, it is not always possible to take advantage of the most efficient computing method even when it is known.

6.4 DIGITAL FILTERS WITH FINITELY MANY NONZERO WEIGHTS

One way to obtain a filter which has characteristics approximating those of an ideal filter, yet has the advantages of filters with finitely many nonzero weights, is to set the weights of the ideal filter equal to zero outside of an interval $-\alpha \le k \le \beta$. This process is called *truncation*. In fact, every filter with finitely many nonzero weights can be viewed as the truncated version of some convolution filter. Consequently, it is important to know how the operation of truncation effects the characteristics of the filter.

As a notational convenience, we will consider only symmetric truncations; that is, we will take $\alpha = \beta = K$. This does not limit the applicability of the discussion to any extent. If $I_{[-K, K]}(k)$ is the indicator function

$$I_{[-K,K]}(k) = \begin{cases} 1, & -K \le k \le K, \\ 0, & \text{otherwise,} \end{cases}$$

the filter weights of the truncated filter are

$$b_k = I_{[-K,K]}(k) a_k, \qquad k = 0, \pm 1, \ldots. \tag{6.27}$$

The transfer function $B_K(\lambda)$ of this filter is the Fourier transform of the weights b_k. However, as was seen in Example 1.2, the Fourier transform of a product is the convolution of the Fourier transforms. Consequently, if $B(\lambda)$ is the transfer function of the filter with weights a_k and $C_K(\lambda)$ is the Fourier transform of $I_{[-K,K]}(k)$, we will have

$$B_K(\lambda) = \frac{1}{2\pi} \int_{-\pi}^{\pi} B(\mu) C_K(\lambda - \mu)\, d\mu.$$

Now, an explicit expression for $C_k(\lambda)$ can be obtained as follows: Note that

$$C_K(\lambda) = \sum_{k=-\infty}^{\infty} I_{[-K,K]}(k) e^{-i\lambda k} = \sum_{k=-K}^{K} e^{-i\lambda k}.$$

Then from expression (1.18) we obtain

$$C_K(\lambda) = \sin((2K+1)\lambda/2)/\sin(\lambda/2). \tag{6.28}$$

Consequently, the transfer function of the truncated filter is

$$B_K(\lambda) = \int_{-\pi}^{\pi} B(\mu) \frac{\sin[(2K+1)(\lambda-\mu)/2]}{2\pi \sin((\lambda-\mu)/2)}\, d\mu \cdot \tag{6.29}$$

That is, at a given frequency λ, $B_K(\lambda)$ is the average of the values of the ideal transfer function $B(\mu)$ with respect to the weight function or *window*, $C_K(\mu)/2\pi$, centered at λ. A graph of $C_K(\lambda)$ is given in Fig. 6.6.

To gain an intuitive understanding of expression (6.29) it is useful to introduce the generalized function $\delta_p(\lambda)$ which is the periodic version of the Dirac delta function defined in Section 4.2. This function has the property

$$\int_{-\pi}^{\pi} B(\lambda - \mu)\delta_p(\mu)\, d\mu = \int_{-\pi}^{\pi} B(\mu)\delta_p(\lambda - \mu)\, d\mu = B(\lambda)$$

for every "well-behaved" periodic function $B(\lambda)$. This delta function is periodic of period 2π and has a (formal) Fourier series expansion with coefficients

$$a_k = \frac{1}{2\pi} \int_{-\pi}^{\pi} \delta_p(\lambda) e^{-i\lambda k}\, d\lambda = \frac{1}{2\pi}, \qquad k = 0, \pm 1, \ldots.$$

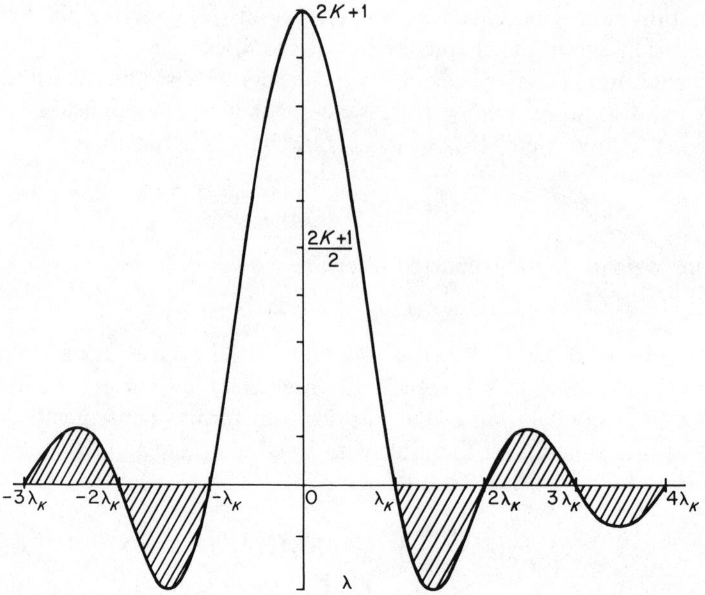

Fig. 6.6 *Graph of the window function* $C_K(\lambda)$. *Side lobes are cross-hatched,* $\lambda_K = 2\pi/(2K+1)$.

Now, if these coefficients are used as the weights of the ideal filter in (6.27), then $B(\lambda) = \delta_p(\lambda)$ and the transfer function $B_K(\lambda)$ of the truncated filter is simply $C_K(\lambda)/2\pi$ from (6.29). Thus, Fig. 6.6 illustrates the effect of truncating the filter whose transfer function has a unit "spike" at $\lambda = 0$ and is zero elsewhere in $(-\pi, \pi)$. The "spike" is broadened into a peak, called the *main lobe*, of width $4\pi/(2K+1)$. Thus, the sharp features of the original transfer function are smoothed and spread into neighboring frequencies.

A more troublesome effect of truncation is the introduction of ripples or *side lobes* on either side of the main lobe of $C_K(\lambda)$. The side lobes are hatched in Fig. 6.6. We will illustrate the problems caused by the side lobes by looking at the truncated version of a filter discussed in Section 6.3. If $B(\lambda)$ is the transfer function of the ideal symmetric low-pass filter with cutoff frequency λ_0, then the transfer function of the truncated filter is easily shown to be

$$B_K(\lambda) = \frac{1}{2\pi} \int_{\lambda-\lambda_0}^{\lambda+\lambda_0} C_K(\mu)\, d\mu. \tag{6.30}$$

A graph of $B_K(\lambda)$ superimposed on $B(\lambda)$ is given in Fig. 6.7 for $K = 5$.

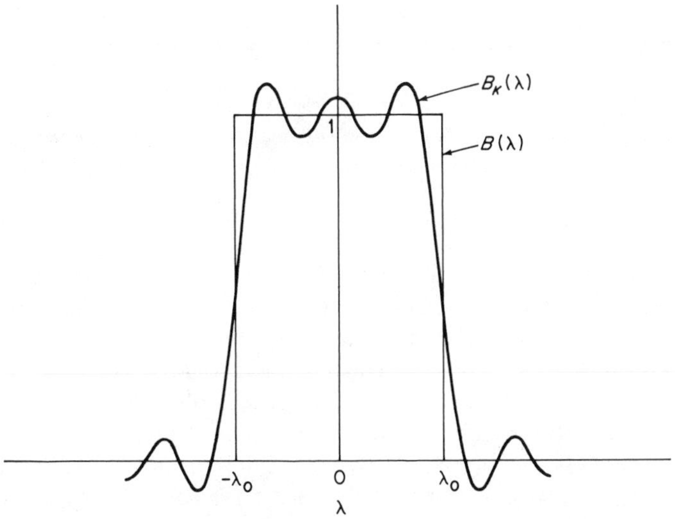

Fig. 6.7 *Graphs of the transfer functions $B(\lambda)$ and $B_K(\lambda)$ of the ideal symmetric low-pass filter and the filter obtained by truncating the weights of the ideal filter at $k = \pm K$, $\lambda_0 = \pi/2$, $K = 5$.*

From expression (6.30) and Fig. 6.6 it can be seen that the distortion and the size of the side lobes of $B_K(\lambda)$ are accounted for by the large side lobes of $C_K(\lambda)$, which are up to 20% as large as the peak. Thus, instead of a sharp cutoff of power at $\pm\lambda_0$, the truncated filter allows through a sizeable amount of power from frequencies outside of the range $[-\lambda_0, \lambda_0]$. This is called *filter leakage*. Filter leakage can be a problem when the filter is used to preprocess data for spectrum analysis. The power allowed through the filter for frequencies $|\lambda| > \lambda_0$ can seriously bias spectral estimates in the range $|\lambda| < \lambda_0$.

Filtering and Decimation

Another standard use of low-pass filters is to process data in preparation for the operation of *decimation*. This term is used for the resampling of a time series at a sampling interval $\Delta t'$ which is some integer multiple of the original sampling interval Δt,

$$\Delta t' = s\,\Delta t.$$

One effect of this procedure is to reduce the original number of data points N to

$$N' = N/s.$$

This is sometimes necessary for data-handling purposes. From the theory of

Chapter 3, it is evident that the Nyquist folding frequency is also made smaller according to the relationship

$$\lambda_N' = \lambda_N/s.$$

Thus, in order to avoid aliasing, it is necessary to low-pass filter the time series before decimation by means of a filter with cutoff frequency satisfying the inequality

$$\lambda_0 \leq \lambda_N'.$$

Any power left outside of the interval $|\lambda| \leq \lambda_0$ due to filter leakage will contribute to the aliasing problem (Section 3.2). Thus, it is desirable to use a low-pass filter with side lobes as small as possible.

We will consider methods for reducing the leakage of low-pass filters next. First, however, it should be noted that leakage problems are not peculiar to low-pass filters, but rather are present to some degree in all (interesting) filters with finitely many nonzero weights. Consequently, although we have concentrated on low-pass filters, the discussion is equally valid for filters of any type.

Methods for Improving Filter Characteristics

The most obvious way to improve the characteristics of a truncated-weights filter is to increase the sample size N so that K can also be made larger. Since by increasing K the weights of the filter become "untruncated," it is reasonable that the output of the truncated-weights filter should tend to the output of the corresponding ideal filter. Thus, for example, if we link K and N by making $K = \alpha N$, $0 < \alpha < \frac{1}{2}$, as in the last section, then an increase in sample size will be coupled with an output closer to that of the ideal filter and with useable length $(1 - 2\alpha)N$ which also increases with N. However, the sample size is not always at our disposal and it is important to have other ways to improve the characteristics of a truncated-weights filter. This requires an analysis of the distortion introduced by truncation.

The amount of distortion evident in the transfer function of Fig. 6.7 is actually due to two causes. As indicated above, the large side lobes of $C_K(\lambda)$ are a principal source of distortion. However, the sharp cutoff of the ideal low-pass filter at $\pm\lambda_0$ also contributes to the problem. To see this, if we go to the extreme of using a filter with the smoothest possible transfer function—namely, a simple amplification filter with constant transfer function

$$B(\lambda) = c, \qquad -\pi < \lambda \leq \pi,$$

then, since the integral of $C_K(\lambda)/2\pi$ is equal to one,

$$B_K(\lambda) = \int_{-\pi}^{\pi} (B(\mu)C_K(\lambda - \mu)/2\pi)\, d\mu = c.$$

Thus, no distortion is introduced by truncation in this case. As the sharpness of the features of the transfer function progresses from the minimum represented by a constant function to the right angles of an ideal low-pass filter, the distortion increases from zero to the degree represented in Fig. 6.7.

This suggests that to improve the fidelity of filters with finitely many weights, one possibility is to require the transfer function of the underlying filter to be smoother. Thus, instead of attempting to approximate rectangular transfer functions, one might instead approximate rectangles with tapered edges in which the dropoff from one to zero is more gradual. An important filter of this type is the *Ormsby filter* which is available as a subroutine in the biomedical computer program package (Dixon, 1969, p. 208). The Ormsby filter is the truncated version of a symmetric low-pass filter with transfer function of trapezoidal shape. Band-pass Ormsby filters with easily selected pass-bands are also available in this reference. Further details concerning these filters are given by Enochson and Otnes (1968).

It is also possible to reduce the distortion due to the side lobes of $C_K(\lambda)$ by another means. The key observation required is that the large side lobes of $C_K(\lambda)$ are due to the "square" shape of the indicator function $I_{[-K,K]}(k)$. If a truncation function is chosen which drops off from one to zero more gradually in the neighborhood of $\pm K$, a window function with smaller side-lobes would result. Such a truncation function is called a *taper*. Here we will apply tapers to the filter weights. However, the process of tapering is also applied to the time series data itself to improve certain properties of spectral estimates as we will see in Chapter 9.

Tapering Filter Weights

The effect of tapering the filter weights will be illustrated by means of a convenient and popular tapering function. To conform with the terminology associated with this function through its use in spectrum analysis we will call it the *Hanning taper*. Let

$$h_K(k) = \begin{cases} \frac{1}{2}(1 + \cos(\pi k/K)), & k = 0, \pm 1, \ldots, \pm K, \\ 0, & |k| > K. \end{cases} \tag{6.31}$$

Then, if the filter weights are defined by

$$b_k = h_K(k) a_k, \qquad k = 0, \pm 1, \ldots,$$

where $a_0, a_{\pm 1}, \ldots$ are the weights of the filter with transfer function $B(\lambda)$, the transfer function of the tapered-weights filter is

$$B_K(\lambda) = \int_{-\pi}^{\pi} (B(\mu) H_K(\lambda - \mu)/2\pi) \, d\mu,$$

where

$$H_K(\lambda) = \sum_{k=-\infty}^{\infty} h_K(k) e^{-i\lambda k}.$$

This function is called the *Hanning window*. It is easily shown that

$$H_K(\lambda) = \tfrac{1}{2}C_K(\lambda) + \tfrac{1}{4}C_K(\lambda - (\pi/K)) + \tfrac{1}{4}C_K(\lambda + (\pi/K)), \qquad (6.32)$$

where $C_K(\lambda)$ is given by (6.28). Compare the graph of this function in Fig. 6.8 with Fig. 6.6. Note that the side lobes of this window are substantially reduced in size relative to the height of the main lobe at the expense of doubling the width of the main lobe. Consequently, the sharp features of the underlying

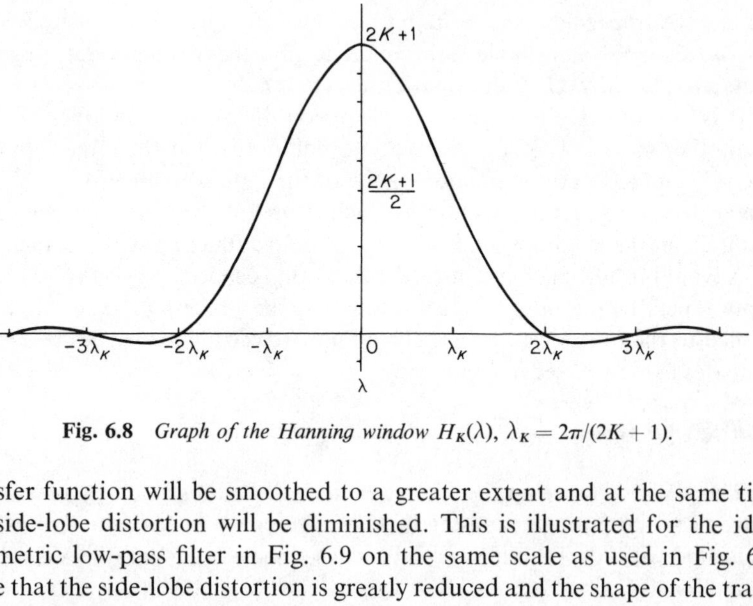

Fig. 6.8 *Graph of the Hanning window* $H_K(\lambda)$, $\lambda_K = 2\pi/(2K+1)$.

transfer function will be smoothed to a greater extent and at the same time the side-lobe distortion will be diminished. This is illustrated for the ideal symmetric low-pass filter in Fig. 6.9 on the same scale as used in Fig. 6.7. Note that the side-lobe distortion is greatly reduced and the shape of the transfer function substantially improved. Consequently, from the viewpoint of preventing filter leakage, this filter would be preferable to the truncated weights filter of Fig. 6.7. Hanning-tapered, ideal symmetric low-pass filters are available as subroutines in the biomedical computer program package (Dixon, 1969).

There is one sense in which the truncated-weights filters are optimal. We define the *integrated squared error* of approximating an ideal filter with transfer function $B(\lambda)$ and weights a_k by a filter with transfer function $B_K(\lambda)$ and weights b_k to be

$$\mathscr{I}_K = \int_{-\pi}^{\pi} |B_K(\lambda) - B(\lambda)|^2 \, d\lambda.$$

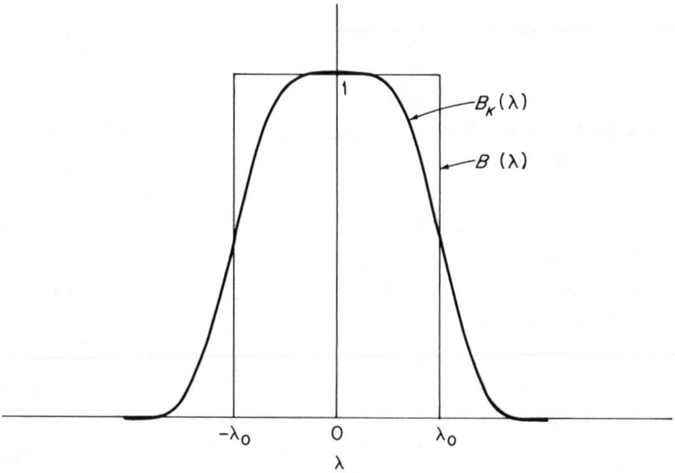

Fig. 6.9 *Graphs of the transfer functions $B(\lambda)$ and $B_K(\lambda)$ of the ideal symmetric low-pass filter and the filter obtained by tapering the weights of the ideal filter using the Hanning taper,* $\lambda_0 = \pi/2$, $K = 5$.

Then if $b_K = 0$ for $|k| > K$, by the Parseval relation,

$$\mathscr{I}_K = 2\pi \sum_{k=-K}^{K} (a_k - b_k)^2 + 2\pi \sum_{|k|>K} a_k^{\,2}.$$

Thus, the minimum integrated squared error is achieved by taking $b_k = a_k$ for $|k| \leq K$. This is the truncated-weights filter. Any other filter with zero weights for $|k| > K$, including the tapered-weights filters, will have larger integrated squared error. By an argument similar to the one given in Section 6.3, if the input time series $X(t)$ has spectral density function satisfying $m \leq f_X(\lambda) \leq M$ and if $Y(t)$ and $Y_K(t)$ are the outputs of the filters with transfer functions $B(\lambda)$ and $B_K(\lambda)$, respectively, then the mean-square error of approximating $Y(t)$ by $Y_K(t)$ satisfies the inequality

$$m\mathscr{I}_K \leq E\big(Y(t) - Y_K(t)\big)^2 \leq M\mathscr{I}_K.$$

Thus, although tapering improves the side-lobe characteristics of the truncated weights filter, it need not yield an output as close "on the average" to that of the ideal filter as does the truncated weights filter. Despite this, the suppression of side lobes is of sufficient importance that it is one of the principal goals of digital filter design. For an excellent discussion of the design of filters with side lobes controlled to be smaller than a prescribed magnitude, see Granger and Hatanaka (1964). This reference also has a good general discussion of digital filtering with some important applications to economics.

6.5 DIGITAL FILTERS OBTAINED BY COMBINING SIMPLE FILTERS

Another approach to designing filters, which has the practical advantage of great flexibility, is to combine a number of simple filters by the methods of Section 4.3 to form more complex filters with the desired shape and side-lobe characteristics. This method requires only the most elementary computer programming ability and the filters can be constructed on a "cut-and-try" basis. For this reason, most of the time series processing done in the past has utilized this technique and it is safe to predict that much of it will be done this way in the future even though more sophisticated methods are available.

Recall that *if* L_1 *and* L_2 *are linear filters with transfer functions* $B_1(\lambda)$ *and* $B_2(\lambda)$, *then the linear combination filter,* $\alpha L_1 + \beta L_2$, *has transfer function* $\alpha B_1(\lambda) + \beta B_2(\lambda)$ *and the sequential filter* $L_1 L_2$ *has transfer function* $B_1(\lambda)B_2(\lambda)$. Starting with low-pass filters such as the averaging filters of Example 6.2, the transfer functions can be manipulated algebraically by forming linear combinations and products to arrive at an acceptable transfer function for the desired filtering operation. Automatic plotting equipment is useful for graphing the gain and phase functions at the various stages of filter construction. The sequence of operations leading to the transfer function determines the sequence of filters to be applied to the data.

When the desired sequence of filters has been decided upon, the data can be processed either by carrying out the operations step by step in several passes through the data or the filter weights of the overall filter can be computed and the filtering performed in one pass through the data. These weights can be determined by the following operations which are the discrete analogs of the results for convolution filters given in Section 4.3: *If* L_1 *has weights* a_k *and* L_2 *has weights* b_k, *then the weights of the filter* $\alpha L_1 + \beta L_2$ *are*

$$c_k = \alpha a_k + \beta b_k$$

and the weights of $L_1 L_2$ *are the discrete convolutions*

$$d_k = \sum_{j=-\infty}^{\infty} a_j b_{k-j} = \sum_{j=-\infty}^{\infty} a_{k-j} b_j .$$

This method also allows us to calculate the transfer functions and weights of filters which are recognized to be the result of algebraically combining filters with known transfer functions. We now give some examples of these ideas.

Example 6.5 *An Alternate Form of the Backward Shift Operator*

If I denotes the "do-nothing" filter with weights $a_0 = 1$, $a_j = 0, j \neq 0$, and Δ is the difference filter of Example 6.1, then the weights of $I - \Delta$ can be calculated by the rules given above:

$$c_0 = 1 - 1 = 0,$$
$$c_1 = 0 - (-1) = 1,$$
$$c_j = 0 \qquad \text{otherwise.}$$

An application of these weights yields $(I - \Delta)(X(t)) = X(t-1)$. It follows that $I - \Delta = B$, the backward shift operator defined in Section 6.2. The transfer function of the filter $I - \Delta$ is $B(\lambda) = 1 - (1 - e^{-i\lambda}) = e^{-i\lambda}$. This computation would also verify that $I - \Delta = B$, since linear filters are uniquely determined by their transfer functions.

Example 6.6 *Symmetric Averaging Filter with R Weights; The Case of R, An Even Number*

When R is an even number, the averaging filter of Example 6.2 must be modified in order to retain symmetry, since symmetry requires an odd number of weights. The simplest procedure is to split one of the extreme weights in half and apply it at both ends of the weight sequence. We then obtain

$$L(X(t)) = \sum_{k=-R/2}^{R/2} c_k X(t-k),$$

where

$$c_k = \begin{cases} 1/R, & k = 0, \pm 1, \ldots, \pm(\tfrac{1}{2}R - 1), \\ 1/2R, & k = \pm\tfrac{1}{2}R, \\ 0, & \text{otherwise.} \end{cases}$$

By observing that these are the sums of the weights of two filters L_1 and L_2, where L_1 has weights

$$a_k = \begin{cases} 1/2R, & k = 0, \pm 1, \ldots, \pm(\tfrac{1}{2}R - 1), \\ 0, & \text{otherwise,} \end{cases}$$

and L_2 has weights

$$b_k = \begin{cases} 1/2R, & k = 0, \pm 1, \ldots, \pm\tfrac{1}{2}R, \\ 0, & \text{otherwise,} \end{cases}$$

we can calculate the transfer function of L from the relation $B(\lambda) = B_1(\lambda) + B_2(\lambda)$. However, $B_1(\lambda)$ and $B_2(\lambda)$ can be obtained from Example 6.2;

$$B_1(\lambda) = \sin(\lambda(R-1)/2)/2R\sin(\lambda/2),$$
$$B_2(\lambda) = \sin(\lambda(R+1)/2)/2R\sin(\lambda/2).$$

Thus,

$$B(\lambda) = \frac{\cos(\lambda/2)\sin(\lambda R/2)}{R\sin(\lambda/2)},$$

where we have used the trigonometric identity $\sin(\alpha+\beta)+\sin(\alpha-\beta) = 2\sin\alpha\cos\beta$.

This same device can be used to construct and calculate the transfer function of filters with triangular and trapezoidal filter weight profiles. For example, the filter with weights

$$c_k = \begin{cases} 1/(2R+K+1), & k = 0, \pm 1, \ldots, \pm(2R+1), \\ (K+1-l)/(K+1)(2R+K+1), & k = \pm(2R+1+l), \\ & l = 1, 2, \ldots, K, \\ 0, & \text{otherwise,} \end{cases}$$

where R and K are nonnegative integers, is the sum of $K+1$ filters L_j with weights $S = 1/[(K+1)(2R+K+1)]$ over the ranges $-(R+j)$ to $(R+j)$ for $j = 0, 1, \ldots, K$, and zero otherwise. These weights can be shown to sum to one. The transfer function is then

$$B(\lambda) = S\sum_{k=0}^{K}\sum_{r=-(R+j)}^{R+j} e^{-i\lambda r} = S\sum_{j=0}^{K}\sin(R+j+\tfrac{1}{2})\lambda/\sin(\lambda/2).$$

The profile of the weights is triangular when $R = 0$. Such weights can be viewed as tapered rectangular filter weights and will correspond to low-pass filters with smaller side lobes than the corresponding rectangular filters.

Example 6.7 *Improving Side-Lobe Characteristics by Applying Filters Sequentially*

Two low-pass filters can be used sequentially to obtain a filter with reasonable side-lobe size by applying the first filter to approximate the desired low-pass characteristics then using the second filter to reduce the side lobes of the first. Thus, for example, if a simple symmetric average L_1 with R weights (R odd) is applied first, the transfer function

$$B_1(\lambda) = \frac{\sin\lambda R}{R\sin(\lambda/2)}$$

has its first side lobes symmetrically placed in the frequency range $\pi \le |\lambda R/2| < 3\pi/2$. Restricting attention to positive frequencies, the maximum height of the side lobe will occur at approximately the midpoint of this range, which is

$$\lambda = 5\pi/4R.$$

Now it is reasonable to use a second symmetric average L_2 with R' weights, where R' is to be selected so that the first zero of $B_2(\lambda)$ falls at approximately the point where the first side lobe of $B_1(\lambda)$ is largest. Since the first zero of

$B_2(\lambda)$ is at $2\pi/R'$, equating this with the above frequency we obtain the following expression for determining R',

$$R' \cong 8R/5.$$

A graphical illustration of the effect of this procedure is given in Fig. 6.10. Actually, the side lobes will be substantially reduced in size simply by repeating L_1. However, this changes the shape of the main lobe of $B_1(\lambda)$ more than does the operation $L_2 L_1$.

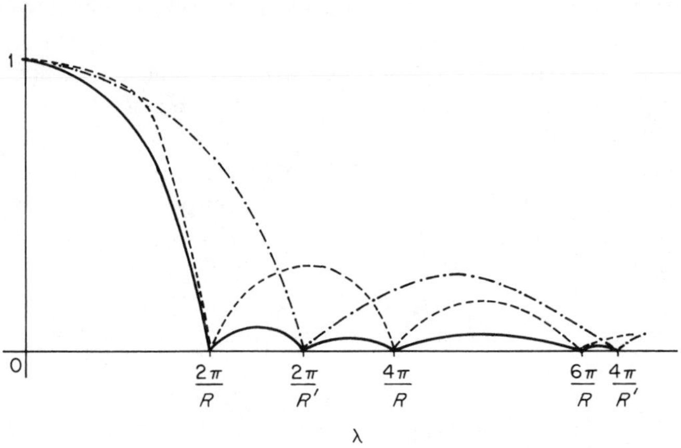

Fig. 6.10 *Illustration of the suppression of side lobes by the sequential application of two simple averages L_1 and L_2 with transfer functions $B_1(\lambda)$ and $B_2(\lambda)$, where* – – –: $|B_1(\lambda)|$; – · –: $|B_2(\lambda)|$; —: $|B_1(\lambda)B_2(\lambda)|$.

Example 6.8 *Families of Band-Pass Filters—Binomial Filters*

An interesting method for generating band-pass filters which decompose the power of a time series into a number of different pass-bands is the following. Let L denote a low-pass filter with transfer function $B(\lambda)$ which is nonzero except at isolated points in $(-\pi, \pi)$. Then if I denotes the do-nothing filter, by the binomial theorem and our "algebra of filters," for any positive integer n we have

$$I^n = (L + (I - L))^n = \sum_{k=0}^{n} \binom{n}{k} L^k (I - L)^{n-k}, \tag{6.33}$$

where $\binom{n}{k}$ is the binomial coefficient, $n!/(k!(n-k)!)$. The filter $\binom{n}{k} L^k (I - L)^{n-k}$ is the sequential filter obtained by applying L, the low-pass filter, k times and $I - L$, the high pass filter, $n - k$ times, then multiplying by the factor $\binom{n}{k}$. The result is a band-pass filter with transfer function $B_k(\lambda) = \binom{n}{k} B^k(\lambda) \times$

$(1 - B(\lambda))^{n-k}$. Moreover, if we define $Y_k(t) = \binom{n}{k} L^k (I - L)^{n-k}(X(t))$ for an input series $X(t)$, then (6.33) implies that

$$X(t) = \sum_{k=0}^{n} Y_k(t).$$

Thus, in this sense, the band-pass filters decompose $X(t)$ into different pass-bands. However, because the pass-bands overlap, in general, it is not true that the power of $X(t)$ is the sum of the power magnitudes of the $Y_k(t)$'s. Thus, the $Y_k(t)$'s do not represent a spectral decomposition of $X(t)$, which entails the partitioning of the power of $X(t)$ into disjoint frequency intervals. However, if the band-pass filters have reasonably well-defined pass-bands which do not overlap too much, a crude spectrum analysis can be achieved by using the power of $Y_k(t)$ as an estimate of the power of $X(t)$ over the pass-band of $\binom{n}{k} L^k (I - L)^{n-k}$.

Blackman and Tukey (1959) use a simple variant of this idea to perform a preliminary analysis of a time series for the purpose of designing appropriate prefilters for a more careful spectrum analysis. They use the simple summation filter $S(X(t)) = X(t) + X(t-1)$ for L in conjunction with the high-pass filter $\Delta(X(t)) = X(t) - X(t-1)$. It is easily seen that $\Delta = 2I - S$. Consequently, by the binomial theorem,

$$2^n I^n = (S + \Delta)^n = \sum_{k=0}^{n} \binom{n}{k} S^k \Delta^{n-k}.$$

Thus, the band-pass filters $L_k = (1/2^n)\binom{n}{k} S^k \Delta^{n-k}$ can be used as described above. An algorithm for carrying out the spectral calculations is given by Blackman and Tukey (1959, p. 135).

Finally, to illustrate another aspect of the use of simple filters to produce more complex filters—and to point out one of the dangers of this method—we give the following example which is also of independent historical interest.

Example 6.9 *The Slutsky Effect*

Slutsky (1927) demonstrated both theoretically and by computations on simulated time series that by applying a sequence of simple averaging and differencing operations to a discrete white noise process it was possible to obtain a nearly perfect sinusoid. This "effect" had a profound influence on the development of time series analysis since it pointed to a possible mechanism for producing cycles and periodicities in data other than that described by simple almost periodic functions with white noise residuals which had been the model considered up to that time. This opened the way for new time series models and, in fact, the important moving average processes to be studied in Chapter 7 were the outgrowth of the work of Slutsky and of Yule whose studies date from about the same time.

The Slutsky effect would seem to contradict the statement that linear transformations cannot transform processes with continuous spectra into processes with discrete spectra. However, the pure sinusoid is the limiting result of a sequence of linear operations and, in fact, the output of a finite number of these operations still has a continuous spectrum although the spectrum is sharply peaked about the frequency of the limiting sinusoid.

The Slutsky effect has also, justifiably, been the object of concern among practicing time series analysts. As was mentioned earlier, it is often necessary or desirable to filter a time series before analyzing it. It is now possible that cycles detected in the analysis are actually a result of the filtering and are not characteristic of the underlying phenomenon at all! There is the possibility that some of the cycles noted in certain time series in economics and geophysics were due to filtering. Fishman (1969, p. 45) gives an example from economics. This problem can be largely avoided by one simple expedient; namely, *always calculate the transfer function of the filter*. This is one of the most important steps in analyzing time series. Since virtually every filter is composed of very simple filters combined by the two basic operations given above, the gain and phase shift functions are easily obtained. Then, potential difficulties with the filter can be detected and avoided.

It is of some interest to explain, in elementary terms, another phenomenon noted by Slutsky in his simulation studies. When a sufficient amount of filtering had been done so that the gain function of the filter was sharply peaked about a frequency λ_0, say, he observed that cycles of frequencies near λ_0 would appear and disappear then reappear with different phase, etc. This kind of phenomenon is often observed in real time series and is the basis for the term *cyclic data*.

With a sharply peaked spectrum, only the frequencies in a small neighborhood $(\lambda_0 - \varepsilon, \lambda_0 + \varepsilon)$ are prominent in the data and the various harmonic components with these frequencies, although having nearly the same amplitude, can have quite different phases. The time series consists of a "sum" of these harmonic components. Take a simple situation in which only two terms of the sum are considered and suppose they are $C\cos(\lambda_0 - \varepsilon)t$ and $C\cos(\lambda_0 + \varepsilon)t$. Now, from the identity $\cos(x - y) + \cos(x + y) = 2\cos x \cos y$, we obtain

$$C\cos(\lambda_0 - \varepsilon)t + C\cos(\lambda_0 + \varepsilon)t = 2C\cos(\varepsilon t)\cos(\lambda_0 t).$$

This is a periodic function of frequency λ_0 *modulated* by a periodic amplitude factor with frequency ε. Thus, a cycle with frequency λ_0 will seem to appear and disappear at intervals of π/ε time units. Since all components with frequencies in $(\lambda_0 - \varepsilon, \lambda_0 + \varepsilon)$ interact in this way, a varied and complex cyclic time series can result. Virtually the only way such a time series can be "sorted out" is by spectrum analysis.

6.6 FILTERS WITH GAPPED WEIGHTS AND RESULTS CONCERNING THE FILTERING OF SERIES WITH POLYNOMIAL TRENDS

In this section some miscellaneous topics of interest are considered. A number of methods for suppressing or revealing seasonal (periodic) trends are based on the simple idea of filters with gapped weights. This is considered first. Then we will look at a method for removing polynomial trends—the variate difference method of Tintner (1940)—and at a criterion that a low-pass filter must satisfy in order to pass polynomial trends due to Brillinger (1965).

Filters with Gapped Weights

If $a_j, j = 0, \pm 1, \ldots$, are the weights of a digital filter with transfer function $B(\lambda)$, then by separating these weights by K indices and assigning the intervening indices zero weights we obtain the corresponding *filter with gapped weights* of gap K,

$$b_j = \begin{cases} a_{j/K}, & j = 0, \pm K, \pm 2K, \ldots, \\ 0, & \text{otherwise.} \end{cases}$$

The transfer function of the gapped weight filter is

$$B_K(\lambda) = B(K\lambda). \tag{6.34}$$

This is easily verified: By the change of index $l = j/K$ we obtain

$$\begin{aligned} B_K(\lambda) &= \sum_{j=-\infty}^{\infty} b_j e^{-i\lambda j} \\ &= \sum_{l=-\infty}^{\infty} a_l e^{-i\lambda(Kl)} = B(K\lambda). \end{aligned}$$

Thus, $B_K(\lambda)$ is a periodic function of period $2\pi/K$ which repeats the gain and phase characteristics of the original filter on $(-\pi, \pi)$ over each interval $\big((2j-1)\pi/K, (2j+1)\pi/K\big)$ for $j = 0, \pm 1, \pm 2, \ldots$.

Example 6.10 *The Buys–Ballot Filter and Its Uses*

The gapped version of the simple symmetric averaging filter with R weights (R odd) is seen to have transfer function

$$B_K(\lambda) = \frac{\sin(\lambda KR/2)}{R \sin(K\lambda/2)}$$

from Example 6.2 and expression (6.34). The following argument will be based on this rather than the more commonly used nonsymmetric filter for ease of exposition. From the spectral representation of the $X(t)$ process we can write

$$Y(t) = \int_{-\pi}^{\pi} e^{i\lambda t} B_K(\lambda) Z_X(d\lambda)$$

$$= \int_{-\pi}^{\pi} e^{i\lambda t} \frac{\sin(\lambda KR/2)}{R \sin(\lambda K/2)} Z_X(d\lambda).$$

As $R \to \infty$, the transfer functions $\sin(\lambda KR/2)/R \sin(\lambda K/2)$ tend to zero for all frequencies λ except those of the form $2\pi l/K$, $l = 0, \pm 1, \ldots,$ at which they are unity. Thus, by the type of argument introduced in Section 2.10, it is seen that $Y(t)$ has the mean-square limit

$$W(t) = \sum_{k=-[(K-1)/2]}^{[K/2]} e^{i2\pi kt/K} Z_X(\{2\pi k/K\}). \tag{6.35}$$

Thus, in the limit, the discrete spectral terms of the time series at the frequencies $\lambda_k = 2\pi k/K$ are isolated by this filter.

For a finite length of data, $X(1), X(2), \ldots, X(N)$, if R is chosen as large as possible, a good approximation to this periodic limit is obtained. By varying K, an elementary sort of harmonic analysis can be carried out. This is one of the earliest methods of harmonic analysis originally done laboriously by hand in tabular form. The resulting tables, called Buys–Ballot tables after the originator, are discussed by Wold (1938, p. 23).

Now consider the nonsymmetric filter

$$Y(t) = \frac{1}{R} \sum_{j=1}^{R} X(t - Kj), \qquad t = 1, 2, \ldots, K,$$

where K is fixed and $R = [N/K]$. This filter, which we will call the Buys–Ballot filter, also has gain function $|B_K(\lambda)|$. A graph of this gain function for a variety of values of K is given in Fig. 6.11. Moreover, the phase shift is zero at the frequencies λ_k. Consequently, since all other frequency components are averaged to zero as $R \to \infty$, this filter has the same limit (6.35) as the symmetric filter and provides a well-resolved estimate of $W(t)$ when R is large.

Because of the limitation of data length and the selection of R, the given filter characteristics are valid only for the output segment $Y(1), Y(2), \ldots, Y(K)$. However, since the limit (6.35) is strictly periodic of period K, it is customary to extend the output periodically to the original data length N by repeating this segment over and over, i.e., define the periodic extension $Y_p(t)$ of $Y(t)$ to be

$$Y_p(t) = Y(k)$$

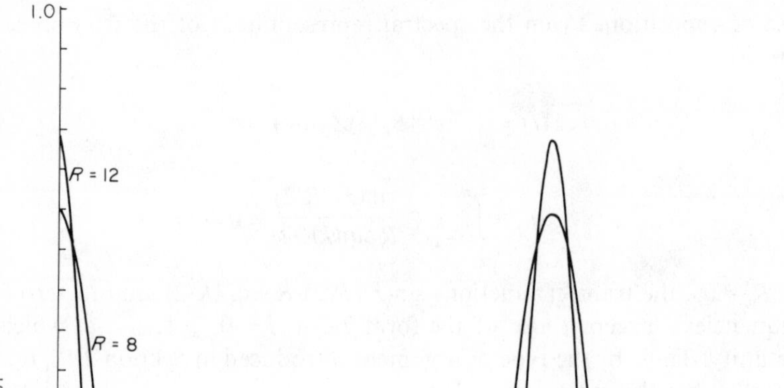

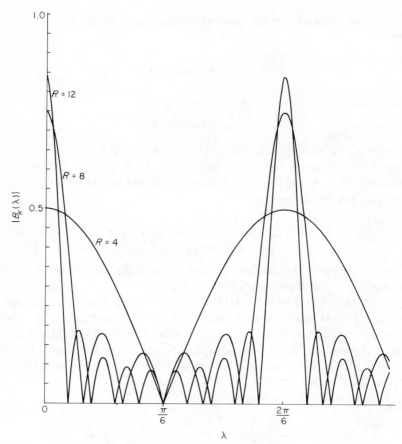

Fig. 6.11 *Gain function of two cycles of the Buys–Ballot filter for K = 6 and R = 4, 8, 12.*

if $t = k + Kl$ for $1 \leq k < K$ and $0 < l \leq [N/K]$. Then, the Buys–Ballot filter can be used to suppress the periodic components in the original time series at frequencies λ_k by forming the residual time series

$$U(t) = X(t) - Y_p(t), \qquad t = 1, 2, \ldots, N.$$

An Application of the Buys–Ballot Filter to Rainfall Runoff Data

The most common use of this filter is to isolate and/or eliminate the seasonal components of time series for which these components are strong and well defined. This situation occurs reasonably frequently in geophysical data and in time series studied in economics. An excellent example of this

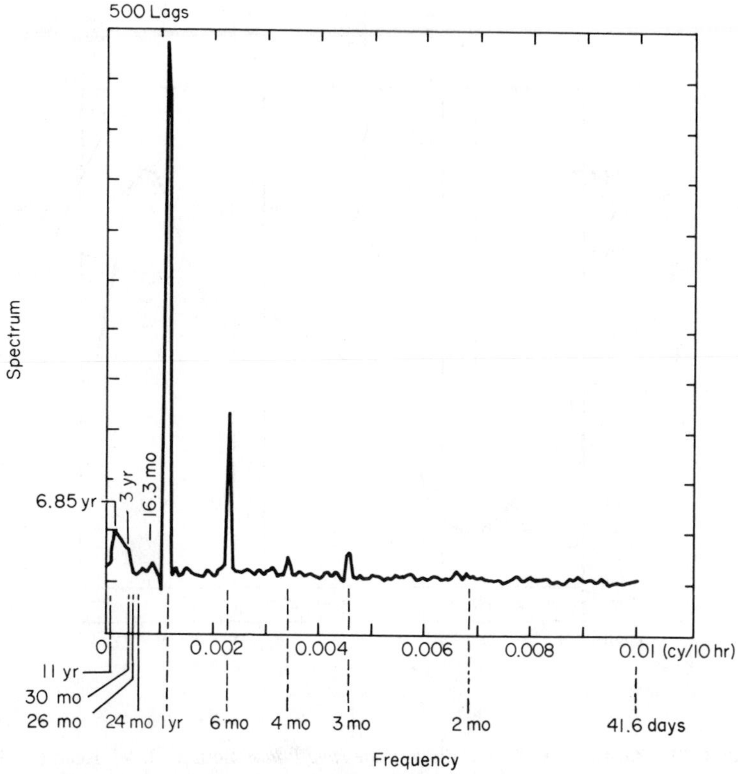

Fig. 6.12 *Rio Chagres runoff spectrum.* Source: J. W. Reed (1971).

type of time series is provided by Reed (1971) in a study of low-frequency cycles in the daily runoff data of the Rio Chagres River in Panama. The spectrum of this data, given in Fig. 6.12, shows how the yearly periodicity and its harmonics at 6, 4, and 3 months dominate the spectrum making the activity below a one-year period difficult to resolve. The daily averages for each day of the year over a 57-year period correspond to the application of a Buys–Ballot filter with $K = 365$ days and $R = 57$ values per average. The filtered series was used to locate the times of maximum and minimum runoff and to study the general characteristics of the seasonal periodic components of the series. A graph of the daily averages, smoothed by a symmetric averaging filter with $R = 20$, is given in Fig. 6.13. The daily averages were then repeated periodically and subtracted from the runoff series to remove the yearly cycle and its harmonics. The degree of success in suppressing this periodic component is seen by comparing Fig. 6.12 with Fig. 6.14 which is a graph of the spectrum of the residual series.

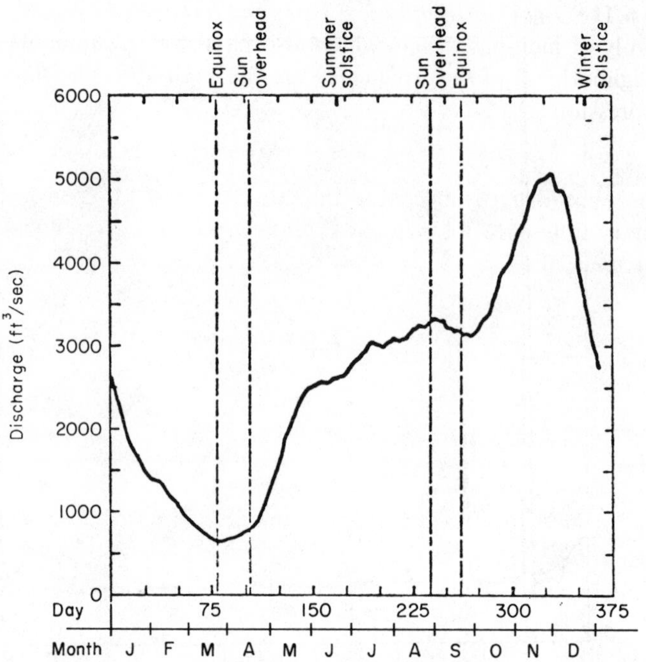

Fig. 6.13 *Rio Chagres River daily average runoff data.* Source: J. W. Reed (1971).

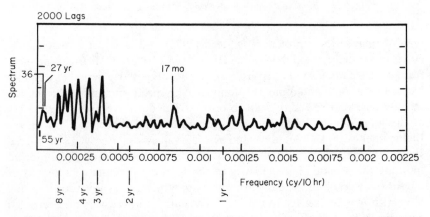

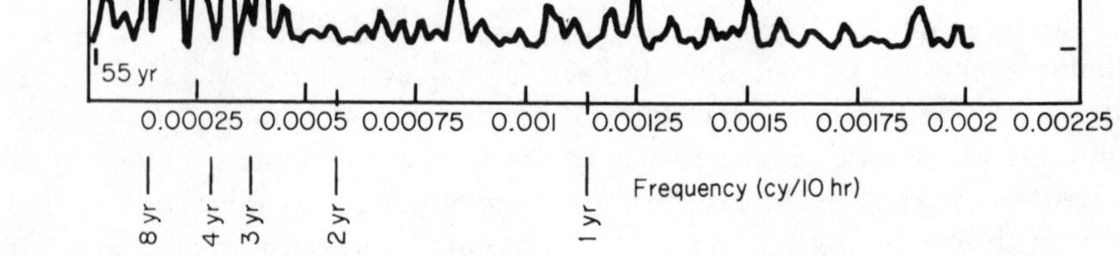

Fig. 6.14 *Spectrum of Rio Chagres runoff data with annual cycle and its harmonics removed.* Source: J. W. Reed (1971).

Example 6.11 *The Quasi-Difference Filter and Its Gapped Version*

A family of high-pass filters which also pass selected amounts of power in the neighborhood of zero frequency are the *quasi-difference filters* defined by the expression

$$Y(t) = X(t) - \alpha X(t-1),$$

where α is a parameter restricted to the range $0 < \alpha \leq 1$. When $\alpha = 1$, this is the ordinary difference filter already studied. The transfer function of the quasi-difference filter is

$$B_\alpha(\lambda) = 1 - \alpha e^{-i\lambda}$$

with gain and phase functions

$$|B_\alpha(\lambda)| = (1 - 2\alpha \cos \lambda + \alpha^2)^{1/2} \qquad \text{and} \qquad \vartheta_\alpha(\lambda) = \text{Arctan}\left(\frac{\alpha \sin \lambda}{1 - \alpha \cos \lambda}\right).$$

A graph of the gain function of this filter is given in Fig. 6.15.

The quasi-difference filter is often used to balance the spectrum of a time series with a large power peak at low frequencies. The balancing process is called *prewhitening* by Blackman and Tukey (1959). This filter has an advantage over the difference filter in that, since the power near zero frequency is not completely eliminated, good estimates of the spectrum in this range can be obtained by multiplying the estimates of the balanced spectrum at frequency λ by $1/|B_\alpha(\lambda)|^2$.

By gapping the quasi-differencing filter, the periodically repeated "valleys" in the gain function can be located over bothersome periodic components in the spectrum allowing these components to be suppressed to a selectable degree. Thus, the filter

$$Y(t) = X(t) - \alpha X(t - K)$$

has gain function

$$|B_{\alpha, K}(\lambda)| = (1 - 2\alpha \cos K\lambda + \alpha^2)^{1/2}$$

which has minima at the frequencies $\lambda_k = 2\pi k/K$, $k = 0, \pm 1, \ldots$. Because these minima are rather broad, the suppression of periodicities is not as selective as with the use of the Buys–Ballot filter. However, this filter is very simple and has frequently been applied to data for which the spectral peaks to be suppressed are not well defined. Such data occur rather commonly in business and economics series [see Fishman (1969)]. This kind of filter is called a *seasonal adjustment filter*. For a discussion of the construction and use of a variety of seasonal adjustment filters in economics, see Granger and Hatanaka (1964).

We next look at the use of filters to modify data with polynomial trends.

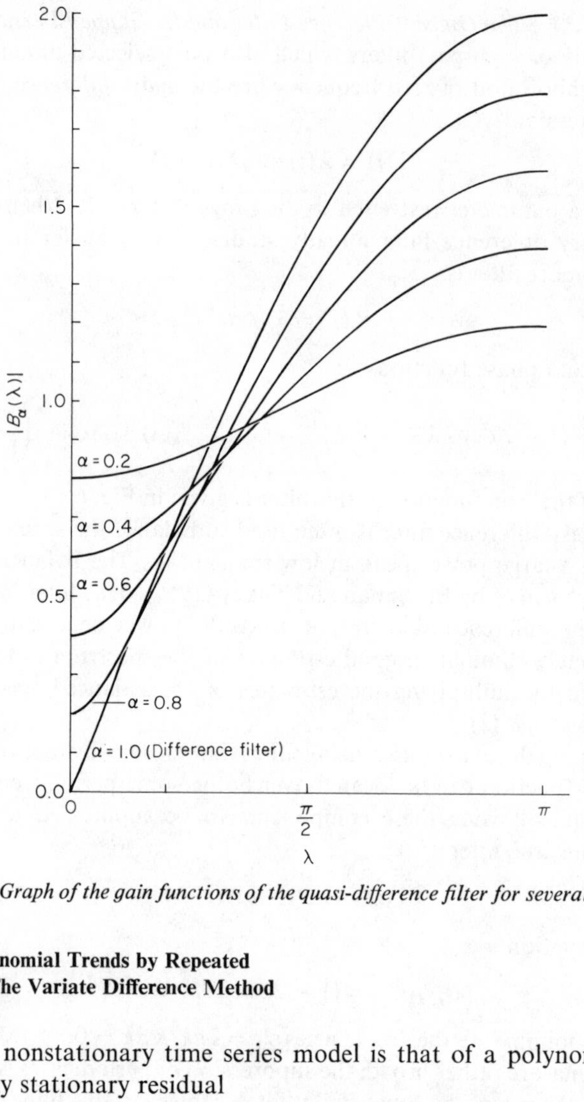

Fig. 6.15 *Graph of the gain functions of the quasi-difference filter for several values of α.*

Removing Polynomial Trends by Repeated Differencing; The Variate Difference Method

A useful nonstationary time series model is that of a polynomial trend with a weakly stationary residual

$$Y(t) = m(t) + X(t), \qquad t = 0, \pm 1, \ldots,$$

where $m(t)$ is a polynomial of degree n. It is assumed that the polynomial has known degree but unknown real-valued coefficients, and that $X(t)$ is weakly stationary with continuous spectrum and spectral density function $f_X(\lambda)$. The problem we wish to consider is the removal of the trend term by digital filtering to obtain the spectrum of the "uncontaminated" residual process. This can be accomplished by means of the *variate difference method* (Tintner, 1940).

The method is based on the important observation that *any polynomial of degree n is eliminated by the repeated difference filter* Δ^{n+1}. To illustrate this fact, consider the special case of a polynomial of degree 2;

$$m(t) = at^2 + bt + c.$$

Then

$$\begin{aligned}\Delta m(t) &= at^2 + bt + c - (a(t-1)^2 + b(t-1) + c)\\ &= 2at - a + b,\\ \Delta^2 m(t) &= (2at - a + b) - (2a(t-1) - a + b)\\ &= 2a,\\ \Delta^3 m(t) &= 2a - 2a = 0.\end{aligned}$$

This property is easily seen to hold for any n. Consequently,

$$\begin{aligned}Z(t) = \Delta^{n+1}(Y(t)) &= \Delta^{n+1}(m(t)) + \Delta^{n+1}(X(t))\\ &= \Delta^{n+1}(X(t)).\end{aligned}$$

The gain function of Δ^{n+1} is $|2\sin(\lambda/2)|^{n+1}$ from Example 6.1 and the theory of repeated linear filters. Thus, $Z(t)$ is a weakly stationary process with continuous spectrum and spectral density function $f_Z(\lambda) = |2\sin(\lambda/2)|^{2(n+1)} f_X(\lambda)$.

We can solve for the spectral density of $X(t)$ in this equation if the spectrum of $Z(t)$ is known. In general it will not be known, thus will have to be estimated. Then this equation can be used to obtain an estimate of the spectrum of $X(t)$. If $\hat{f}_Z(\lambda)$ denotes an estimate of the spectral density of $Z(t)$, then

$$\hat{f}_X(\lambda) = [1/|2\sin(\lambda/2)|^{2(n+1)}]\,\hat{f}_Z(\lambda)$$

will be a reasonable estimate of $f_X(\lambda)$. Note that this estimate will be somewhat unreliable near $\lambda = 0$ since $\sin(\lambda/2)$ is close to zero there. (See the discussion of correcting filter bias in Section 4.4.)

If n is also unknown but the polynomial model seems reasonable, it is common practice to plot the time series after each application of the difference filter. When no obvious trend remains, the differencing operations are terminated and the spectrum analysis is carried out as before.

Filters Which Pass Polynomial Trends

Occasionally it is of interest to construct a low-pass filter which passes a polynomial trend without change. Brillinger (1965) has shown that filters which do this can be characterized in terms of the derivatives of their transfer functions at $\lambda = 0$. He showed that *if* $B(\lambda)$ *is the transfer function of the filter, then a polynomial of degree n will be passed through the filter unchanged if and only if* $B(0) = 1$ *and* $B^{(r)}(0) = 0$ *for* $1 \leq r \leq n$, *where* $B^{(r)}(\lambda)$ *denotes the rth derivative of the transfer function at* λ.

The proof of this result is simple and elegant: First note that if the monomials t^r are passed for $1 \leq r \leq n$, then any polynomial of degree n will be passed because of the linearity of the filter. If $a_j, j = 0, \pm 1, \ldots$, are the filter weights, then t^r is passed if

$$\sum_{j=-\infty}^{\infty} a_j(t-j)^r = t^r \quad \text{for all } t. \tag{6.36}$$

However, by the binomial theorem,

$$(t-j)^r = \sum_{l=0}^{r} \binom{r}{l}(-j)^{r-l} t^l.$$

Thus,

$$\sum_{j=-\infty}^{\infty} a_j(t-j)^r = \sum_{l=0}^{r} \binom{r}{l} \left[\sum_{j=-\infty}^{\infty} (-j)^{r-l} a_j \right] t^l,$$

and identity (6.36) will hold if and only if

$$\binom{r}{l} \left[\sum_{j=-\infty}^{\infty} (-j)^{r-l} a_j \right] = \begin{cases} 0, & \text{for } 0 \leq l \leq r-1, \\ 1, & \text{for } l = r. \end{cases}$$

Equivalently, since (6.36) is to hold for $1 \leq r \leq n$,

$$\sum_{j=-\infty}^{\infty} (-j)^r a_j = \begin{cases} 0, & 1 \leq r \leq n, \\ 1, & r = 0. \end{cases} \tag{6.37}$$

However, $B(\lambda) = \sum_{j=-\infty}^{\infty} a_j e^{-i\lambda j}$, and if the derivatives of order r exist for $1 \leq r \leq n$, [which will be true if $B^{(n)}(\lambda)$ exists], they can be taken inside the sum to obtain

$$B^{(r)}(\lambda) = \sum_{j=-\infty}^{\infty} (-j)^r a_j e^{-i\lambda j}.$$

Thus, (6.37) is equivalent to $B(0) = 1$ and $B^{(r)}(0) = 0$, $1 \leq r \leq n$, as was to be shown.

From the above discussion of the variate difference method it is evident that the filter

$$L_n = I - \Delta^{n+1}$$

passes polynomials of degree n. It is easy to check that Brillinger's conditions are satisfied by the transfer function of this filter.

APPENDIX TO CHAPTER 6

A6.1 The Convolution Form of Digital Filters

The discussion of linear filters given in Section A4.1 carries over to discrete processes and, in fact, the theory is somewhat simplified in that the transformations U_t for integer values of t are all obtained from a single unitary transformation U by the relation

$$U_t = U^t, \qquad t = 0, \pm 1, \ldots .$$

Thus, U is the forward shift transformation which takes $X(t)$ into $X(t+1)$ for all t. The backward shift transformation defined in Section 6.2 is, then,

$$B = U^{-1}.$$

The general time-domain representation of a linear filter was shown to be (A4.3). In the discrete case this represents any linear filter L as a limit of convolution filters L_n with finitely many nonzero weights,

$$L_n(X(t)) = \sum_{j \in J_n} a_{n,j} X(t - t_{n,j})$$

$$= \sum_{k=-\infty}^{\infty} b_{n,k} X(t-k),$$

where $b_{n,k} = a_{n,j}$ when $k = t_{n,j}$, $j \in J_n$ and $b_{n,k} = 0$ otherwise. Even so there is no guarantee that L is a convolution filter on $X(t)$, since the $b_{n,k}$'s need not converge to a square summable set of filter weights for L. However, we will now show that *if $X(t)$ has a continuous spectrum and a spectral density function which is bounded from above and away from zero,*

$$0 < m \le f_X(\lambda) \le M, \tag{A6.1}$$

then not only does every convolution filter match $X(t)$ but every linear filter matching $X(t)$ is of convolution type.

Let L denote a filter which matches $X(t)$ and let $B(\lambda)$ be its transfer function. Then $B(\lambda) \in \mathscr{L}_2(-\pi, \pi)$, since

$$\int_{-\pi}^{\pi} |B(\lambda)|^2 \, d\lambda \le \frac{1}{m} \int_{-\pi}^{\pi} |B(\lambda)|^2 f_X(\lambda) \, d\lambda < \infty.$$

However, as was indicated in Example 1.2, every element of $\mathscr{L}_2(-\pi, \pi)$ has a Fourier series expansion with square summable coefficients which converges in mean-square. That is, there exists a unique set of weights c_j such that

$$\lim_{n\to\infty} \int_{-\pi}^{\pi} \left| \sum_{j=-n}^{n} c_j e^{-i\lambda j} - B(\lambda) \right|^2 d\lambda = 0, \tag{A6.2}$$

and

$$\sum_{j=-\infty}^{\infty} c_j^2 < \infty.$$

Now, consider the sequence of convolution filters

$$L_n(X(t)) = \sum_{j=-n}^{n} c_j X(t-j).$$

Then,

$$\begin{aligned} E|L_n(X(t)) - L(X(t))|^2 &= E\left|\int_{-\pi}^{\pi} e^{i\lambda t}\left\{\sum_{j=-n}^{n} c_j e^{-i\lambda j} - B(\lambda)\right\} Z_X(d\lambda)\right|^2 \\ &= \int_{-\pi}^{\pi} \left|\sum_{j=-n}^{n} c_j e^{-i\lambda j} - B(\lambda)\right|^2 f_X(\lambda)\, d\lambda \\ &\le M \int_{-\pi}^{\pi} \left|\sum_{j=-n}^{n} c_j e^{-i\lambda j} - B(\lambda)\right|^2 d\lambda. \end{aligned}$$

This last term tends to zero as $n \to \infty$ by (A6.2). It follows that the partial sums $L_n(X(t))$ have a mean-square limit and this limit is equal to $L(X(t))$. Thus, L has the convolution representation

$$L(X(t)) = \sum_{j=-\infty}^{\infty} c_j X(t-j).$$

The limitations imposed by condition (A6.1) can be better appreciated by the following argument. If the time series $X(t)$ has spectral density $f_X(\lambda)$ which is equal to zero on a set A of positive measure, then it is possible to construct transfer functions $B(\lambda)$ for linear filters such that

$$\int_A |B(\lambda)|^2\, d\lambda = \infty \qquad \text{yet} \qquad \int_{-\pi}^{\pi} |B(\lambda)|^2 f_X(\lambda)\, d\lambda < \infty.$$

Thus, these filters will match $X(t)$ but will not be of convolution type. If $f_X(\lambda) \le M$, then all filters of convolution type will match $X(t)$ as was argued in the text. When $f_X(\lambda)$ is unbounded this need not be the case since examples can be easily constructed for which

$$\int_{-\pi}^{\pi} |B(\lambda)|^2\, d\lambda < \infty \qquad \text{but} \qquad \int_{-\pi}^{\pi} |B(\lambda)|^2 f_X(\lambda)\, d\lambda = \infty.$$

A6.2 Nonnegative Definite Filters

From the definition of the transfer function $B(\lambda)$ and phase shift $\vartheta(\lambda)$ of a linear filter it is easily seen that $\vartheta(\lambda) = 0$ for all λ if and only if $B(\lambda)$ is real-valued *and nonnegative* for all λ. Symmetric filters have real-valued but not necessarily nonnegative transfer functions, which leads to the possibility that $\vartheta(\lambda) = \pm\pi$ for some frequencies.

A digital filter with (real-valued) filter weights $\{c_k\colon k = 0, \pm 1, \ldots\}$ is said to be *nonnegative definite* if for every positive integer n and complex numbers a_k, $k = 0, \pm 1, \ldots, \pm n$, we have

$$\sum_{j=-n}^{n} \sum_{k=-n}^{n} a_j \bar{a}_k c_{j-k} \geq 0.$$

(See Section A2.1 for the definition of continuous-time nonnegative definite functions. This definition and the theory given here can be adapted to give the comparable results for continuous-time convolution filters.) We now show that the transfer functions of nonnegative definite filters are, indeed, real-valued and nonnegative and conversely. Thus, in order for a filter not to shift phases, it is necessary and sufficient that it be nonnegative definite.

First, taking $a_0 = 1$, $a_j = i$, and $a_k = 0$ otherwise, the above inequality becomes $2c_0 + i(c_j - c_{-j}) \geq 0$. However, this sum must be real-valued, which can happen only if $c_j = c_{-j}$. Thus, as is not surprising, nonnegative definite filters are always symmetric. It follows that the transfer function is real-valued.

If $B(\lambda)$ is the transfer function of the filter with the given (square-summable) weight sequence, it follows from expression (6.4) that

$$\sum_{j=-n}^{n} \sum_{k=-n}^{n} a_j \bar{a}_k c_{j-k} = \frac{1}{2\pi} \int_{-\pi}^{\pi} B(\lambda) \left| \sum_{j=-n}^{n} a_j e^{i\lambda j} \right|^2 d\lambda \geq 0. \tag{A6.3}$$

Clearly, this inequality holds if $B(\lambda) \geq 0$ a.e. Thus, the filter is nonnegative definite if $\vartheta(\lambda) = 0$ a.e. Now, suppose that $B(\lambda) < 0$ on a set $A = B \cup (-B)$ of positive measure in $(-\pi, \pi)$ and let $I_A(\lambda)$ be the set characteristic function of this set. Since this is a square integrable function, it has a Fourier series representation

$$I_A(\lambda) = \sum_{j=-\infty}^{\infty} a_j e^{i\lambda j}.$$

The identity $I_A(\lambda) = |I_A(\lambda)|^2$ implies the alternate representation

$$I_A(\lambda) = \left| \sum_{j=-\infty}^{\infty} a_j e^{i\lambda j} \right|^2.$$

Moreover, if

$$I_{A,n}(\lambda) = \left| \sum_{j=-n}^{n} a_j e^{i\lambda j} \right|^2,$$

it is easily argued that

$$\lim_{n\to\infty}\int_{-\pi}^{\pi} B(\lambda)I_{A,n}(\lambda)\,d\lambda = \int_{-\pi}^{\pi} B(\lambda)I_A(\lambda)\,d\lambda.$$

Since each term of the sequence of integrals is nonnegative because of (A6.3), it follows that $\int_{-\pi}^{\pi} B(\lambda)I_A(\lambda)\,d\lambda \geq 0$. However, this is impossible if $B(\lambda) < 0$ on A. Thus, necessarily $B(\lambda) \geq 0$ a.e. in $(-\pi, \pi)$.

It is easily seen that all of the ideal symmetric filters discussed in the text are nonnegative definite, since the transfer functions were chosen to be nonnegative. It is of interest to note that this property is lost by truncation as can be observed, for example, from the negative side lobes of the transfer function of the truncated weights filter in Fig. 6.7. When the filter is modified to minimize the side lobes as described in Section 6.4, then the power transmitted by the filter in frequency ranges for which $\vartheta(\lambda) \neq 0$ is extremely small. Thus, for practical purposes, properly designed symmetric filters have actual phase shift characteristics which are nearly identical to those of nonnegative definite filters.

A6.3 The Error of the Fast Fourier Transform Filtering Method

By substituting the spectral representation for the process $X(t)$ in (6.24) and using the resulting expression in (6.25) we obtain

$$W(t) = \int_{-\pi}^{\pi}\left\{\frac{1}{N}\sum_{s=1}^{N} e^{i\lambda s}H(t-s)\right\}Z(d\lambda),$$

where

$$H(r) = \sum_{\nu=-[(N-1)/2]}^{[N/2]} C(\lambda_\nu)e^{i\lambda_\nu r}.$$

Now,

$$C(\lambda) = \sum_{k=-\infty}^{\infty} e^{-i\lambda k}b_k,$$

thus,

$$H(r) = \sum_{k=-\infty}^{\infty} b_k \sum_{\nu=-[(N-1)/2]}^{[N/2]} e^{i\lambda_\nu(r-k)}.$$

However, since $\lambda_\nu = 2\pi\nu/N$,

$$\sum_{\nu=-[(N-1)/2]}^{[N/2]} e^{i\lambda_\nu m} = \begin{cases} N, & \text{if } m = pN, \quad p = 0, \pm 1, \dots, \\ 0, & \text{otherwise.}\end{cases}$$

This implies

$$H(r) = N \sum_{p=-\infty}^{\infty} b_{r+pN} = N\tilde{b}_r .$$

Finally,

$$W(t) = \int_{-\pi}^{\pi} \left\{ \sum_{s=1}^{N} e^{i\lambda s} \tilde{b}_{t-s} \right\} Z(d\lambda)$$

$$= \int_{-\pi}^{\pi} e^{i\lambda t} \left\{ \sum_{k=t-N}^{t-1} e^{-i\lambda k} \tilde{b}_k \right\} Z(d\lambda).$$

Then (6.26) is obtained from this expression and (6.22) exactly as was expression (6.20) from (6.19).

CHAPTER

7

Finite Parameter Models, Linear Prediction, and Real-Time Filtering

7.1 INTRODUCTION

In the early years of the twentieth century, time series studies were based on an implicit model consisting of an almost periodic (nonstochastic) trend term with a white noise residual. This model was used to search for "hidden periodicities," primarily in geophysical data such as the Wolfer sunspot series. A graph of this historic series is given in Fig. 7.1. At first glance a model of pure sinusoids would seem reasonable. However, closer inspection revealed properties of the data that could not be readily accounted for by such a model. Observations of this nature caused early criticism of the "scheme of hidden periodicities" and prompted the search for models which better described the observed data.

Yule (1921) considered the consecutive differences of a purely random series (the values of a series of independent, uniformly distributed random variables) and noted a "tendency toward regularity" in the resulting data similar to that exhibited in Fig. 7.1. Continuing this work, Slutsky (1927) studied sums and differences of purely random series and formulated the observed regular behavior discussed in Example 6.9 as one of the first probabilistic limit theorems for stochastic processes. He called the series resulting from the summing and differencing operations *processes of moving summation*. The more recent terminology, apparently dating from the definitive study of discrete models by Wold (1938), is *moving average processes*. This class of

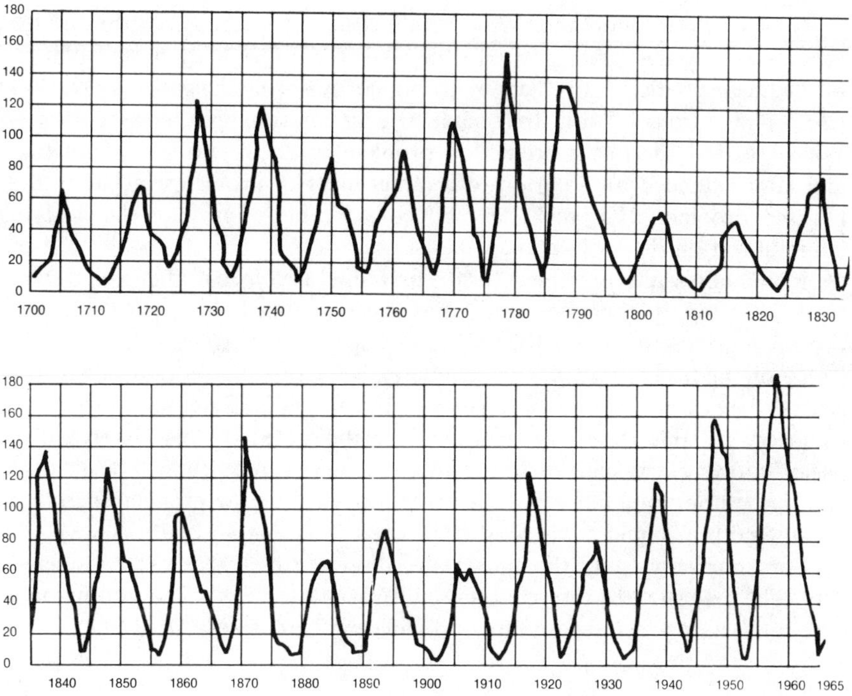

Fig. 7.1 *Graph of Wolfer sunspot data. Annual mean sunspot numbers observed over a period of* 265 *years.*

processes remains one of the most important collections of finite parameter models and will be our first object of study in this chapter.

In a study of the Wolfer sunspot series, Yule (1927) used techniques of regression analysis to approximate the value of the series at a given time as a linear function of a fixed number of previous values. The scheme implicitly defined by this procedure was termed the *scheme of linear autoregression* by Wold. *Finite autoregressive processes*, as they are more commonly called today, constitute the second important class of finite parameter models to be considered in this chapter.

Autoregressive processes appeared in economics during the 1930s in the work of Frisch (1933) and Tinbergen (1937), who first used stochastic elements in dynamic models of economic systems. Because of the great interest in forecasting future values of economic variables, the unique predictive feature of autoregressive processes (which we will illustrate in Section 7.3) played a significant role in the applications of these models.

In his 1938 monograph, Wold established a key decomposition of the general weakly stationary process which led to the formulation and solution of the linear prediction problem by Kolmogorov during the period 1939–1941 [see Kolmogorov (1941b)]. Independently, during roughly the same period, Wiener solved the linear prediction problem in an important special case and later extended his solution to a larger class of problems including the filtering problem. His work has had considerable impact on the field of communication theory [see Lee (1964)].

Kolmogorov also presented, for the first time, a geometric interpretation of weakly stationary processes which has greatly unified the subject and has proved most useful in the theoretical development of time series analysis. [see Kolmogorov (1941a)]. This geometric view of time series is one of the central themes of this book and some of the more significant applications of geometry to time series problems will be illustrated in this chapter. The basic features of the solution to the linear prediction problem will be detailed and the solution of a more realistic version of the filtering problem than that considered in Chapter 5 will be sketched.

For simplicity, only the univariate, discrete-time theory is considered here. The extension to discrete-time multivariate processes and continuous-time univariate processes is amply provided by Hannan (1970).

7.2 MOVING AVERAGES

Recall that a white noise process is a sequence of uncorrelated random variables $\xi(t)$, $t = 0, \pm 1, \ldots$, with common mean $E\xi(t) = 0$ and variance $E\xi^2(t) = \sigma^2$. Then, a *moving average process* is defined by the expression

$$X(t) = \sum_{j=-m}^{n} a_j \xi(t-j), \qquad t = 0, \pm 1, \ldots, \tag{7.1}$$

where m and n are nonnegative integers and the a_j's are real constants. This is, more precisely, a *finite moving average*. By letting n and/or m tend to infinity, with the added condition that

$$\sum_{j=-\infty}^{\infty} a_j^2 < \infty, \tag{7.2}$$

an *infinite moving average* is obtained, i.e.,

$$X(t) = \sum_{j=-\infty}^{\infty} a_j \xi(t-j). \tag{7.3}$$

Then (7.1) is simply a special case of (7.3) with coefficients $a_j = 0$ for $j < -m$ and $j > n$.

From the theory of digital filters given in Chapter 6, $X(t)$ is the output of the linear filter with transfer function

$$B(\lambda) = \sum_{j=-\infty}^{\infty} a_j e^{-i\lambda j} \tag{7.4}$$

to the white noise input. Since the spectrum of the white noise process is continuous with spectral density

$$f_\xi(\lambda) = \sigma^2/2\pi, \qquad -\pi \le \lambda \le \pi,$$

the matching condition is satisfied by virtue of the Parseval relation (Example 1.2) and condition (7.2):

$$\int |B(\lambda)|^2 f_\xi(\lambda)\, d\lambda = \sigma^2 \sum_{j=-\infty}^{\infty} a_j^2 < \infty.$$

Thus, (7.3) defines a weakly stationary process with continuous spectrum and spectral density

$$\begin{aligned} f_X(\lambda) &= |B(\lambda)|^2 f_\xi(\lambda) \\ &= \frac{\sigma^2}{2\pi} \left| \sum_{j=-\infty}^{\infty} a_j e^{-i\lambda j} \right|^2. \end{aligned} \tag{7.5}$$

This process has zero mean and variance $\sigma^2 \sum_{j=-\infty}^{\infty} a_j^2$.

The finite moving average model has a useful intuitive interpretation. At time t, "nature" produces a random shock or *innovation* $\xi(t)$ which is unrelated to the shocks at other times. The observed quantity $X(t)$ at time t is then an "average" of the random shocks over times close to t. If $m > 0$ and $n > 0$ in (7.1), then $X(t)$ depends on innovations from the past, which corresponds to terms $\xi(t-j)$ for $j = 1, 2, \ldots, n$, the present $\xi(t)$, and the future, which is determined by the terms $\xi(t-j)$ for $j = -1, -2, \ldots, -m$. With this interpretation it would be more realistic to consider models which depend only on the present and past, i.e., for which $m = 0$. This restriction, extended to (7.3), leads to the class of *one-sided moving averages*

$$X(t) = \sum_{j=0}^{\infty} a_j \xi(t-j),$$

which will play an important role in prediction theory. By the change of index from j to $j - m$ in (7.1), we obtain

$$X(t) = \sum_{j=-m}^{n} a_j \xi(t-j) = \sum_{j=0}^{m+n} a_{j-m} \xi(t-j+m).$$

Now, $\eta(t) = \xi(t + m)$, $t = 0, \pm 1, \ldots$, is another white noise process with the same mean and variance as $\xi(t)$. Thus, the finite moving average (7.1) can always be put in the form of a one-sided moving average

$$X(t) = \sum_{j=0}^{p} b_j \eta(t - j) \tag{7.6}$$

with $p = m + n$ and $b_j = a_{j-m}$, $j = 0, \ldots, p$. This is the form most often found in the literature and we will adopt this definition in later sections. However, the applicability of finite moving average processes as models for weakly stationary time series is best illustrated in terms of the two-sided representation (7.1).

The Approximation of Processes with Continuous Spectra by Finite Moving Averages

Adopting $E(X - Y)^2$ as the measure of (squared) distance between random variables as before, we will see that *if* $Y(t)$ *is any weakly stationary stochastic process with continuous spectrum and if* ε *is any positive number, then there is a two-sided, finite moving average process* $X(t)$ *given by* (7.1) *such that if m and n are sufficiently large, we will have*

$$E(X(t) - Y(t))^2 < \varepsilon \qquad \textit{for} \quad \textit{all } t.$$

That is, every weakly stationary process with continuous spectrum can be approximated arbitrarily closely in mean-square by a finite moving average process. This fact is a simple consequence of the even more remarkable fact that *every weakly stationary time series with continuous spectrum can be represented as a (possibly) infinite moving average.* In Section 2.6 we saw that every process with a discrete spectrum is a stochastic almost periodic function. This provided a concrete representation of such processes as linear functions of the elementary sinusoidal functions. The expression of processes with continuous spectra as moving averages provides a comparable concrete representation of these time series as linear functions of elementary white noise processes.

Every Weakly Stationary Process with Continuous Spectrum Has an Infinite Moving Average Representation

This representation is relatively easy to establish and we carry out the derivation here for the case in which the spectral density function $f_Y(\lambda)$ is everywhere strictly greater than zero. We will show how to remove this restriction in the appendix to this chapter.

Let σ^2 be a positive number. A linear filter for $Y(t)$ is completely determined by specifying its transfer function, provided the matching condition is satisfied (see Section A4.1). Consider the linear filter determined by the transfer function

$$A(\lambda) = \frac{\sigma}{(2\pi)^{1/2}} \frac{1}{(f_Y(\lambda))^{1/2}}.$$

Since $f_Y(\lambda)$ is strictly positive, there is no difficulty with the division by $(f_Y(\lambda))^{1/2}$. This filter does indeed match $Y(t)$, since

$$\int_{-\pi}^{\pi} |A(\lambda)|^2 f_Y(\lambda)\, d\lambda = \sigma^2.$$

Thus,

$$\xi(t) = \int_{-\pi}^{\pi} e^{i\lambda t} A(\lambda) Z_Y(d\lambda)$$

is a well-defined, weakly stationary stochastic process. Moreover, the spectral density function of $\xi(t)$ is

$$f_\xi(\lambda) = |A(\lambda)|^2 f_Y(\lambda) = \sigma^2/2\pi, \qquad -\pi \le \lambda \le \pi.$$

That is, $\xi(t)$ is a white noise process with variance σ^2.

The random spectral measure of this process is

$$Z_\xi(d\lambda) = A(\lambda) Z_Y(d\lambda).$$

Thus, since $A(\lambda)$ is never zero, $Y(t)$ can be recovered from $\xi(t)$ by inverting the filter (Section 4.4):

$$Z_Y(d\lambda) = \frac{1}{A(\lambda)} Z_\xi(d\lambda) = \frac{(2\pi)^{1/2}}{\sigma} (f_Y(\lambda))^{1/2} Z_\xi(d\lambda).$$

Now, $((2\pi)^{1/2}/\sigma)(f_Y(\lambda))^{1/2}$ is a square-integrable periodic function and by the theory of Example 1.2 it has a Fourier series expansion

$$\frac{(2\pi)^{1/2}}{\sigma} (f_Y(\lambda))^{1/2} = \sum_{j=-\infty}^{\infty} a_j e^{-i\lambda j}.$$

The a_j's are real constants and $\sum_{j=-\infty}^{\infty} a_j^2 < \infty$. Then,

$$\begin{aligned} Y(t) &= \int_{-\pi}^{\pi} e^{i\lambda t} \left(\sum_{j=-\infty}^{\infty} a_j e^{-i\lambda j} \right) Z_\xi(d\lambda) \\ &= \sum_{j=-\infty}^{\infty} a_j \int_{-\pi}^{\pi} e^{i\lambda(t-j)} Z_\xi(d\lambda) = \sum_{j=-\infty}^{\infty} a_j \xi(t-j). \end{aligned} \tag{7.7}$$

This is the desired moving average representation of $Y(t)$.

This representation is by no means unique, however. Suppose that $C(\lambda)$ is a complex-valued function, nonvanishing for $-\pi \leq \lambda \leq \pi$ and with the property

$$f_Y(\lambda) = (\sigma^2/2\pi)|C(\lambda)|^2. \tag{7.8}$$

Then, taking $A(\lambda) = 1/C(\lambda)$, the construction of a white noise process and the representation of $Y(t)$ as a moving average with weights equal to the Fourier coefficients of $C(\lambda)$ can be carried out exactly as before. Thus, there are at least as many different moving average representations as there are different factorizations of $f_Y(\lambda)$ of form (7.8).

For example, if $\vartheta(\lambda)$ is any valid phase shift function,

$$C(\lambda) = \frac{(2\pi)^{1/2}}{\sigma}(f_Y(\lambda))^{1/2}e^{i\vartheta(\lambda)}$$

satisfies (7.8). Since the class of phase shift functions is quite large, so is the class of moving average representations of $Y(t)$.

An important question for prediction theory is whether *one-sided* moving average representations exist for $Y(t)$. If, in addition to (7.8), $C(\lambda)$ satisfies the condition

$$C(\lambda) = \sum_{j=0}^{\infty} a_j e^{-i\lambda j}, \qquad \sum_{j=0}^{\infty} a_j^2 < \infty, \tag{7.9}$$

i.e., if $C(\lambda)$ has a one-sided Fourier series expansion, then the moving average will depend only on the past and present of the corresponding white noise process. Moreover, every one-sided moving average has a spectral density function of the form (7.8) with transfer function $C(\lambda)$ satisfying (7.9). Consequently, the construction of one-sided moving average representations for $Y(t)$ is equivalent to finding functions $C(\lambda)$ which satisfy conditions (7.8) and (7.9). As we will see, the construction of these functions can be framed as a problem in complex analysis. An elegant solution to this problem exists and will be discussed in Section 7.3 in the context of prediction theory.

We now return to the justification of the use of finite moving averages as models for weakly stationary processes. If $Y(t)$ is any weakly stationary process with zero mean and continuous spectrum and if

$$Y(t) = \sum_{j=-\infty}^{\infty} a_j \xi(t-j)$$

is a moving average representation of this process, then a finite moving average

$$X(t) = \sum_{j=-m}^{n} a_j \xi(t-j),$$

using the same coefficients and white noise process, can be made arbitrarily close to $Y(t)$ in mean-square by proper selection of m and n. To see this, note that since the $\xi(t)$'s are uncorrelated,

$$E(Y(t) - X(t))^2 = E\left(\sum_{j<-m} a_j \xi(t-j) + \sum_{j>n} a_j \xi(t-j)\right)^2$$
$$= \left(\sum_{j<-m} a_j^2 + \sum_{j>n} a_j^2\right)\sigma^2.$$

However, since $\sum_{j=-\infty}^{\infty} a_j^2 < \infty$, it follows that $\sum_{j<-m} a_j^2$ and $\sum_{j>n} a_j^2$ tend to zero as m and n approach infinity. Thus, $E(Y(t) - X(t))^2$ can be made as small as desired for all t by selecting m and n large enough.

There is an extensive literature on the statistical problem of fitting time series data by finite moving average schemes. We will not pursue this topic in this book. The interested reader is referred to the excellent treatments by Hannan (1970) and Anderson (1971). For practical applications of these techniques to problems of the forecasting and control of random processes, see Box and Jenkins (1970).

7.3 AUTOREGRESSIVE PROCESSES

Let $\xi(t)$, $t = 0, \pm 1, \ldots,$ be a white noise process with zero mean and variance σ^2 and let $b_0, b_1, \ldots, b_q$ be real constants such that $b_0 = 1$ and $b_q \neq 0$. Then if $X(t)$ is a weakly stationary process which satisfies the "difference" equation

$$X(t) + b_1 X(t-1) + \cdots + b_q X(t-q) = \xi(t), \qquad t = 0, \pm 1, \ldots, \tag{7.10}$$

and for which

$$EX(s)\xi(t) = 0 \qquad \text{for all } s \le t-1, \quad t = 0, \pm 1, \ldots, \tag{7.11}$$

then $X(t)$ is said to be a (finite) *autoregressive process* (*autoregression*). The number q is called the *order* of the autoregression.

Similarly, an *infinite* (*order*) *autoregression* is required to satisfy conditions (7.11) and, in addition, equations of the form

$$\sum_{k=0}^{\infty} b_k X(t-k) = \xi(t), \qquad t = 0, \pm 1, \ldots, \tag{7.12}$$

where $b_0 = 1$ and $\sum_{k=0}^{\infty} b_k^2 < \infty$.

Existence of Finite Autoregressions

Equation (7.10) is the discrete analog of the stochastic differential equations of Section 4.4 and the theory of inverting linear filters can again be applied to determine conditions under which a stationary solution exists. Recall that this theory was based on the characteristic equation of the differential equation. The corresponding characteristic equation for the difference equation is

$$\mathscr{H}(z) = z^q + b_1 z^{q-1} + \cdots + b_q = 0.$$

It is customary to deal with a variant of this equation which is defined in terms of the z-transform of the linear filter determined by $b_0, \ldots, b_q$. In Chapter 6 we defined the *z-transform* of a linear filter $V(t) = \sum_{j=-\infty}^{\infty} a_j U(t-j)$, $\sum_{j=-\infty}^{\infty} a_j^2 < \infty$, to be the function

$$\mathscr{A}(z) = \sum_{j=-\infty}^{\infty} a_j z^j.$$

This is viewed as a complex-valued function of the complex variable z. Note that the transfer function of the filter is obtained by replacing z by $e^{-i\lambda}$, i.e., it is the "value" of the z-transform on the unit circle,

$$A(\lambda) = \mathscr{A}(e^{-i\lambda}).$$

The z-transform of the linear filter (7.10) which transforms $X(t)$ into $\xi(t)$ is

$$\mathscr{B}(z) = 1 + b_1 z + \cdots + b_q z^q.$$

This is a well-defined function over the entire complex plane. It is related to the characteristic polynomial $\mathscr{H}(z)$ by the expression

$$\mathscr{B}(z) = z^q \mathscr{H}(1/z).$$

Since we have assumed $b_q \neq 0$, $z = 0$ is not a solution of either of the equations $\mathscr{B}(z) = 0$ or $\mathscr{H}(z) = 0$. Consequently, it follows that whenever z' is a root of $\mathscr{B}(z) = 0$, then $1/z'$ is a root of $\mathscr{H}(z) = 0$, i.e., any zero of $\mathscr{H}(z)$ inside the unit circle ($z' = re^{i\vartheta}$ for $r < 1$) corresponds to a zero of $\mathscr{B}(z)$ outside the unit circle $[1/z' = (1/r)e^{-i\vartheta}]$, and conversely. Since some of the literature concerning autoregressive processes deals with the characteristic equation while the rest is based on the z-transform, it is worthwhile keeping this relationship in mind.

Since $\mathscr{B}(z)$ is a polynomial, if the transfer function $B(\lambda) = \mathscr{B}(e^{-i\lambda}) \neq 0$ for $-\pi < \lambda < \pi$ [i.e., $\mathscr{B}(z)$ has no zeros on the unit circle], then $B(\lambda)$ is actually bounded away from zero. Writing $|B(\lambda)| \geq M > 0$, the linear filter with this transfer function can be inverted, since

$$\int_{-\pi}^{\pi} |1/B(\lambda)|^2 f_\xi(\lambda)\, d\lambda = (\sigma^2/2\pi) \int_{-\pi}^{\pi} |1/B(\lambda)|^2\, d\lambda \leq \sigma^2/M^2.$$

Thus, $X(t) = \int_{-\pi}^{\pi} e^{i\lambda t}(1/B(\lambda)) Z_\xi(d\lambda)$ is a weakly stationary solution to (7.10). Moreover, since $1/B(\lambda)$ is square integrable, it has a (unique) Fourier series expansion (Example 1.2);

$$1/B(\lambda) = \sum_{j=-\infty}^{\infty} a_j e^{-i\lambda j} \qquad \text{with} \qquad \sum_{j=-\infty}^{\infty} a_j^2 < \infty.$$

It follows that the solution to (7.10) can be put in the form of a moving average

$$X(t) = \sum_{j=-\infty}^{\infty} a_j \xi(t-j).$$

Not every such solution satisfies (7.11), however, thus not every polynomial $\mathscr{B}(z)$ of the above description leads to a finite autoregression. In order for the moving average to satisfy this condition it is necessary and sufficient that

$$EX(s)\xi(t) = \sum_{j=-\infty}^{\infty} a_j E\xi(s-j)\xi(t) = a_{s-t}\sigma^2 = 0$$

for $s = t-1,\ t-2, \ldots$. That is, $a_{-1} = a_{-2} = \cdots = 0$. Moreover, by the same computation,

$$EX(t)\xi(t) = a_0\sigma^2.$$

However, from (7.10) and (7.11),

$$\begin{aligned} EX(t)\xi(t) &= E\left\{\left[\sum_{k=1}^{q}(-b_k)X(t-k) + \xi(t)\right]\xi(t)\right\} \\ &= \sum_{k=1}^{q}(-b_k)EX(t-k)\xi(t) + E\xi^2(t) \\ &= E\xi^2(t) = \sigma^2. \end{aligned}$$

It follows that $a_0 = 1$. Thus, we see that the above Fourier series expansion must be one-sided and have $a_0 = 1$ in order for (7.10) to have a weakly stationary solution satisfying (7.11). When this is the case, the solution is the one-sided moving average

$$X(t) = \sum_{j=0}^{\infty} a_j \xi(t-j).$$

It is convenient to be able to state the condition or conditions for the existence of an autoregression in terms of the z-transform $\mathscr{B}(z)$ alone. One simple and appealing condition suffices which will be seen to be a special case of a more general result to be stated in Section A7.2. However, this condition follows easily from elementary properties of analytic functions and power series and the argument will be given here.

Since $\mathscr{B}(z)$ is a polynomial, if it does not vanish on the unit circle it will be nonzero in a region $1/\rho < |z| < \rho$ for some $\rho > 1$. Thus $1/\mathscr{B}(z)$ will be analytic and have a series expansion in positive and negative powers of z (Laurent expansion) in this region (Titchmarsh, 1939, p. 89). When $z = e^{-i\lambda}$, this series necessarily coincides with the Fourier series expansion of $1/B(\lambda)$. Thus, by the above argument, the coefficients of the negative powers of z must be zero. However, this corresponds to the case in which $1/\mathscr{B}(z)$ is analytic for $|z| < \rho$. Equivalently, $\mathscr{B}(z)$ is nonzero for this region.

In summary, we find that *in order for* (7.10) *to have a weakly stationary solution satisfying* (7.11), *i.e., for a finite autoregression to exist, it is sufficient that all of the zeros of* $\mathscr{B}(z)$ *lie outside of the unit circle. The solution will be a one-sided moving average of the defining white noise process with coefficients equal to those of the power series expansion of* $1/\mathscr{B}(z)$.

The equivalent condition that all of the zeros of $\mathscr{H}(z)$ lie inside the unit circle is reminiscent of the condition leading to one-sided (realizable) solutions of the linear differential equations with random forcing functions considered in Section 4.4. In fact, this is not surprising since the methods of proof used in the two cases are quite similar.

It is known that this condition is also necessary for the existence of a stationary autoregression [see Pagano (1973)]. Thus, the difference equation (7.10) can be treated by other methods and nonstationary solutions will be obtained if some of the roots of $\mathscr{B}(z)$ lie on the unit circle. These solutions have received considerable attention in economics and form an interesting class of nonstationary processes. These processes are treated extensively by Box and Jenkins (1970). We will return to this topic briefly in Section 7.5.

Infinite Autoregressions

The conditions under which an infinite autoregression exists are somewhat less simple than those for a finite autoregression. If $\mathscr{B}(z)$ is the z-transform of the filter defined by (7.12),

$$\mathscr{B}(z) = \sum_{k=0}^{\infty} b_k z^k,$$

then, except in the case where this reduces to a finite polynomial, this function is no longer defined over the entire complex plane. In fact, the domain of convergence of the infinite series is a circle of finite radius centered at $z = 0$.

If, for example, the domain of convergence is $\mathscr{D} = \{z : |z| < \rho\}$ with $\rho > 1$ and if $\mathscr{B}(z) \neq 0$ in this set, then $1/\mathscr{B}(z)$ will have a power series expansion

$$1/\mathscr{B}(z) = \sum_{j=0}^{\infty} a_j z^j$$

valid for $|z| < \rho$ with $a_0 = 1/\mathscr{B}(0) = 1/b_0 = 1$. In particular, $1/\mathscr{B}(z)$ will be bounded on the unit circle which implies that the linear filter with transfer function $1/B(\lambda)$, $B(\lambda) = \mathscr{B}(e^{-i\lambda})$, matches the white noise input. Then the weakly stationary, one-sided moving average

$$X(t) = \int_{-\pi}^{\pi} e^{i\lambda t}(1/B(\lambda))Z_\xi(d\lambda) = \int_{-\pi}^{\pi} e^{i\lambda t}\left(\sum_{j=0}^{\infty} a_j e^{-i\lambda j}\right) Z_\xi(d\lambda)$$
$$= \sum_{j=0}^{\infty} a_j \xi(t-j)$$

solves (7.10) as before. Now condition (7.11) is automatically satisfied.

This is an excessively restrictive situation, however, since in order for the domain of convergence of $\mathscr{B}(z)$ to be $\mathscr{D}$ it is easy to show that the coefficients b_k must satisfy inequalities of the form $|b_k| \le Ml^k$, $k = 0, 1, 2, \ldots$, for some l such that $1/\rho \le l < 1$. This is a substantially more demanding assumption than is the condition $\sum_{k=0}^{\infty} b_k{}^2 < \infty$. The a_k's will satisfy the same inequalities (possibly with a different value of M) which implies, in particular, that $\sum_{k=0}^{\infty} a_k{}^2 < \infty$, as required by the theory. The best condition known for the existence of an infinite autoregression is quite close to this but requires a more delicate analysis. For a statement of this condition, the reader is referred to Section A7.2.

Since autoregressive processes, finite or infinite, are expressible as linear transformations of white noise processes, they have continuous spectra with spectral densities of the form

$$f_X(\lambda) = \frac{\sigma^2}{2\pi}\frac{1}{|B(\lambda)|^2} = \frac{\sigma^2}{2\pi}\frac{1}{|\sum_{k=0}^{\infty} b_k e^{-i\lambda k}|^2}. \tag{7.13}$$

Note that the z-transform $\mathscr{A}(z)$ of the one-sided moving average solution is always related to the z-transform $\mathscr{B}(z)$ of the autoregression by the equation

$$\mathscr{A}(z) = 1/\mathscr{B}(z). \tag{7.14}$$

Thus, one means of obtaining the coefficients of the moving average solution is to explicitly expand $1/\mathscr{B}(z)$ in a power series for $|z| < 1$. This can occasionally be done rather simply. An example will be given shortly.

Autoregressions with Nonzero Means

It is sometimes of interest to consider autoregressive processes which satisfy (7.10) or (7.12) and (7.11) but for which the white noise process $\xi(t)$ has a nonzero mean, m. The above results remain valid up to the point where $X(t)$ is expressed as a one-sided moving average. Rewrite $\xi(t)$ in the form

$$\xi(t) = m + \eta(t),$$

where $\eta(t)$ is a white noise process with zero mean. Then, $X(t)$ can be written in the form

$$X(t) = m \sum_{k=0}^{\infty} a_k + \sum_{k=0}^{\infty} a_k \eta(t-k)$$

provided both terms on the right make sense. The second term is well defined by the previous theory. However, the convergence of $\sum_{k=0}^{\infty} a_k$ is not guaranteed by the condition $\sum_{k=0}^{\infty} a_k{}^2 < \infty$. If the stronger condition

$$\sum_{k=0}^{\infty} |a_k| < \infty$$

holds, then both terms are well defined and $X(t)$ becomes a standard autoregression with a constant trend $m \sum_{k=0}^{\infty} a_k$. This condition is, in fact, satisfied for all finite autoregressions. It is also satisfied for the infinite autoregressions considered in the special case above. An example in which these considerations are important will be given below.

Finite-order autoregressions are quite useful as models for observed time series and, as in the case of moving averages, a great deal of effort has gone into devising schemes for fitting them to time series data. The most popular method is based on the Yule–Walker equations, which comprise a set of linear equations relating the coefficients of the autoregression to its autocovariances. By estimating the autocovariances using observations from the given time series, the coefficients of the autoregression which best fits this series can be obtained by solving the equations with the estimated autocovariances in place of the real ones. An example of this procedure will be given in conjunction with the so-called autoregressive method of spectral estimation in Chapter 9. We now derive these equations.

The Yule–Walker Equations

From expression (7.10) and the fact that $E\xi(t)X(t-l) = 0$ for $l = 1, 2, \ldots, q$, we obtain

$$EX(t-l)\left[\sum_{k=0}^{q} b_k X(t-k)\right] = \sum_{k=0}^{q} b_k EX(t-l)X(t-k) = 0$$

for $l = 1, 2, \ldots, q$. However, by stationarity of the autocovariance $C(\tau)$,

$$EX(t-l)X(t-k) = C(l-k).$$

Thus, using the fact that $b_0 = 1$, these equations become

$$\sum_{k=1}^{q} b_k C(l-k) = -C(l). \qquad l = 1, 2, \ldots, q. \tag{7.15}$$

These are the Yule–Walker equations.

One additional equation is useful. Earlier we showed that

$$EX(t)\xi(t) = \sigma^2.$$

However,

$$EX(t)\xi(t) = EX(t)\left[\sum_{k=0}^{q} b_k X(t-k)\right] = \sum_{k=0}^{q} b_k C(k).$$

Thus,

$$\sigma^2 = \sum_{k=0}^{q} b_k C(k). \tag{7.16}$$

This equation makes it possible to estimate the variance of the white noise process from the estimates of the autoregressive coefficients and autocovariances. A method for fitting the degree of the autoregression to the data based on this equation is given in the above mentioned discussion in Chapter 9.

Physical Interpretation of the Autoregressive Model

With the reassignment of parameters, $c_k = -b_k$, for $k = 1, 2, \ldots, q$, expression (7.10) can be put in the form

$$X(t) = \sum_{k=1}^{q} c_k X(t-k) + \xi(t). \tag{7.17}$$

Thus, the value of the autoregression at any time t consists of a linear function of the values of the process for the past q observation times plus a random shock or *innovation* $\xi(t)$, which is unrelated to the past values of the process [condition (7.11)]. This is, perhaps, the simplest possible model for a stochastic process with a memory of fixed length. If $X(t-1), \ldots, X(t-q)$ were fixed, known functions, then (7.17) would be what is known as a regression model in statistics. However, the process is regressed on its own past values; hence the name, autoregression.

Autoregressive models and simple variants of them appeared in some of the earliest work on the construction of stochastic models for economic systems. The economics student will find the paper by Wold (1959) a most interesting and useful survey of this area. An example of a simple economics model, which is taken from Wold's paper, is the following.

Example 7.1 *A Simple Model of Supply and Demand*

The (logarithmic) available supply $s(t)$, demand $d(t)$, and price $p(t)$ of a given commodity for the time period t are assumed to satisfy the following relations:

$$d(t) = \alpha_1 - \beta_1 p(t-1) + u(t),$$
$$s(t) = \alpha_2 + \beta_2 p(t) + w(t).$$

Here, $u(t)$ and $w(t)$ are unpredictable random disturbances "independent" of the past. These equations reflect the assumptions that demand is adversely affected by the price over the previous time period and that the remaining stock will depend directly on the price charged for the commodity during the same time period. It is also assumed that supply and demand are in "instantaneous equilibrium"

$$d(t) = s(t) = q(t),$$

where the common value $q(t)$ can be thought of as the quantity of the commodity on hand during the time period t.

Now, one of the principal objectives of this type of model is to forecast future values of $p(t)$ and $q(t)$ from values already observed. To do this, the above equations are reduced algebraically to a form in which the unknown quantities are written explicitly in terms of the known quantities (the prices up to time $t-1$, say) and the random quantities $u(t)$ and $w(t)$. We obtain, after a little algebra,

$$p(t) = \frac{\alpha_1 - \alpha_2}{\beta_2} - \frac{\beta_1}{\beta_2} p(t-1) + \frac{u(t) - w(t)}{\beta_2}, \qquad (7.18)$$
$$q(t) = \alpha_1 - \beta_1 p(t-1) + u(t).$$

Now, if $\alpha_1 = \alpha_2$ and if we assume that $u(t)$ and $w(t)$ are zero-mean white noise processes, uncorrelated with each other and with the past of the price time series, then $p(t)$ has the form of a first-order autoregression

$$p(t) = bp(t-1) + \xi(t),$$

where $b = \beta_1/\beta_2$, $\xi(t) = (u(t) - w(t))/\beta_2$. Equation $\mathscr{B}(z) = 0$ is, in this case,

$$1 - bz = 0.$$

Consequently, $p(t)$ will be a well-defined weakly stationary process if the root $z = 1/b$ lies outside of the unit circle. That is, we must have $|b| < 1$, thus $|\beta_1| < |\beta_2|$. In fact, the z-transform of the one-sided moving average is easily obtained from (7.14) in this case by means of the well-known formula for the sum of a geometric series;

$$a(z) = 1/(1 - bz) = \sum_{k=0}^{\infty} b^k z^k.$$

It follows that

$$p(t) = \sum_{k=0}^{\infty} b^k \xi(t-k). \qquad (7.19)$$

It is of interest to note that if $\alpha_1 \neq \alpha_2$, then the first equation in (7.18) can be put in the form

$$p(t) = bp(t-1) + (\xi(t) + m),$$

where $m = (\alpha_1 - \alpha_2)/\beta_2$ may be viewed as a constant trend or nonzero mean in the random innovation. Now, repeating the steps leading to (7.19), we obtain

$$p(t) = m \sum_{k=0}^{\infty} b^k + \sum_{k=0}^{\infty} b^k \xi(t-k).$$

Since $\sum_{k=0}^{\infty} b^k$ is finite, this is a well-defined weakly stationary process with mean

$$Ep(t) = m \sum_{k=0}^{\infty} b^k.$$

That is, the price time series now has a dc trend. It follows that this series has a mixed spectrum; a discrete term with power $(m \sum_{k=0}^{\infty} b^k)^2$ at $\lambda = 0$ and a continuous term with spectral density function

$$\begin{aligned} f_p(\lambda) &= \frac{\sigma^2}{2\pi} \frac{1}{|1 - be^{-i\lambda}|^2} \\ &= \frac{\sigma^2}{2\pi} \frac{1}{1 + b^2 - 2b\cos\lambda}, \qquad -\pi \le \lambda \le \pi, \end{aligned}$$

where $\sigma^2 = (\sigma_u{}^2 + \sigma_w{}^2)/\beta_2{}^2$ and $\sigma_u{}^2$, $\sigma_w{}^2$ are the variances of $u(t)$ and $w(t)$, respectively.

To obtain price and quantity predictions at time $t + 1$ under the assumption that prices are known up to and including time t, the following intuitive reasoning can be used. First, advance the time index by one in (7.18),

$$\begin{aligned} p(t+1) &= \frac{\alpha_1 - \alpha_2}{\beta_2} - \frac{\beta_1}{\beta_2} p(t) + \frac{u(t+1) - w(t+1)}{\beta_2}, \\ q(t+1) &= \alpha_1 - \beta_1 p(t) + u(t+1). \end{aligned}$$

Now the values of $u(t+1)$ and $w(t+1)$ are completely unpredictable from values of $p(s)$, $u(s)$, and $w(s)$ for $s \le t$, since these random variables are completely uncorrelated with the present and past. Consequently, the best one can do is to estimate these quantities by their mean values $Eu(t+1) = Ew(t+1) = 0$. Thus, the forecasts or predictions $\hat{p}(t+1)$ and $\hat{q}(t+1)$ of $p(t+1)$ and $q(t+1)$ are taken to be

$$\begin{aligned} \hat{p}(t+1) &= \frac{\alpha_1 - \alpha_2}{\beta_2} - \frac{\beta_1}{\beta_2} p(t), \\ \hat{q}(t+1) &= \alpha_1 - \beta_1 p(t). \end{aligned}$$

These are extremely simple and convenient predictors.

The same reasoning for the general qth order autoregression in the form given by expression (7.17) would lead to the predictor

$$\hat{X}(t+1) = \sum_{k=1}^{q} c_k X(t+1-k).$$

Again, this is an extremely simple predictor and it is reasonable to ask whether it is not possible to do better by taking more complicated functions of the past observations. In fact, without some additional mathematical structure this question cannot be answered, since no means for evaluating predictors has been given. We now turn to the job of introducing a simple and elegant geometric structure which will make it possible to deal with this question.

7.4 THE LINEAR PREDICTION PROBLEM

Let $X(t)$, $t = 0, \pm 1, \ldots$, be a weakly stationary process with $EX(t) = 0$. The random variables of the process are elements of the Hilbert space of real-valued random variables X for which $EX = 0$ and $EX^2 < \infty$. (This is the real $L_2(P)$ space discussed in Section 1.4.) The inner product for this space is $\langle X, Y \rangle = EXY$. Thus $X \perp Y$ if $EXY = 0$ and $E(X - Y)^2$ is the squared distance between the elements X and Y.

The *linear subspace* $\mathcal{M}^X$ *generated by the process* is the collection of all finite linear combinations of elements of the process and all limits of Cauchy sequences of these finite linear combinations (Section 1.3). We will call any random variable defined in this manner a *linear function* of the process. Thus $\mathcal{M}^X$ is the class of all such linear functions.

We will also be interested in the linear subspace $\mathcal{M}_t^X$ generated by the elements $X(s)$ for $s \le t$. Thus $\mathcal{M}_t^X$ represents the (linear) past of the process up to time t in the sense that $\mathcal{M}_t^X$ is the class of linear functions of the observations $X(s)$, for $s \le t$.

This geometric setting provides a natural framework for stating (and solving) the problem of predicting future values of the process from values obtained in the past. Suppose a realization of the process for times $s \le t$ is available and the value of $X(t + v)$ for some $v \ge 1$ is to be predicted from this information. Since we will never know what particular realization is being observed, it is reasonable to select a function of the past observations, $f(X(t), X(t-1), \ldots)$, which is good "on the average." The distance function of the Hilbert space can be utilized for this purpose. That is, f can be selected to minimize the squared distance

$$E\big(X(t+v) - f(X(t), X(t-1), \ldots)\big)^2.$$

With this measure of goodness and with the restriction that only linear functions are to be considered, the *linear prediction problem* can be stated as follows: *Find the element* $\hat{X}_\nu(t)$ *in* $\mathcal{M}_t^X$ *which minimizes the squared distance* $E(X(t+\nu) - Y)^2$ *among all elements* Y *of* $\mathcal{M}_t^X$. *The solution (if one exists), is called the best* ν*-step predictor of the process.*

The advantage of the Hilbert space setting now becomes apparent. The theory summarized in Section 1.3 guarantees the existence of a unique solution of the prediction problem, namely, the orthogonal projection of $X(t+\nu)$ on the subspace $\mathcal{M}_t^X$. However, in order to be a useful solution it is necessary to be able to obtain an explicit expression for the coefficients of the predictor in terms of the "parameters" of the process. The reason for restricting attention to linear functions is that the representation for the best linear predictor then depends only on a knowledge of the covariances or, equivalently, of the spectrum. When nonlinear functions are allowed, the problem becomes much more difficult. A good elementary discussion of nonlinear prediction is given by Breiman (1969). See Hannan (1970) for a more advanced treatment.

We can extend the statement of the linear prediction problem to include the following requirements: *Assuming that the spectrum* (or *autocovariance function*) *of the process is known, give an explicit expression for the linear predictor* $\hat{X}_\nu(t)$ *and evaluate the mean-square prediction error* $E(X(t+\nu) - \hat{X}_\nu(t))^2$. The solution to this problem will be sketched for the case when the prediction error is greater than zero—which is the most important case in any event. We will not always be able to fulfill the first requirement in a completely satisfactory manner from a practical standpoint. A convenient explicit representation of the predictor need not exist. However, such a representation does exist for autoregressive processes and we look at the solution of the prediction problem for them first.

The Best Linear Predictor for Autoregressive Processes

Write the general autoregression (7.12) (with index advanced one time unit) in the form

$$X(t+1) = \sum_{k=1}^{\infty} c_k X(t+1-k) + \xi(t+1), \tag{7.20}$$

by setting $c_k = -b_k$, $k = 1, 2, \ldots$. It is assumed that the c_k's are known which is seemingly a stronger condition than the knowledge of the spectral density function. However, as we will see later, the coefficients are actually uniquely determined by the spectral density function.

The criterion for determining orthogonal projections given in Section 1.3 allows us to establish easily that the best (one-step) linear predictor of $X(t+1)$ is

$$\hat{X}_1(t) = \sum_{k=1}^{\infty} c_k X(t+1-k). \tag{7.21}$$

Since this is clearly an element of $\mathcal{M}_t^X$, in order to verify that it is the projection of $X(t+1)$ on $\mathcal{M}_t^X$ it is sufficient to show that $X(t+1) - \hat{X}_1(t)$ is orthogonal to all of the generators $X(s)$, $s \le t$. Now,

$$X(t+1) - \hat{X}_1(t) = \xi(t+1)$$

by (7.20) and since $\xi(t+1) \perp X(s)$ for all $s \le t$ by (7.11) we are finished!

The autoregressive form (7.20) clearly contains the most useful and explicit representation of the one-step predictor as a function of the past observations and the parameters of the time series. Thus, a weakly stationary process which has an autoregressive representation is in the most desirable form possible from the viewpoint of one-step prediction. *The one-step prediction error is also immediately available, since*

$$E(X(t+1) - \hat{X}_1(t))^2 = E\xi^2(t+1) = \sigma^2, \tag{7.22}$$

the innovation variance.

Note that this argument can also be used to justify the form of the one-step predictor obtained heuristically in Example 7.1 even when the random innovations have nonzero means.

The criterion for determining projections is extremely useful and we will further demonstrate its utility by establishing that the following construction leads to the best v-step predictor for $v > 1$ in the autoregressive case.

First, we obtain the best two-step predictor $\hat{X}_2(t)$. Advance the time index in (7.20) to $t+2$ and write the expression as

$$X(t+2) = c_1 X(t+1) + \sum_{k=2}^{\infty} c_k X(t+2-k) + \xi(t+2).$$

The heuristic reasoning of Example 7.1 is extended as follows: Replace $\xi(t+2)$ by its expectation as before and then substitute for the unobservable random variable $X(t+1)$ its best prediction $\hat{X}_1(t)$ based on the observations $X(s)$, $s \le t$,

$$\hat{X}_2(t) = c_1 \hat{X}_1(t) + \sum_{k=2}^{\infty} c_k X(t+2-k).$$

If this is the best predictor, then necessarily $(X(t+2) - \hat{X}_2(t)) \perp X(s)$ for all $s \le t$. However,

$$\begin{aligned} X(t+2) - \hat{X}_2(t) &= c_1(X(t+1) - \hat{X}_1(t)) + \xi(t+2) \\ &= c_1\xi(t+1) + \xi(t+2). \end{aligned}$$

Now, by (7.11), $\xi(t+1) \perp X(s)$, $s \leq t$ and $\xi(t+2) \perp X(s)$, $s \leq t+1$. Thus, $[c_1\xi(t+1) + \xi(t+2)] \perp X(s)$, $s \leq t$. It follows that $\hat{X}_2(t)$ is indeed the projection of $X(t+2)$ on $\mathscr{M}_t^X$. Note that the prediction error for predicting two steps ahead is

$$E\big(X(t+2) - \hat{X}_2(t)\big)^2 = E\big(c_1\xi(t+1) + \xi(t+2)\big)^2 \\ = (c_1^2 + 1)\sigma^2.$$

This reasoning can be further extended to obtain the v-step predictor for any $v \geq 1$. The index in (7.20) is advanced to $t + v$ and $\xi(t+v)$ is replaced by zero. Any random variable $X(t+k)$ which is not one of the observables $X(s)$, $s \leq t$, is replaced by its best linear predictor based on the observable random variables. Thus, in general,

$$\hat{X}_v(t) = \sum_{k=1}^{v-1} c_k \hat{X}_{v-k}(t) + \sum_{k=v}^{\infty} c_k X(t+v-k). \tag{7.23}$$

This is a useful expression when predictors for several different steps are desired. The v-step prediction error is easily computed to be

$$E\big(X(t+v) - \hat{X}_v(t)\big)^2 = \left(1 + \sum_{k=1}^{v-1} c_k^2\right)\sigma^2.$$

The Best Predictor as the Output of a Linear Filter

By using (7.23) inductively in v, it is possible to express the predictors $\hat{X}_v(t)$ as explicit linear combinations of the random variables $X(s)$, $s \leq t$;

$$\hat{X}_v(t) = \sum_{k=0}^{\infty} d_k^{(v)} X(t-k). \tag{7.24}$$

Consequently, for fixed v, the predictor $\hat{X}_v(t)$, viewed as a time series in t, is the output of a linear filter with input $X(t)$ and transfer function

$$D_v(\lambda) = \sum_{k=0}^{\infty} d_k^{(v)} e^{-i\lambda k}.$$

This function necessarily matches the input, since the power in the series $\hat{X}^v(t)$ is the norm square of the projection of $X(t+v)$ on $\mathscr{M}_t^X$ and, as was indicated in Section 1.3, projections satisfy the inequality $\|\mathscr{P}(x \mid \mathscr{M})\| \leq \|x\|$.

That the projection is time-invariant is not peculiar to autoregressive processes but is a consequence of two features of the general structure under consideration. First, $X(t+v)$ is projected on a linear subspace which varies with t so as to remain exactly v steps behind this element. Second, $X(t)$, $t = 0, \pm 1, \ldots$, is a weakly stationary process and, because of this, the prediction error $E\big(X(t+v) - \hat{X}_v(t)\big)^2$ does not depend on t. Consequently, if the linear function $f(X(t), X(t-1), \ldots)$ produces the minimum prediction error at time t, the minimum error at time $t+\tau$ will be achieved by $f(X(t+\tau), X(t+\tau-1), \ldots)$. Thus, the best predictor will always be expressible as the output of a linear filter.

This suggests that an alternative method for solving the prediction problem would be to determine the transfer function $D_\nu(\lambda)$ of the best predictor, since a linear filter is completely and uniquely determined by its transfer function. This solution would be valid even in situations where the best predictor does not have an explicit representation of the form (7.24). It is actually in this form that the general prediction problem must be solved, since representation (7.24) is not universally valid, but, in fact, depends upon the possibility of representing the stochastic process under consideration as an autoregression. Unfortunately, this possibility cannot always be realized as we will see later.

We will now show that the transfer function can be determined whenever the process has a one-sided moving average representation. As was seen in Section 7.3, every autoregression has such a representation. Consequently, we will first illustrate the construction of the transfer function in the autoregressive context.

The Transfer Function of the Best Predictor in the Autoregressive Case

By the one-sided moving average representation for an autoregression, every linear function of the random variables $X(s)$ for $s \leq t$ can be expressed as a linear function of the random variables $\xi(s)$ for $s \leq t$. That is, $\mathscr{M}_t^X \subset \mathscr{M}_t^\xi$, where $\mathscr{M}_t^\xi$ is the linear subspace generated by $\xi(s)$, $s \leq t$. Conversely, because of the defining expressions (7.10) or (7.12), every linear function of the random variables $\xi(s)$ can be written as a linear function of $X(s)$, $s \leq t$. It follows that

$$\mathscr{M}_t^X = \mathscr{M}_t^\xi$$

for all t. As a consequence of this, $\hat{X}_\nu(t)$ can be computed as the projection of $X(t+\nu)$ on $\mathscr{M}_t^\xi$.

Since the generators $\xi(s)$ of $\mathscr{M}_t^\xi$ are orthogonal, this projection is quite easy to construct. In fact, if

$$\begin{aligned} X(t+\nu) &= \sum_{k=0}^{\infty} a_k \xi(t+\nu-k) \\ &= \sum_{s=-\nu}^{\infty} a_{\nu+s} \xi(t-s) \end{aligned}$$

is the one-sided moving average representation derived in Section 7.3, then by (1.14),

$$\hat{X}_\nu(t) = \sum_{s=0}^{\infty} a_{\nu+s}\xi(t-s).$$

This expresses the projection as the output of a linear filter with input $\xi(t)$ and transfer function

$$A_\nu(\lambda) = \sum_{s=0}^{\infty} a_{\nu+s} e^{-i\lambda s} = e^{i\lambda\nu} \sum_{k=\nu}^{\infty} a_k e^{-i\lambda k}. \tag{7.25}$$

It follows that the random spectral measures of the time series $\xi(t)$ and $\hat{X}_\nu(t)$ are related by the expression

$$Z_{\hat{X}_\nu}(d\lambda) = A_\nu(\lambda) Z_\xi(d\lambda).$$

From the one-sided moving average representation of $X(t)$ we have

$$Z_X(d\lambda) = A(\lambda) Z_\xi(d\lambda), \tag{7.26}$$

where

$$A(\lambda) = \sum_{k=0}^{\infty} a_k e^{-i\lambda k}. \tag{7.27}$$

Moreover, by the definition of $D_\nu(\lambda)$,

$$Z_{\hat{X}_\nu}(d\lambda) = D_\nu(\lambda) Z_X(d\lambda).$$

Thus, eliminating $Z_\xi(d\lambda)$ in the first two expressions and equating the result with the third, we obtain

$$D_\nu(\lambda) = A_\nu(\lambda)/A(\lambda). \tag{7.28}$$

This expresses the desired transfer function uniquely in terms of the coefficients of the one-sided moving average representation of $X(t)$.

It is now quite easy to obtain an independent check that the filter with transfer function $D_\nu(\lambda)$ matches the input $X(t)$. By virtue of expression (7.5) for the spectral distribution of a moving average and by the Parseval relation of Example 1.2,

$$\int_{-\pi}^{\pi} |D_\nu(\lambda)|^2 F_X(d\lambda) = \int_{-\pi}^{\pi} \left|\frac{A_\nu(\lambda)}{A(\lambda)}\right|^2 \frac{\sigma^2}{2\pi} |A(\lambda)|^2 \, d\lambda$$

$$= \frac{\sigma^2}{2\pi} \int_{-\pi}^{\pi} |A_\nu(\lambda)|^2 \, d\lambda = \sigma^2 \sum_{k=\nu}^{\infty} a_k{}^2.$$

This quantity is finite, since $\sum_{k=0}^{\infty} a_k{}^2 < \infty$.

The prediction error $E(X(t+v)-\hat{X}_v(t))^2$ can also be computed in terms of the moving average coefficients: Let $X_v(t)$ denote the time series obtained by shifting the time index of $X(t)$ ahead v steps,

$$X_v(t) = X(t+v).$$

This defines $X_v(t)$ as a linear transformation of $X(t)$ with transfer function $e^{i\lambda v}$. (Here we have used the method for computing transfer functions given in Section 6.2.) Thus, the error time series $X_v(t) - \hat{X}_v(t)$ has the spectral representation

$$\begin{aligned} X_v(t) - \hat{X}_v(t) &= \int_{-\pi}^{\pi} e^{i\lambda(t+v)} Z_X(d\lambda) - \int_{-\pi}^{\pi} e^{i\lambda t} D_v(\lambda) Z_X(d\lambda) \\ &= \int_{-\pi}^{\pi} e^{i\lambda t}(e^{i\lambda v} - D_v(\lambda)) Z_X(d\lambda), \end{aligned}$$

i.e., by (7.25)–(7.28),

$$\begin{aligned} Z_{X_v - \hat{X}_v}(d\lambda) &= (e^{i\lambda v} - D_v(\lambda)) Z_X(d\lambda) \\ &= e^{i\lambda v}\left(\frac{A(\lambda) - A_v(\lambda)e^{-i\lambda v}}{A(\lambda)}\right) Z_X(d\lambda) \\ &= e^{i\lambda v}\left(\sum_{k=0}^{v-1} a_k e^{-i\lambda k}\right) Z_\xi(d\lambda). \end{aligned}$$

Thus, by expression (2.31) and the Parseval relation,

$$\begin{aligned} E(X(t+v) - \hat{X}_v(t))^2 &= \int_{-\pi}^{\pi} F_{X_v - \hat{X}_v}(d\lambda) \\ &= \int_{-\pi}^{\pi} \left|\sum_{k=0}^{v-1} a_k e^{-i\lambda k}\right|^2 (\sigma^2/2\pi)\, d\lambda \\ &= \sigma^2 \sum_{k=0}^{v-1} a_k^2. \end{aligned}$$

It follows that the transfer function $D_v(\lambda)$ and the prediction error can be obtained as soon as $A(\lambda)$ is available. Note that if we set $v = 1$ in this expression for the prediction error, the result must coincide with (7.22). It follows that $a_0^2 = 1$. Of the two possibilities, we will take $a_0 = 1$ hereafter.

The Transfer Function of the Best Predictor in the General Case

Using the same technique, the transfer function of the best predictor can be obtained for any weakly stationary process $X(t)$ possessing a one-sided moving average representation in terms of a white noise process $\xi(t)$ for which

$$\mathscr{M}_t^{\xi} = \mathscr{M}_t^{X}, \qquad t = 0, \pm 1, \ldots. \tag{7.29}$$

We will retain the name, *innovation process*, for a white noise process with this property. The only use made of the autoregressive representation was to establish (7.29). A key theorem due to Wold (1938) shows that, in fact, all weakly stationary processes which are "interesting" from the viewpoint of the prediction problem can, after suitable preprocessing, be represented as one-sided moving averages satisfying (7.29). We consider this next.

The Wold Decomposition Theorem

The Wold decomposition theorem can be stated as follows.

Theorem *Let $X(t)$, $t = 0, \pm 1, \ldots$, be a zero-mean, weakly stationary stochastic process. Then $X(t)$ can be expressed as the sum of two zero-mean, weakly stationary processes,*

$$X(t) = U(t) + V(t), \tag{7.30}$$

such that:

(i) *the process $U(t)$ is uncorrelated with the process $V(t)$;*
(ii) *$U(t)$ has a one-sided moving average representation*

$$U(t) = \sum_{k=0}^{\infty} a_k \xi(t - k)$$

with $a_0 = 1$ and $\sum_{k=0}^{\infty} a_k^2 < \infty$ and the subspace generated by the unique white noise process $\xi(t)$ satisfies

$$\mathscr{M}_t^{\xi} = \mathscr{M}_t^{U}$$

for all t;

(iii) *the $V(t)$ process is completely determined by linear functions of its past values in the sense that $\mathscr{M}_t^V = \mathscr{M}_s^V$ for every pair of integers t and s.*

In order to retain the continuity of the discussion, the proof of this theorem will be given in the Appendix at the end of the chapter. To see that (iii) implies what it says, note that, in particular, $\mathscr{M}_{t+v}^V = \mathscr{M}_t^V$ for all t and v. Thus $V(t + v)$, which is an element of $\mathscr{M}_{t+v}^V$, is also in $\mathscr{M}_t^V$. Consequently, the projection of $V(t + v)$ on $\mathscr{M}_t^V$ will be $V(t + v)$ itself. It follows that once the process $V(s)$ has been observed for times $s \le t$ for *any* value of t, the remaining values $V(t + 1)$, $V(t + 2), \ldots$ can be obtained without error as linear functions of the observed random variables. Such a process is called *deterministic.*

The moving average process $U(t)$ is said to be *nondeterministic.* Its variance is $\sigma^2 \sum_{k=0}^{\infty} a_k^2$ by Parseval's relation, where σ^2 is the variance of the innovation process $\xi(t)$. The only way that $U(t)$ can have zero variance, and, thus, equal 0 for all t, is for $\sigma^2 = 0$. In this case, $X(t) = V(t)$ and the original process is deterministic.

When $\sigma^2 > 0$, the $X(t)$ process is said to be *regular*. In this case, either the nondeterministic component $U(t)$ is present alone or both components are present. Suppose both are present. Then, as we show in Section A7.3, the condition (7.30) and property (i) imply that

$$\hat{X}_\nu(t) = \hat{U}_\nu(t) + \hat{V}_\nu(t),$$

where $\hat{X}_\nu(t)$ is the projection of $X(t+\nu)$ on $\mathscr{M}_t{}^X$, $\hat{U}_\nu(t)$ is the projection of $U(t+\nu)$ on $\mathscr{M}_t{}^U$, and $\hat{V}_\nu(t)$ is the projection of $V(t+\nu)$ on $\mathscr{M}_t{}^V$ [thus is equal to $V(t+\nu)$]. That is, the best ν-step predictor of each component can be obtained separately and the results added to obtain the optimal ν-step predictor for the original process. This result has practical value only if it is possible to separate the two components of the $X(t)$ process so that they can be dealt with individually. This can be done in theory—and to a great extent in practice—when the process is regular because of the following consideration.

The spectral distribution of $X(t)$ can be evaluated in terms of the spectral distributions of $U(t)$ and $V(t)$. Because of (7.30) and property (i),

$$F_X(d\lambda) = F_U(d\lambda) + F_V(d\lambda). \tag{7.31}$$

However, the spectrum of $U(t)$ is continuous because $U(t)$ is a moving average process. In fact,

$$F_U(d\lambda) = \frac{2\pi}{\sigma^2}\left|\sum_{k=0}^{\infty} a_k e^{-i\lambda k}\right|^2 d\lambda$$

by the theory of Section 7.2. Now, it can be shown [see, e.g., Doob (1953, p. 569) or Hannan (1970, p. 140)] *that* (7.31) *is actually the Lebesgue decomposition of* F_X *into continuous and discrete components, respectively*. Thus, in all cases of practical interest, the deterministic component is simply a stochastic almost periodic function. This term can be estimated without error by the technique outlined in Section 2.10 if the $X(t)$ process is available from the infinite past. More realistically, it can be estimated with increasing reliability as more and more data is accumulated from the finite past as time progresses. In either case, the deterministic term can be (essentially) removed from $X(t)$ and the more interesting problem of predicting the $U(t)$ process can then be dealt with.

We will obtain an expression for the innovation variance of $X(t)$ shortly and it will be apparent that regular processes are by far the more important and commonly occurring from the viewpoint of modeling physical time series. For nonregular processes, linear prediction can, in theory, be carried out without error. Nature is seldom so accommodating as to permit this to occur.

Hereafter, we assume that $X(t)$ is regular and that the $V(t)$ term is missing in (7.30). Then $X(t) = U(t)$ has a one-sided moving average representation which satisfies condition (7.29). It follows that if we can determine the coefficients of $A(\lambda)$ in the moving average representation of $X(t)$ and if the innovation variance σ^2 can be determined, then the construction of the transfer function of the ν-step predictor (7.28) can be carried out exactly as before. Also, the ν-step prediction error can be computed. Thus, the solution of the general linear prediction problem hinges on the purely mathematical problem of calculating the coefficients of the moving average representation and of determining σ^2 from the spectrum of $X(t)$. We discuss this next.

Determination of the Moving Average Transfer Function $A(\lambda)$

Depending on the application, $A(\lambda)$ can be determined either by direct construction or by means of the following consideration: From the above discussion, it follows that $X(t)$ has a continuous spectrum with spectral density function

$$f_X(\lambda) = (\sigma^2/2\pi)|A(\lambda)|^2, \tag{7.32}$$

where $A(\lambda) = \sum_{k=0}^{\infty} a_k e^{-i\lambda k}$, $a_0 = 1$, and $\sum_{k=0}^{\infty} a_k{}^2 < \infty$. In some problems it is possible to obtain the class of all functions satisfying these conditions. This will be the case, for example, for processes with rational spectral densities to be discussed in the next section. Then it suffices to be able to select $A(\lambda)$ from among the functions in this class. This requires an additional characterization of $A(\lambda)$ which, as we will see, depends on the z-transform of the sequence $\{a_k\}$. We first characterize this class of functions.

Let $\{b_k : k \geq 0\}$ be a sequence of numbers and consider its z-transform

$$\mathscr{B}(z) = \sum_{k=0}^{\infty} b_k z^k. \tag{7.33}$$

Then

$$B(\lambda) = \mathscr{B}(e^{-i\lambda})$$

will satisfy the above conditions if

(a) $\mathscr{B}(z)$ is analytic in the region $\mathscr{D} = \{z\colon |z| < 1\}$ with power series given by expression (7.33). Moreover, it will be necessary that $\mathscr{B}(z)$ can be extended to the boundary of $\mathscr{D}$ in such a way that (7.33) remains valid there (in some sense). [This will yield the one-sided Fourier expansion of $B(\lambda)$. One condition under which this holds is given in Section A7.2.]

(b) $\sum_{k=0}^{\infty} b_k{}^2 < \infty$,

(c) $f_X(\lambda) = (\sigma^2/2\pi)\,|\mathscr{B}(e^{-i\lambda})|^2$.

The condition $b_0 = 1$ is equivalent to the requirement

(d) $\mathscr{B}(0) = 1$.

We will gain some idea of the additional properties to be satisfied by the function $\mathscr{A}(z)$ which determines $A(\lambda)$ by looking at a construction which will be shown to produce $\mathscr{A}(z)$ in an important special case. The derivation is due to Whittle (1963, p. 26).

Example 7.2 *The Construction of $A(\lambda)$ in a Special Case*

Suppose that $\mathscr{F}(z)$ is a complex-valued function such that

$$f_X(\lambda) = \mathscr{F}(e^{-i\lambda})$$

and assume that $\log \mathscr{F}(z)$ is analytic in a region $\mathscr{R} = \{z\colon \rho < |z| < 1/\rho\}$ for some $\rho < 1$. Then, $\log \mathscr{F}(z)$ has a Laurent expansion

$$\log \mathscr{F}(z) = \sum_{k=-\infty}^{\infty} c_k z^k,$$

valid in $\mathscr{R}$, where by the change of variables, $z = e^{-i\lambda}$,

$$\begin{aligned} c_k &= \frac{1}{2\pi i}\int_{|z|=1} z^{-k-1} \log \mathscr{F}(z)\, dz \\ &= \frac{1}{2\pi}\int_{-\pi}^{\pi} e^{i\lambda k} \log f_X(\lambda)\, d\lambda. \end{aligned} \tag{7.34}$$

Since the c_k's are the Fourier coefficients of an even, real-valued function they are also real-valued and even. Thus,

$$\mathscr{F}(z) = \exp\left\{\sum_{k=-\infty}^{\infty} c_k z^k\right\} = e^{c_0}\mathscr{A}(z)\mathscr{A}(z^{-1}),$$

where

$$\mathscr{A}(z) = \exp\left\{\sum_{k=1}^{\infty} c_k z^k\right\}. \tag{7.35}$$

Now, $\mathscr{A}(z)$ is analytic in the region $\mathscr{R}' = \{z\colon |z| < 1/\rho\}$ and $\mathscr{A}(0) = 1$. Thus, $\mathscr{A}(z)$ has a power series expansion

$$\mathscr{A}(z) = \sum_{k=0}^{\infty} a_k z^k, \tag{7.36}$$

with $a_0 = 1$. Since this power series converges in $\mathcal{R}'$, the coefficients satisfy inequalities of the form $|a_k| \le Ml^k$ for some $M > 0$ and $\rho \le l < 1$. In particular,

$$\sum_{k=0}^{\infty} a_k^2 < \infty.$$

Moreover,

$$\begin{aligned} f_X(\lambda) = \mathscr{F}(e^{-i\lambda}) &= e^{c_0}\mathscr{A}(e^{-i\lambda})\mathscr{A}(e^{i\lambda}) \\ &= e^{c_0}|A(\lambda)|^2. \end{aligned} \tag{7.37}$$

Thus, $\mathscr{A}(z)$ will satisfy properties (a)–(d) if it can be shown that $e^{c_0} = \sigma^2/2\pi$, where σ^2 is the variance of the innovation process guaranteed to exist by the Wold decomposition theorem. To show this, we establish that an autoregressive representation for $X(t)$ exists in this special case.

Note that $\mathscr{A}(z)$ is nonzero for $z \in \mathcal{R}'$ because of (7.35). Thus, $1/\mathscr{A}(z)$ is analytic in this region and has a power series expansion,

$$1/\mathscr{A}(z) = \sum_{k=0}^{\infty} u_k z^k, \qquad u_0 = 1.$$

In particular,

$$1/A(\lambda) = \sum_{k=0}^{\infty} u_k e^{-i\lambda k} \tag{7.38}$$

and

$$\sum_{k=0}^{\infty} u_k^2 < \infty.$$

Now, the construction given in Section 7.2 can be carried out to obtain the white noise process

$$\xi(t) = \int_{-\pi}^{\pi} e^{i\lambda t} \frac{1}{A(\lambda)} Z_X(d\lambda). \tag{7.39}$$

The corresponding moving average representation is one-sided because of (7.36). Thus, we obtain

$$X(t) = \sum_{k=0}^{\infty} a_k \xi(t-k).$$

Note that this implies

$$X(s) \perp \xi(t) \qquad \text{for} \quad s \le t-1, \tag{7.40}$$

since

$$\langle X(s), \xi(t)\rangle = \sum_{k=0}^{\infty} a_k \langle \xi(s-k), \xi(t)\rangle = 0.$$

Substituting (7.38) into (7.39) we obtain

$$\xi(t) = \sum_{k=0}^{\infty} u_k \int_{-\pi}^{\pi} e^{i\lambda(t-k)} Z_X(d\lambda)$$

$$= \sum_{k=0}^{\infty} u_k X(t-k). \tag{7.41}$$

This, along with condition (7.40), yields the autoregressive representation of $X(t)$.

Now, by the argument given in Section 7.3, it follows that

$$\hat{X}_1(t) = \sum_{k=1}^{\infty} (-u_k) X(t+1-k) \tag{7.42}$$

is the projection of $X(t+1)$ on $\mathscr{M}_t^X$. Thus, since the white noise process satisfies the equation

$$\xi(t+1) = X(t+1) - \hat{X}_1(t),$$

it is precisely the innovation process of the time series $X(t)$ guaranteed by the Wold decomposition theorem.

Identifying the expression for the spectral density of a moving average process with (7.37) we obtain

$$e^{c_0} = \sigma^2/2\pi. \tag{7.43}$$

Thus, the function $\mathscr{A}(z)$ given by (7.35) satisfies all of the conditions (a)–(d) and, moreover, it determines the coefficients of the one-sided moving average representation of $X(t)$ as a function of its innovation process, i.e., $A(\lambda) = \mathscr{A}(e^{-i\lambda})$.

The Construction of $A(\lambda)$ in the General Case

The construction of $A(\lambda)$ in the general prediction problem is actually very similar to the one given in this example and some of the details will be sketched without proof. If we substitute the value of c_0 from (7.34) into (7.43), the following expression for the innovation variance is obtained:

$$\sigma^2 = 2\pi \exp\left\{\frac{1}{2\pi} \int_{-\pi}^{\pi} \log f_X(\lambda)\, d\lambda\right\}. \tag{7.44}$$

As was shown in (7.22) and again above, the innovation variance is the same as the one-step prediction error in the autoregressive case. This is true in general and (7.44) is a key expression for this error due to Szegö (1939) and

Kolmogorov [see Kolmogorov (1941b)]. By means of this expression, it is seen that a *sufficient condition for* $X(t)$ *to be regular* is

$$\int_{-\pi}^{\pi} \log f_X(\lambda)\, d\lambda > -\infty. \tag{7.45}$$

This is the condition upon which the solution to the general prediction problem rests. It is weaker than the analyticity of $\log \mathscr{F}(z)$, assumed in the example, but with proper interpretation and substantially more mathematical labor, many of the details of the example can be preserved. For instance, expressions (7.35) and (7.36), properly interpreted, lead to the correct one-sided moving average solution to the prediction problem. However, (7.45) allows for the possibility that $f_X(\lambda)$ has zeros in $(-\pi, \pi)$. Thus $\mathscr{F}(z)$ can have zeros on the unit circle in which case $\log \mathscr{F}(z)$ cannot be analytic in a region containing the unit circle as was assumed in the example. Then, $\mathscr{A}(z)$ will be analytic and nonzero for the restricted domain $|z| < 1$ but the boundary value $A(\lambda) = \mathscr{A}(e^{-i\lambda})$ must be defined through a limiting process (see Section A7.2). This would rule out a one-sided expansion of $1/A(\lambda)$ such as (7.38) and thus, an autoregressive representation of $X(t)$ will not always exist. The following example illustrates this situation.

Example 7.3 *A Nondeterministic Process Which Does Not Have an Autoregressive Representation*

Let $\xi(t)$ be a white noise process with variance σ^2 and define the process $X(t)$ by

$$X(t) = \xi(t) - \xi(t-1). \tag{7.46}$$

This process is already in the form of a one-sided moving average with $A(\lambda) = 1 - e^{-i\lambda}$ so it is not surprising that (7.45) is satisfied. The (unique) autoregressive representation would be of the form

$$\xi(t) = \sum_{k=0}^{\infty} u_k X(t-k).$$

However, by substituting this into expression (7.46) it is easily seen that the only solution has coefficients

$$u_k = 1, \qquad k = 0, 1, 2, \ldots .$$

This does not satisfy the necessary condition $\sum_{k=0}^{\infty} {u_k}^2 < \infty$ for the existence of a valid (weakly stationary) autoregression.

However, the best one-step predictor $\hat{X}_1(t)$ is a well-defined element of ${\mathscr{M}_t}^X$ as the above theory indicates. Hannan (1970, p. 131) shows that the following limit yields the correct element:

$$\hat{X}_1(t) = \lim_{N\to\infty} \sum_{k=0}^{N} (1 - (k/N))X(t-k).$$

Now, recall that this projection can be viewed as a linear filter taking the time series $X(t)$ into $\hat{X}_1(t)$. Here we have an example, then, of a digital filter which is not of convolution type.

Factoring the Spectral Density

Finally, as stated earlier, in some instances the class of complex-valued functions $\mathscr{B}(z)$ satisfying conditions (a)–(d) will be given, and the solution of the prediction problem then depends on being able to identify the element $\mathscr{A}(z)$ in this class. From expression (7.35) for $\mathscr{A}(z)$, obtained in the example, it is seen that $\mathscr{A}(z)$ is never zero as long as the power series in the exponent converges. In the general case, the domain of convergence of this power series will contain and often equal $\{z: |z| < 1\}$. Thus, along with the properties (a)–(d) the distinguishing feature we seek is that $\mathscr{A}(z)$ is nonvanishing for $|z| < 1$.

In summary, $A(\lambda) = \mathscr{A}(e^{-i\lambda})$ *will be the transfer function of the one-sided moving average representation of* $X(t)$ *in terms of its innovation process provided* $\mathscr{A}(z)$ *satisfies conditions* (a)–(d) *and, in addition,* $\mathscr{A}(z) \neq 0$ *for* $|z| < 1$. $\mathscr{A}(z)$ *is uniquely determined by these properties.* For a proof of this result in the general (univariate) case, see Doob (1953, pp. 569–579).

The determination of the class of functions $\mathscr{B}(z)$ satisfying (a)–(d) is known in prediction theory as the problem of *factoring the spectral density function.* A good treatment of the general factorization problem and its applications is given for multidimensional time series by Hannan (1970, p. 62). We illustrate the use of spectral density factoring in the next section.

7.5 MIXED AUTOREGRESSIVE–MOVING AVERAGE PROCESSES AND RECURSIVE PREDICTION

The equivalence between mixed autoregressive–moving average processes and processes with rational spectral densities will be established in this section. These processes will then be shown to have predictors which can be put in recursive form. Adaptive prediction for a class of nonstationary processes will be discussed briefly.

The Mixed Autoregressive–Moving Average Model

A natural generalization of both finite autoregressions and finite moving averages is to combine the two models by using a one-sided moving average in place of the white noise process as input to the stochastic difference equation (7.10):

$$\sum_{j=0}^{q} d_j X(t-j) = \sum_{k=0}^{p} c_k \xi(t-k), \tag{7.47}$$

where $\xi(t)$ is a given zero-mean, white noise process with variance σ^2. A weakly stationary solution $X(t)$ to (7.47) is called a *mixed autoregressive-moving average process*. It is usually assumed that $c_0 = d_0 = 1$ and that $\mathscr{C}(z)$ has no zeros in $\{z\colon |z| < 1\}$ and $\mathscr{D}(z)$ has no zeros in $\{z\colon |z| \le 1\}$, where $\mathscr{C}(z) = \sum_{k=0}^{p} c_k z^k$ and $\mathscr{D}(z) = \sum_{j=0}^{q} d_j z^j$. The reason for adopting these conditions will now be considered.

Viewing $X(t)$ as the output of a linear filter with input $\xi(t)$, the transfer function $A(\lambda)$ can be obtained by substituting $A(\lambda)e^{i\lambda t}$ for $X(t)$ and $e^{i\lambda t}$ for $\xi(t)$ in (7.47) and solving for $A(\lambda)$. The result is

$$A(\lambda) = \sum_{k=0}^{p} c_k e^{-i\lambda k} \Big/ \sum_{j=0}^{q} d_j e^{-i\lambda j}. \tag{7.48}$$

In order for this filter to match the white noise input, it is necessary and sufficient that none of the roots of the equation $\mathscr{D}(z) = 0$ lie on the unit circle. With this restriction alone, $A(\lambda)$ has a Fourier series expansion

$$A(\lambda) = \sum_{j=-\infty}^{\infty} a_j e^{-i\lambda j}$$

with $\sum_{j=-\infty}^{\infty} a_j^2 < \infty$. This leads to the representation of a solution of (7.47) as a two-sided moving average in terms of the original white noise process,

$$X(t) = \sum_{j=-\infty}^{\infty} a_j \xi(t-j).$$

The process clearly has a continuous spectrum and the spectral density function is

$$f_X(\lambda) = \frac{\sigma^2}{2\pi} |A(\lambda)|^2 = \frac{\sigma^2}{2\pi} \left| \frac{\sum_{k=0}^{p} c_k e^{-i\lambda k}}{\sum_{j=0}^{q} d_j e^{-i\lambda j}} \right|^2. \tag{7.49}$$

If all of the zeros of $\mathscr{D}(z)$ lie outside of the region $\{z : |z| \le 1\}$, then $\mathscr{A}(z) = \mathscr{C}(z)/\mathscr{D}(z)$ is analytic, thus, has a power series expansion in a region containing this set. It follows that the above moving average is one-sided and depends only on the past and present of the $\xi(t)$ process. This is a reasonable property for the model to have, since the $\xi(t)$ process can again be viewed as a time series of unobserved shocks produced by the underlying physical process as time progresses.

This does not imply, however, that this is the innovation process for $X(t)$. That is, the pasts of the $\xi(t)$ and $X(t)$ processes need not coincide.

From our discussion of prediction theory, the given white noise process will be the innovation process for $X(t)$ only if $\mathscr{A}(z)$ is analytic and non-vanishing for $|z| < 1$. This requires the additional condition that the zeros of $\mathscr{C}(z)$ lie outside of the set $\{z: |z| < 1\}$. Moreover, to obtain the normalization $a_0 = 1$, it is necessary that $c_0 = d_0 = 1$. Thus, the conditions imposed on (7.47) are designed to guarantee that a weakly stationary solution of the difference equation always exists and that the given white noise process is the innovation process of the solution.

Finally, note that if $\mathscr{C}(z)$ has no zeros on $|z| = 1$, then $1/\mathscr{A}(z)$ has a power series expansion which converges for $|z| \le 1$ and, thus, the $X(t)$ process has an (infinite) autoregressive representation as well.

Processes with Rational Spectral Densities

A model for time series which has received considerable attention is that of a weakly stationary process $X(t)$ with continuous spectrum and rational spectral density

$$f_X(\lambda) = \sum_{j=-p}^{p} \gamma_j e^{-i\lambda j} \Big/ \sum_{k=-q}^{q} \delta_k e^{-i\lambda k}. \tag{7.50}$$

It is assumed that the numerator and denominator have no factors in common. The numbers p and q are nonnegative integers and the quantities γ_j and δ_k are real-valued parameters such that $\gamma_{-j} = \gamma_j$ and $\delta_{-k} = \delta_k$. In order for $f_X(\lambda)$ to be integrable it is necessary that the denominator of (7.50) not vanish for $-\pi < \lambda \le \pi$.

The importance of this model results from the fact that any continuous spectral density function can be approximated arbitrarily closely by a rational function such as (7.50) by proper choice of p, q and the other parameters. While this is also true of the spectral densities of finite moving averages and finite autoregressions, a model with a rational spectral density will almost invariably lead to a better fit with fewer parameters. Box and Jenkins (1970) call this property *parsimony*. The parameters of the process are usually estimated by statistical techniques which increase rapidly in computational effort with increasing numbers of parameters. Consequently, it is not only convenient but necessary to have relatively parsimonious models. We will not cover the statistical methods for estimating the parameters of a rational spectral density in this book. Excellent treatments giving a variety of methods and applications are provided by Anderson (1971), Box and Jenkins (1970), and Hannan (1970).

It is easily seen that every mixed autoregressive–moving average process has a rational spectral density. The converse is also true in the sense that *for every process $X(t)$ with rational spectrum there exists a white noise process $\xi(t)$ and parameters d_j, $j = 0, \ldots, q$ and c_k, $k = 0, \ldots, p$, satisfying the conditions* *specified for the parameters of a mixed autoregressive–moving average model such that* (7.47) *holds.*

From our discussion of prediction theory and the fact that the condition imposed on (7.50) is sufficient to guarantee that the regularity condition (7.45) is valid, the innovation process $\xi(t)$ and the function $\mathscr{A}(z)$ which provides the parameters of the one-sided moving average representation of $X(t)$ are computable from the above theory. However, to show that $X(t)$ satisfies the difference equation (7.47), it is convenient in this instance to display the possible factors of the spectral density $f_X(\lambda)$ and select $\mathscr{A}(z)$ from among them. The reason for this is that the factors are easily and explicitly obtainable in the form (7.48). The details of the factorization will only be sketched. An excellent treatment providing the missing mathematical steps can be found in the book by Hannan (1970, p. 62).

The numerator and denominator of (7.50) are of the form

$$R(\lambda) = \sum_{j=-l}^{l} r_j e^{-i\lambda j}$$

where the r_j's are real numbers for which $r_{-j} = r_j$ and $r_l \neq 0$. If

$$\mathscr{R}(z) = \sum_{j=-l}^{l} r_j z^j$$

is the z-transform of these coefficients, then it is easy to see that

$$\mathscr{R}(z) = \mathscr{R}(1/z)$$

for all z. In particular, if z_0 is a number such that $\mathscr{R}(z_0) = 0$, then $\mathscr{R}(1/z_0) = 0$ as well.

Now, let $\Phi(z) = z^l \mathscr{R}(z)$. Then by a change of index of summation,

$$\Phi(z) = \sum_{k=0}^{2l} r_{k-l} z^k.$$

This is a polynomial of degree $2l$. Hence, it has $2l$ zeros. Moreover, by its relationship to $\mathscr{R}(z)$, whenever z_0 is a root of $\Phi(z) = 0$, so is $1/z_0$. Thus, the $2l$ roots can be grouped into l pairs, $(z_j, 1/z_j)$, $j = 1, 2, \ldots, l$. Now, $\Phi(z)$ can be written in the form

$$\Phi(z) = \text{const} \prod_{j=1} (z - z_j)(zz_j - 1).$$

Thus, by associating one factor of $1/z$ with each term $(z - z_j)(zz_j - 1)$, $\mathscr{R}(z)$ can be written in the form

$$\begin{aligned}\mathscr{R}(z) &= \text{const} \prod_{j=1}^{l} (z - z_j)((1/z) - z_j) \\ &= \text{const}\, \mathscr{S}(z)\mathscr{S}(1/z),\end{aligned}$$

where $\mathscr{S}(z) = \prod_{j=1}^{l} (z - z_j)$ contains one element from each of the l pairs of roots. The constants indicated above are different and will depend on the selection of roots, etc. Finally, by expanding the product, we can write

$$\mathscr{S}(z) = \sum_{j=0}^{l} s_j z^j. \tag{7.51}$$

Thus,

$$R(\lambda) = \mathscr{R}(e^{-i\lambda}) = \text{const} \left| \sum_{j=0}^{l} s_j e^{-i\lambda j} \right|^2.$$

The constant can be adjusted so as to make $s_0 = 1$.

Applying this result to the numerator and denominator of (7.50) we obtain

$$f_X(\lambda) = \text{const}\,|B(\lambda)|^2 \tag{7.52}$$

where

$$B(\lambda) = \mathscr{B}(e^{-i\lambda}) \qquad \text{and} \qquad \mathscr{B}(z) = \mathscr{C}(z)/\mathscr{D}(z)$$

with $\mathscr{C}(z)$ and $\mathscr{D}(z)$ both of form (7.51). A variety of functions $\mathscr{B}(z)$ can be obtained according to which roots of the various pairs $(z_j, 1/z_j)$ are selected to make up $\mathscr{C}(z)$ and $\mathscr{D}(z)$. By the initial condition on (7.50), $\mathscr{D}(z)$ will have no roots on the unit circle. Consequently, if $\mathscr{D}(z)$ is selected to have all of its roots outside of the unit circle, $\mathscr{B}(z)$ will be analytic in a region containing the set $\{z\colon |z| \leq 1\}$. The collection of functions $\mathscr{B}(z)$ obtained by varying the roots in $\mathscr{C}(z)$ is the collection of z-transforms of one-sided moving averages which satisfy conditions (a)–(d), above. The element $\mathscr{A}(z)$ must, in addition, be nonvanishing for $|z| < 1$. This function corresponds to the polynomial $\mathscr{C}(z)$ with zeros selected to be outside or on the unit circle.

With $\mathscr{A}(z)$ so specified, from the prediction theory discussion we obtain

$$Z_X(d\lambda) = A(\lambda) Z_\xi(d\lambda) = \frac{C(\lambda)}{D(\lambda)} Z_\xi(d\lambda),$$

where

$$C(\lambda) = \sum_{k=0}^{p} c_k e^{-i\lambda k}, \qquad D(\lambda) = \sum_{j=0}^{q} d_j e^{-i\lambda j}$$

and $\xi(t)$ is the innovation process for $X(t)$. Thus,

$$\begin{aligned}
\sum_{j=0}^{q} d_j X(t-j) &= \int_{-\pi}^{\pi} e^{i\lambda t} D(\lambda) Z_X(d\lambda) \\
&= \int_{-\pi}^{\pi} e^{i\lambda t} C(\lambda) Z_\xi(d\lambda) \\
&= \sum_{k=0}^{p} c_k \xi(t-k).
\end{aligned}$$

This is precisely the mixed autoregressive–moving average representation (7.47). Because of (7.49), the constant in (7.52) must be $\sigma^2/2\pi$, where σ^2 is the variance of the innovation process.

Recursive Predictors

For a process with rational spectral density, the best linear predictor can be put in a distinctive and computationally useful form. A predictor will be called *recursive* if it can be represented as a function of a *fixed* number of previous observations and previous predictors,

$$\hat{X}_v(t) = f(\hat{X}_v(t-1), \ldots, \hat{X}_v(t-r); X(t), X(t-1), \ldots, X(t-s)). \quad (7.53)$$

More generally, we could allow this function to depend on previous predictors with different prediction steps $v' \le v$, for example. However, we will restrict attention to this form here. The computational advantage of such an expression is that it is only necessary to store $r + s + 1$ items of data to compute the next predictor at any given time. Contrast this with, say, the one-step autoregressive predictor (7.21) which depends on the infinite past of the process. To use the autoregressive predictor in practice it would be necessary to carry along each value of the process as it is received in order to calculate the next prediction. This is, in fact, never done. The expression is truncated by setting $c_j = 0$ for all "sufficiently large" values of j. The resulting expression is no longer that of the best one-step predictor except in the case of a finite autoregressive process of the appropriate order.

The truncation can be viewed as an attempt to fit the true process by a finite autoregression. As was discussed above, it is almost always preferable to approximate the process by a mixed autoregressive–moving average process. We will now see that a mixed autoregressive–moving average process always has a recursive representation of the form (7.53) in which the function f is linear. A complete treatment of the prediction of processes with rational spectral densities is given by Yaglom (1962).

Construction of the Recursive Predictor for a Mixed Autoregressive–Moving Average Process

Expression (7.28) for the transfer function of the best v-step predictor can be written in the form

$$D_v(\lambda) = e^{i\lambda v} \frac{A(\lambda) - \sum_{k=0}^{v-1} a_k e^{-i\lambda k}}{A(\lambda)}. \quad (7.54)$$

For mixed autoregressive–moving average processes,

$$A(\lambda) = \frac{C(\lambda)}{D(\lambda)} = \sum_{j=0}^{\nu-1} a_j e^{-i\lambda j} + \frac{C_\nu(\lambda)}{D(\lambda)}, \tag{7.55}$$

where $C_\nu(\lambda)$ is the remainder after dividing the trigonometric polynomial $C(\lambda)$ by $D(\lambda)$ for ν terms. The usual algorithm for dividing polynomials can be adapted to yield this result. When this expression is substituted into (7.54), we obtain

$$D_\nu(\lambda) = e^{i\lambda\nu} C_\nu(\lambda)/C(\lambda).$$

Thus, since

$$Z_{\hat{X}_\nu}(d\lambda) = D_\nu(\lambda) Z_X(d\lambda),$$

we have

$$C(\lambda) Z_{\hat{X}_\nu}(d\lambda) = e^{i\lambda\nu} C_\nu(\lambda) Z_X(d\lambda).$$

Now, it is always possible to write $C_\nu(\lambda)$ in the form

$$e^{i\lambda\nu} C_\nu(\lambda) = \sum_{j=0}^{p} c_j^{(\nu)} e^{-i\lambda j}.$$

Then, multiplying both sides of the previous expression by $e^{i\lambda t}$ and integrating, we obtain

$$\sum_{k=0}^{p} c_k \hat{X}_\nu(t-k) = \sum_{j=0}^{p} c_j^{(\nu)} X(t-j).$$

Since $c_0 = 1$, this is equivalent to

$$\hat{X}_\nu(t) = \sum_{k=1}^{p} (-c_k) \hat{X}_\nu(t-k) + \sum_{j=0}^{p} c_j^{(\nu)} X(t-j). \tag{7.56}$$

This is the desired recursive form. The following (somewhat contrived) numerical example will illustrate these ideas.

Example 7.4 *Construction of a Recursive Predictor*

Suppose that $X(t)$ is a weakly stationary process with continuous spectrum and spectral density function

$$f_X(\lambda) = \frac{2e^{-i\lambda} + 5 + 2e^{i\lambda}}{-12e^{-i2\lambda} + 11e^{-i\lambda} + 146 + 11e^{i\lambda} - 12e^{i2\lambda}}.$$

Let $\mathscr{F}(z)$ represent this expression with $e^{-i\lambda}$ replaced by z. Then, normalized so that $c_0 = d_0 = 1$, $\mathscr{F}(z)$ can be written in factored form as

$$\mathscr{F}(z) = \frac{1}{36} \frac{(\frac{1}{2}z + 1)(\frac{1}{2}z^{-1} + 1)}{(1 - \frac{1}{4}z)(1 - \frac{1}{4}z^{-1})(\frac{1}{3}z + 1)(\frac{1}{3}z^{-1} + 1)}.$$

The extent to which this example has been "manufactured" is rather evident. In practice, the factorization of the spectral density function is the most difficult step in the process of determining the coefficients of the best predictor. This step relies on standard techniques for factoring polynomials which are available in texts on algebraic equations.

The factor $\mathscr{A}(z)$ is

$$\mathscr{A}(z) = \frac{\frac{1}{2}z + 1}{(1 - \frac{1}{4}z)(\frac{1}{3}z + 1)},$$

since this factorization puts the zeros of the polynomials in numerator and denominator outside of the unit circle. Also, since

$$f_X(\lambda) = (\sigma^2/2\pi)|\mathscr{A}(e^{-i\lambda})|^2,$$

it follows that the innovation variance is $\sigma^2 = 2\pi/36 = \pi/18$.

Now, $\mathscr{C}(z) = 1 + \frac{1}{2}z$ and $\mathscr{D}(z) = (1 - \frac{1}{4}z)(\frac{1}{3}z + 1) = 1 + \frac{1}{12}z - \frac{1}{12}z^2$. Expression (7.55) can be obtained by dividing $\mathscr{D}(z)$ into $\mathscr{C}(z)$ then making the substitution $z = e^{-i\lambda}$. Carrying out the division for two terms yields

$$\frac{\mathscr{C}(z)}{\mathscr{D}(z)} = 1 + \tfrac{5}{12}z + \frac{\frac{7}{144}z^2 + \frac{5}{144}z^3}{\mathscr{D}(z)}.$$

Thus, setting $z = e^{-i\lambda}$ and comparing this expression with (7.55) we obtain

$$e^{i2\lambda}C_2(\lambda) = \tfrac{7}{144} + \tfrac{5}{144}e^{-i\lambda}.$$

Since $C(\lambda) = 1 + \frac{1}{2}e^{-i\lambda}$, the best two-step recursive predictor is

$$\hat{X}_2(t) = -\tfrac{1}{2}\hat{X}_2(t-1) + \tfrac{7}{144}X(t) + \tfrac{5}{144}X(t-1).$$

To use this predictor in practice, if we begin observing the time series at $t = 1$, say, we would have to "initialize" the prediction process by assigning values to $\hat{X}_2(0)$ and $X(0)$ in some fashion. This is usually done arbitrarily unless "natural" values of these variables are available from some source. After several predictions have been made, the dependence on the initializing values "washes out" or, at least, becomes of negligible importance. The length of time for this to occur will depend on the magnitude of the initial errors so some care is warranted in the selection of the initial values.

Adaptive Prediction

In recent years, progress has been made in extending the design of predictors of recursive type to more realistic prediction problems. Even when the standard weakly stationary prediction model is accurate, the parameters of the model will seldom be known. Then, if prediction is to be carried out in real time, one is faced with the problem of estimating the parameters at the same time the predictions are being made. Thus, the usual prediction error is compounded by the error of estimation. It is desirable that the predictor "adapt itself" to the underlying process so that the error of this *adaptive predictor* tends to the true prediction error as time progresses. Moreover, the *rate* at which this occurs should be as rapid as possible. The current state of investigation of this subject is summarized by Gardner (1962) and Davis and Koopmans (1970).

The term "adaptive prediction" is also used to describe the situation in which the underlying model is nonstationary, but otherwise specified. Then, the predictor "adapts" if it follows the process as it evolves in time. Box and Jenkins (1970) skillfully exploit the use of an interesting nonstationary model which lends itself to adaptive prediction by recursive formulas. The model is an extension of the mixed autoregressive–moving average model and we discuss it here briefly.

A Nonstationary Model and Its Prediction

If B denotes the backward shift operator defined in Section 6.2, then the stochastic equation (7.47) can be represented in the form

$$\mathscr{D}(B)X(t) = \mathscr{C}(B)\xi(t). \tag{7.57}$$

In order for this equation to have a weakly stationary solution it is necessary that $\mathscr{D}(z)$ have no zeros on the unit circle. If this condition is removed and $\mathscr{C}(z)$ and $\mathscr{D}(z)$ are simply required to have no zeros for $|z| < 1$, then this equation will still have a solution for, say, $t \geq 0$, but the solution will now "evolve" as time progresses.

To see this, write

$$\mathscr{D}(z) = \mathscr{U}(z)\mathscr{V}(z),$$

where $\mathscr{U}(z)$ is composed of the factors of $\mathscr{D}(z)$ containing the roots which are strictly greater than one in absolute value, and let $\mathscr{V}(z)$ consist of the factors corresponding to roots lying on the unit circle. Then, if we define

$$Y(t) = \mathscr{V}(B)X(t),$$

it follows that $Y(t)$ is to satisfy the equation

$$\mathscr{U}(B)Y(t) = \mathscr{C}(B)\xi(t). \tag{7.58}$$

This equation has a weakly stationary solution as we have seen. However, we can write

$$\mathscr{V}(B) = \sum_{j=0}^{r} v_j B^j, \qquad v_0 = 1.$$

Thus,

$$Y(t) = \sum_{j=0}^{r} v_j X(t-j)$$

or

$$X(t) = Y(t) + \sum_{j=1}^{r} (-v_j) X(t-j). \tag{7.59}$$

Consequently, once the $X(t)$ process has been initialized through the specification of $X(1), \ldots, X(r)$, its "law of evolution" is completely determined by (7.58) and (7.59). Box and Jenkins call these processes *autoregressive, integrated, moving average* (*ARIMA*) *models.*

One of the most important ARIMA models is that of a stochastic process with weakly stationary kth differences, where the process of kth differences has a rational spectral density. Such a process will satisfy a stochastic difference equation of the form (7.57) with

$$\mathscr{D}(B) = \mathscr{U}(B)\,\Delta^k = \mathscr{U}(B)(1 - B)^k.$$

This model can be expected to fit a large number of practical problems. From our discussion of the variate difference method in Section 6.6, any process with a polynomial trend of degree $k - 1$ and weakly stationary, zero-mean residual with continuous spectrum will have weakly stationary kth differences. Then the spectral density function of the kth difference process can be approximated as closely as desired by a rational function. Consequently, this model will represent a class of processes with polynomial trends. A theoretical account of stochastic processes with stationary differences was first given by Yaglom (1955).

7.6 LINEAR FILTERING IN REAL TIME

In this section we will treat briefly a more realistic version of the filtering problem than the one introduced in Section 5.5. Recall that $X(t)$ and $Y(t)$ were taken to be weakly stationary, stationarily correlated processes, $X(t)$ unobservable and $Y(t)$ observable. The filtering problem was to approximate $X(t)$ as closely as possible by linearly filtering $Y(t)$, assuming that $Y(t)$ has been observed for $-\infty < t < \infty$. More precisely, we determined the linear filter L which minimized $E\big(X(t) - L(Y(t))\big)^2$. If both processes have continuous spectra with spectral densities $f_X(\lambda)$, $f_Y(\lambda)$ and cross-spectral density $f_{XY}(\lambda)$, we showed that the transfer function of the filter L was

$$B(\lambda) = f_{XY}(\lambda)/f_Y(\lambda). \tag{7.60}$$

More realistically, for a given time t, observations on $Y(s)$ will be available only for $s \leq t$. Consequently, we will now require the filter to be restricted to $Y(s)$, $s \leq t$, and we will combine the filtering operation with prediction by asking for the "best" value of $X(t+v)$ for some $v \geq 0$. That is, we will determine the filter L_v which minimizes $E\big(X(t+v) - L_v(Y(t))\big)^2$ among all linear filters restricted to $Y(s)$, $s \leq t$.

The solution to this problem is, clearly, the projection of $X(t+v)$ onto the linear subspace $\mathscr{M}_t^{\,Y}$ generated by $Y(s)$, $s \leq t$. A simple geometric observation will permit us to find this projection by the techniques used to solve the prediction problem. The observation, which is based on expression (1.16), is that if $\mathscr{M}$ and $\mathscr{N}$ are two subspaces with $\mathscr{N} \subset \mathscr{M}$, then the projection of an element $\mathbf{x}$ onto $\mathscr{N}$ can be obtained by first projecting $\mathbf{x}$ onto $\mathscr{M}$, then projecting the resulting element onto $\mathscr{N}$. In our case, we take $\mathscr{N} = \mathscr{M}_t^{\,Y}$ and $\mathscr{M} = \mathscr{M}^Y$, the subspace generated by $Y(t)$, $-\infty < t < \infty$. Thus, L_v is the sequential filter

$$L_v = \mathscr{P}L,$$

where L is the projection of $X(t+v)$ onto $\mathscr{M}^Y$ and $\mathscr{P}$ is the projection of the result on $\mathscr{M}_t^{\,Y}$. The filter L is the one determined in the first version of the filtering problem but operating on $Y_v(t) = Y(t+v)$ rather than on $Y(t)$. It follows that its transfer function is $B(\lambda)e^{i\lambda v}$; where $B(\lambda)$ is given by (7.60). In fact, if $\hat{X}_v(t)$ is the output of this filter, then

$$\hat{X}_v(t) = \int_{-\pi}^{\pi} e^{i\lambda(t+v)} B(\lambda) Z_Y(d\lambda).$$

We will assume that $Y(t)$ has a one-sided moving average representation with respect to a white noise process $\xi(t)$ and that $\mathscr{M}_t^{\,Y} = \mathscr{M}_t^{\,\xi}$. Then,

$$Z_Y(d\lambda) = C(\lambda) Z_\xi(d\lambda)$$

with

$$C(\lambda) = \sum_{j=0}^{\infty} c_j e^{-i\lambda j}, \qquad \sum c_j^2 < \infty. \tag{7.61}$$

It follows that

$$\hat{X}_v(t) = \int_{-\pi}^{\pi} e^{i\lambda(t+v)} D(\lambda) Z_\xi(d\lambda) = \sum_{j=-\infty}^{\infty} d_j \xi(t+v-j),$$

where

$$D(\lambda) = C(\lambda)B(\lambda). \tag{7.62}$$

The filter with transfer function $D(\lambda)$ matches the white noise input. Consequently,

$$D(\lambda) = \sum_{j=-\infty}^{\infty} d_j e^{-i\lambda j}$$

is a valid Fourier series expansion with $\sum_{j=-\infty}^{\infty} d_j^2 < \infty$.

Now, since $\mathscr{M}_t^{\xi} = \mathscr{M}_t^{Y}$, the projection of $\hat{X}_\nu(t)$ onto $\mathscr{M}_t^{Y}$ has the representation

$$\mathscr{P}(\hat{X}_\nu(t)) = \sum_{j=\nu}^{\infty} d_j \xi(t+\nu-j) = \sum_{j=0}^{\infty} d_{j+\nu} \xi(t-j).$$

The process $\tilde{X}_\nu(t) = \mathscr{P}(\hat{X}_\nu(t))$ is $L_\nu(Y(t))$, the solution of the problem, and it only remains to calculate the transfer function of this filter. However, note that

$$Z_{\tilde{X}_\nu}(d\lambda) = D_\nu(\lambda) Z_\xi(d\lambda),$$

where

$$D_\nu(\lambda) = \sum_{j=0}^{\infty} d_{j+\nu} e^{-i\lambda j}. \tag{7.63}$$

It follows that

$$Z_{\tilde{X}_\nu}(d\lambda) = A_\nu(\lambda) Z_Y(d\lambda),$$

where

$$A_\nu(\lambda) = D_\nu(\lambda)/C(\lambda). \tag{7.64}$$

This is the transfer function of L_ν. Thus, in theory, if functions (7.60) and (7.61) are available, a contingency which depends on knowing $f_{XY}(\lambda)$ and $f_Y(\lambda)$, then $A_\nu(\lambda)$ can be calculated from (7.62)–(7.64). If, in addition, $f_Y(\lambda)$ is bounded and $A_\nu(\lambda)$ has a one-sided Fourier series expansion with square-summable coefficients, then the projection will have an explicit representation

$$\tilde{X}_\nu(t) = \sum_{k=0}^{\infty} r_k^{(\nu)} Y(t-k).$$

(See Section A7.2.) By truncating this expression for a "sufficiently large" value of k, a useful and nearly optimal filter can be constructed.

The parameters of this filter can be estimated if records of both the $X(t)$ and $Y(t)$ processes are available for some interval of time. This situation occasionally occurs in practice. For example, if $Y(t) = X(t) + N(t)$, where $N(t)$ is a noise process uncorrelated with $X(t)$, then it is sometimes possible to obtain stretches of record in which either $X(t)$ or $N(t)$ is absent. Then $f_{XY}(\lambda) = f_X(\lambda)$ and $f_Y(\lambda) = f_X(\lambda) + f_N(\lambda)$ can be estimated from the available data.

For certain restricted classes of processes, explicit recursive filters can be constructed which are optimal within these classes. Particularly important work in this direction has been published by Kalman (1960, 1963) and Kalman and Bucy (1961). Kalman filters can be constructed for a large and useful class of multivariate nonstationary processes both in continuous and discrete time. These filters have been used widely in many diverse fields of application. As in the case of recursive predictors, recent work has been devoted to increasing the adaptive features of these filters by allowing for adjustments of parameters as information about the process accumulates [see, for example, Bucy and Follin (1962)]. A good introduction to Kalman filtering is given by Hannan (1970).

APPENDIX TO CHAPTER 7

A7.1 Extension of the Moving Average Representation to Processes for Which the Spectral Density Can Have Zeros

Let $X(t)$ be a continuous-spectrum, weakly stationary process with spectral measure $Z_X(d\lambda)$ and spectral density function $f_X(\lambda)$ and let B denote the set of frequencies for which $f_X(\lambda) = 0$. It is always possible to extend the underlying probability space in order to construct on it a white noise process $W(t)$ with spectral measure $Z_W(d\lambda)$ such that $W(t)$ is uncorrelated with $X(t)$. [See, e.g., Doob (1953, p. 71).] Now, define the process $Y(t)$ to have spectral measure

$$Z_Y(d\lambda) = \begin{cases} Z_X(d\lambda), & \text{if } \lambda \in B^c, \\ Z_W(d\lambda), & \text{if } \lambda \in B. \end{cases}$$

Then, $f_Y(\lambda) > 0$ for all λ and the theory for nonzero spectral densities given in the text yields

$$Y(t) = \sum_{j=-\infty}^{\infty} a_j \xi(t-j),$$

where $\sum_{j=-\infty}^{\infty} a_j^2 < \infty$ and $\xi(t)$ is a white noise process with spectral measure $Z_\xi(d\lambda)$. However,

$$Z_X(d\lambda) = I_{B^c}(\lambda) Z_Y(d\lambda) = I_{B^c}(\lambda) A(\lambda) Z_\xi(d\lambda),$$

where $I_{B^c}(\lambda)$ is the set characteristic function of B^c and $A(\lambda) = \sum_{j=-\infty}^{\infty} a_j e^{-i\lambda j}$. Then, since

$$\int_{-\pi}^{\pi} |I_{B^c}(\lambda) A(\lambda)|^2 \, d\lambda \le \int_{-\pi}^{\pi} |A(\lambda)|^2 \, d\lambda,$$

it follows that $I_{B^c}(\lambda)A(\lambda)$ has a Fourier series expansion with square summable coefficients b_j and

$$\begin{aligned} X(t) &= \int_{-\pi}^{\pi} \left(\sum_{j=-\infty}^{\infty} b_j e^{-i\lambda j} \right) e^{i\lambda t} Z_\xi(d\lambda) \\ &= \sum_{j=-\infty}^{\infty} b_j \xi(t-j). \end{aligned}$$

This is the desired moving average representation of $X(t)$.

A7.2 One-Sided Fourier Series Expansions

The best available condition for a function to have a one-sided Fourier expansion with square summable coefficients is the following: *If $\mathscr{A}(z)$ is an analytic function for $|z| < 1$ with power series expansion $\mathscr{A}(z) = \sum_{j=0}^{\infty} a_j z^j$, a_j real-valued, and if*

$$\lim_{r \uparrow 1} \int_{-\pi}^{\pi} |\mathscr{A}(re^{-i\lambda})|^2 \, d\lambda < \infty, \tag{A7.1}$$

then $A(\lambda) = \mathscr{A}(e^{-i\lambda})$ is well defined as the $\mathscr{L}_2(-\pi, \pi)$ limit of $\mathscr{A}(re^{-i\lambda})$ as $r \to 1$, $0 < r < 1$. Moreover, the power series expansion of $\mathscr{A}(z)$ is valid for $|z| = 1$ with $\sum_{j=0}^{\infty} a_j^2 < \infty$ and *$\sum_{j=0}^{\infty} a_j e^{-i\lambda j}$ is the Fourier series expansion of $A(\lambda)$.*

The proof of this important theorem is given by Grenander and Rosenblatt (1957, p. 288).

The function $A(\lambda)$ defined in this way is called the *radial limit* of $\mathscr{A}(z)$. Consequently, functions which are radial limits have one-sided Fourier series expansions.

This theorem has the following corollary: *If $X(t)$ is a discrete-time weakly stationary process with zero mean and continuous spectrum and if the spectral density function is bounded, then any linear filter L for which the transfer function $A(\lambda)$ is a radial limit can be represented in the form*

$$L(X(t)) = \sum_{j=0}^{\infty} a_j X(t-j),$$

where the coefficients a_j are those of the Fourier series expansion of $A(\lambda)$.

The proof of this corollary follows exactly the verification of the convolution representation of a digital filter given in Section A6.1.

This result has a number of important implications. For example, the condition

$$\int_{-\pi}^{\pi} \log f(\lambda) \, d\lambda > -\infty,$$

encountered in prediction theory, implies the existence of a function $\mathscr{A}(z)$ with a radial limit $A(\lambda)$ for which

$$f(\lambda) = |A(\lambda)|^2.$$

This theorem, due to Szegö (1939), guarantees the existence of the one-sided moving average representation of the process which was basic to the Kolmogorov solution of the prediction problem.

The existence of infinite autoregressive processes satisfying (7.12) is also covered by this result: Apply the theorem to $\mathscr{A}(z) = 1/\mathscr{B}(z)$, where $\mathscr{B}(z)$ is the z-transform of the filter determined by the difference equation (7.12). Thus, if

$$\lim_{r\uparrow 1} \int_{-\pi}^{\pi} |1/\mathscr{B}(re^{-i\lambda})|^2 \, d\lambda < \infty \tag{A7.2}$$

and $1/\mathscr{B}(z) = \sum_{k=0}^{\infty} a_k z^k$, it follows that

$$X(t) = \sum_{k=0}^{\infty} a_k \xi(t-k)$$

will be a valid, zero-mean, weakly stationary process which satisfies (7.12) where $\xi(t)$ is the innovation process. (We implicitly use the fact that the spectral density of a white noise process is bounded.) In order for $\mathscr{A}(z)$ to be analytic for $|z| < 1$ it is necessary that $\mathscr{B}(z)$ be analytic and nonzero in this region. Thus, if $\mathscr{B}(z)$ is analytic and nonzero for $|z| < \rho$, $\rho > 1$, then (A7.2) is certainly satisfied. This contains the condition for finite autoregressions and is the special case for infinite autoregressions considered in Section 7.3.

Finally, this theorem provides the best known condition for the existence of an autoregressive representation of a one-sided moving average process. Let

$$X(t) = \sum_{k=0}^{\infty} b_k \xi(t-k)$$

where $\xi(t)$ is a white noise process and let

$$\mathscr{B}(z) = \sum_{k=0}^{\infty} b_k z^k.$$

Then if $\mathscr{B}(z)$ is analytic and nonzero for $|z| < 1$ and if (A7.2) is satisfied, it follows that $1/B(\lambda) = 1/\mathscr{B}(e^{-i\lambda})$ will have a one-sided Fourier series expansion, $\sum_{j=0}^{\infty} c_j e^{-i\lambda j}$, with $\sum_{j=0}^{\infty} c_j^2 < \infty$. If, in addition, the spectral density of $X(t)$, $f_X(\lambda) = (\sigma^2/2\pi)|B(\lambda)|^2$, is bounded, then $\sum_{j=0}^{\infty} c_j X(t-j)$ will be a well-defined element of $L_2(P)$ for every t. However,

$$\begin{aligned}\sum_{j=0}^{\infty} c_j X(t-j) &= \int_{-\pi}^{\pi} \left(\sum_{j=0}^{\infty} c_j e^{-i\lambda j}\right) e^{i\lambda t} Z_X(d\lambda) \\ &= \int_{-\pi}^{\pi} (1/B(\lambda)) e^{i\lambda t} B(\lambda) Z_\xi(d\lambda) \\ &= \xi(t).\end{aligned}$$

Thus, $X(t)$ satisfies the difference equation of an infinite autoregression.

Note that if $B(\lambda)$ is bounded and bounded away from zero, as is the case when $\mathscr{B}(z)$ is a polynomial with all of its zeros strictly outside of the unit circle, then (A7.2) and the boundedness of $f_X(\lambda)$ are immediate. Thus, a finite-order, one-sided moving average for which the z-transform has all zeros outside of the unit circle can always be represented as an infinite autoregression.

A7.3 Proof of the Wold Decomposition Theorem

We will show that although this theorem is concerned with weakly stationary processes, it is actually purely geometric in character. Let $X(t)$ be a zero-mean, weakly stationary process in discrete time and, as in the text, let $\mathcal{M}_t^X$ represent the subspace of $L_2(P)$ generated by the random variables $X(s)$ for $s \leq t$. In the notation of Section 1.3, let $\hat{X}_1(t-1) = \mathcal{P}(X(t) \mid \mathcal{M}_{t-1}^X)$ and define the stochastic process $\xi(t)$ by

$$\xi(t) = X(t) - \hat{X}_1(t-1).$$

The variance of $\xi(t)$ is, then, the one-step prediction error σ^2. Note that by this construction, $\xi(t) \perp \mathcal{M}_{t-1}^X$.

Now $\xi(t)$ is a process of uncorrelated (orthogonal) random variables, since if $t < s$, then $\xi(s) \perp \mathcal{M}_{s-1}^X$, while $\xi(t) \in \mathcal{M}_{s-1}^X$. Thus, $\xi(t) \perp \xi(s)$. This is the innovation process of $X(t)$ as we now show.

Let $\mathcal{M}_t^\xi$ be the subspace of $L_2(P)$ generated by the random variables $\xi(s)$ for $s \leq t$. Since $\xi(s) \in \mathcal{M}_s^X$ for all s, it follows that $\mathcal{M}_t^\xi \subset \mathcal{M}_t^X$. Let $\mathcal{M}_t^V$ be the orthogonal complement of $\mathcal{M}_t^\xi$ in $\mathcal{M}_t^X$. Then, $\mathcal{M}_t^X = \mathcal{M}_t^\xi \oplus \mathcal{M}_t^V$ and, by expression (1.15), we have

$$X(t) = U(t) + V(t), \tag{A7.3}$$

where

$$U(t) = \mathcal{P}(X(t) \mid \mathcal{M}_t^\xi) \qquad \text{and} \qquad V(t) = \mathcal{P}(X(t) \mid \mathcal{M}_t^V).$$

This is statement (7.30) of the theorem.

Since the generators of $\mathcal{M}_t^\xi$ are orthogonal, it follows from (1.14) that the projection of $X(t)$ on $\mathcal{M}_t^\xi$ must be of the form

$$U(t) = \sum_{k=0}^{\infty} a_k \xi(t-k),$$

where

$$a_k = \frac{\langle X(t), \xi(t-k) \rangle}{\|\xi(t-k)\|^2} = \frac{1}{\sigma^2} \langle X(0), \xi(-k) \rangle$$

for $k \geq 0$ and

$$\sum_{k=0}^{\infty} a_k^2 < \infty.$$

Note that $X(0) = \hat{X}_1(-1) + \xi(0)$. Thus, since $\xi(0) \perp \hat{X}_1(-1)$,

$$a_0 = (1/\sigma^2)\langle \xi(0), \xi(0) \rangle = 1.$$

This establishes statement (ii) of the theorem.

Now, if $\mathcal{M}_t^U$ denotes the linear subspace spanned by the random variables $U(s)$ for $s \le t$, then $\mathcal{M}_t^U = \mathcal{M}_t^\xi$ for all t. To see this, note that $\mathcal{M}_t^U \subset \mathcal{M}_t^\xi$ by the one-sided representation of $U(t)$ given above. To establish the reverse inclusion, take $W \in \mathcal{M}_t^\xi$. Since $\mathcal{M}_t^\xi \subset \mathcal{M}_t^X$, W is the limit of a Cauchy sequence of finite linear combinations; $W = \lim_n \sum_j c_{j,n} X(t - t_{j,n})$, $t_{j,n} \ge 0$. But then,

$$W = \mathcal{P}(W \mid \mathcal{M}_t^\xi) = \lim_n \sum_j c_{j,n} \mathcal{P}(X(t - t_{j,n}) \mid \mathcal{M}_t^\xi)$$
$$= \lim_n \sum_j c_{j,n} U(t - t_{j,n}) \in \mathcal{M}_t^U.$$

(We have used the continuity of projections discussed in Section 1.3.) In the remainder of the proof we will replace $\mathcal{M}_t^\xi$ by $\mathcal{M}_t^U$.

Next, if $s \le t$, then $\xi(s) \in \mathcal{M}_t^U$. However, $V(t) \perp \mathcal{M}_t^U$, thus $\xi(s) \perp V(t)$. Similarly, if $s > t$, then $\xi(s) \perp \mathcal{M}_t^X$. But $V(t) \in \mathcal{M}_t^X$ and again, $\xi(s) \perp V(t)$. It follows that $\xi(s) \perp V(t)$ for all s, t. Thus $U(s) \perp V(t)$ for all s, t. That is, the processes $U(t)$ and $V(t)$ are uncorrelated. This is statement (i) of the theorem. Again by (1.15),

$$\mathcal{P}(X(t) \mid \mathcal{M}_{t-1}^X) = \mathcal{P}(X(t) \mid \mathcal{M}_{t-1}^U) + \mathcal{P}(X(t) \mid \mathcal{M}_{t-1}^V).$$

However, $U(t) \perp \mathcal{M}_{t-1}^V$ and $V(t) \perp \mathcal{M}_{t-1}^U$, which implies that $\mathcal{P}(U(t) \mid \mathcal{M}_{t-1}^V) = \mathcal{P}(V(t) \mid \mathcal{M}_{t-1}^U) = 0$. Then, by (A7.3), we obtain

$$\hat{X}_1(t-1) = \mathcal{P}(X(t) \mid \mathcal{M}_{t-1}^X) = \mathcal{P}(U(t) \mid \mathcal{M}_{t-1}^U) + \mathcal{P}(V(t) \mid \mathcal{M}_{t-1}^V).$$

This expression is valid with the subscripts $t-1$ replaced by s for any $s \le t$ by exactly the same argument. We needed this fact at one point in the text of this chapter.

Now, by an application of expression (1.14) we obtain

$$\mathcal{P}(U(t) \mid \mathcal{M}_{t-1}^U) = \sum_{k=1}^{\infty} a_k \xi(t-k).$$

Thus,

$$\xi(t) = X(t) - \hat{X}_1(t-1)$$
$$= U(t) - \sum_{k=1}^{\infty} a_k \xi(t-k) + V(t) - \mathcal{P}(V(t) \mid \mathcal{M}_{t-1}^V)$$
$$= \xi(t) + V(t) - \mathcal{P}(V(t) \mid \mathcal{M}_{t-1}^V).$$

From this we obtain

$$V(t) = \mathcal{P}(V(t) \mid \mathcal{M}_{t-1}^V)).$$

It follows that $\mathcal{M}_t^V \subset \mathcal{M}_{t-1}^V$. However, since $\mathcal{M}_{t-1}^V \subset \mathcal{M}_t^V$, this can only happen if $\mathcal{M}_s^V = \mathcal{M}_t^V$ for all s, t and the theorem is proved.

CHAPTER

8

The Distribution Theory of Spectral Estimates with Applications to Statistical Inference

8.1 INTRODUCTION

The statistical analysis of time series actually predates the introduction of the models we have considered in previous chapters of this book. Early investigators, beginning with Schuster in the late nineteenth century, were interested in looking for periodicities in geophysical and economics data. The "tool" adopted for such studies was the periodogram which is, essentially, the squared absolute value of the finite Fourier transform of the time series. Thus, Fourier analysis is not only one of the most important forms of analysis for time series, it is also one of the oldest. The distribution theory of the periodogram based on a trigonometric polynomial regression function with a white noise residual was derived by Fisher (1929) and was used as the basis for testing for the existence of periodicities.

With the introduction of moving average and autoregressive processes by Yule in the 1920s, interest shifted away from the frequency domain analysis of time series to the time domain. This shift was accelerated after the time series models of Chapter 2 were introduced in the 1930s because the periodogram, which had demonstrated its value for locating periodicities, proved to be an erratic and unfaithful estimator of the newly introduced power spectral density. A considerable body of time domain techniques were developed during the period 1920–1950. This theory is summarized in detail by Anderson (1971).

Interest in the frequency domain analysis of time series was reawakened by a suggestion of Daniell (1946) to the effect that by averaging the periodogram in neighborhoods of each frequency of interest, well-behaved estimators of the spectral density function could be obtained. This idea was pursued by Bartlett (1948) and then by Tukey (1949) who introduced the idea of the spectral window and investigated the properties of windowed estimators. These developments ushered in the modern era of time series spectrum analysis which has witnessed important contributions by a number of investigators. We will summarize their work in this and the next chapter.

The distribution theory for standard estimators of the spectral density of a one-dimensional time series will be considered first. Then the theory for multivariate time series will be outlined and the distributions of the important spectral parameters considered in Chapter 5 will be given. The theory will be applied to the calculation of confidence intervals and the testing of hypotheses for these parameters. These results will be applied to the important problem of designing time series experiments in Chapter 9.

8.2 DISTRIBUTION OF THE FINITE FOURIER TRANSFORM AND THE PERIODOGRAM

Throughout this chapter we will assume that the time series under consideration are discrete, zero-mean, stationary Gaussian processes with continuous spectra. The restriction to Gaussian processes simplifies the theory substantially and makes it possible to present a rather intuitive yet precise account of results. These results are actually valid for a much broader class of time series. That is, the theory is reasonably robust against deviations from the Gaussian assumption. For details of the more general theory the reader is referred to Grenander and Rosenblatt (1957) and Hannan (1970).

In this section, $X(t)$, $t = 0, \pm 1, \ldots$, will be a univariate process with spectral density function $f(\lambda)$. We assume that the process has been observed for times $1 \leq t \leq N$. Thus, N is the *sample size* and is one of the important parameters upon which the properties of spectral estimates will depend. We will be interested in the finite Fourier transform of the data. For reasons to be discussed presently, we take the following normalized version:

$$z_\nu^{(N)} = \frac{1}{(2\pi N)^{1/2}} \sum_{t=1}^{N} X(t) e^{-i\lambda_\nu t}, \tag{8.1}$$

where $\lambda_\nu = 2\pi\nu/N$, $-[(N-1)/2] \leq \nu \leq [N/2]$. As before, $[x]$ denotes the largest integer less than or equal to x. [Compare (8.1) with expression (3.28).]

Since linear combinations of normal random variables are normal, the random variables $z_\nu^{(N)}$ have a multivariate complex normal distribution with zero means. We will also want to calculate the variance $E|z_\nu^{(N)}|^2$ of $z_\nu^{(N)}$. To do this we substitute the spectral representation for $X(t)$ in (8.1):

$$z_\nu^{(N)} = \frac{1}{(2\pi N)^{1/2}} \sum_{t=1}^{N} e^{-i\lambda_\nu t} \int_{-\pi}^{\pi} e^{-i\lambda t} Z(d\lambda)$$

$$= \int_{-\pi}^{\pi} H_N(\lambda - (2\pi\nu/N)) Z(d\lambda), \tag{8.2}$$

where

$$H_N(\lambda) = \frac{1}{(2\pi N)^{1/2}} \sum_{t=1}^{N} e^{i\lambda t}. \tag{8.3}$$

Then,

$$E|z_\nu^{(N)}|^2 = \int_{-\pi}^{\pi}\int_{-\pi}^{\pi} H_N(\lambda - (2\pi\nu/N))\overline{H_N(\mu - (2\pi\nu/N))} EZ(d\lambda)\overline{Z(d\mu)}$$

$$= \int_{-\pi}^{\pi} |H_N(\lambda - (2\pi\nu/N))|^2 f(\lambda)\, d\lambda, \tag{8.4}$$

by the properties of random spectral measures given in Chapter 2.

Now, by (8.3),

$$|H_N(\lambda)|^2 = \frac{1}{2\pi N} \sum_{t=1}^{N}\sum_{s=1}^{N} e^{i\lambda(t-s)}$$

$$= \frac{1}{2\pi N}\left\{N + \sum_{\substack{t=1\\ t\neq s}}^{N}\sum_{s=1}^{N} e^{i\lambda(t-s)}\right\}.$$

It follows that

$$\int_{-\pi}^{\pi} |H_N(\lambda)|^2\, d\lambda = \frac{1}{2\pi N}\left\{2\pi N + \sum_{\substack{t=1\\ t\neq s}}^{N}\sum_{s=1}^{N} \int_{-\pi}^{\pi} e^{i\lambda(t-s)}\, d\lambda\right\}$$

$$= 1.$$

Thus, the normalization of (8.1) by $(2\pi N)^{-1/2}$ was chosen to make the integral of the weight function $|H_N(\lambda)|^2$ equal to unity. Now, by expression (1.18), it is easily seen that

$$|H_N(\lambda)|^2 = \frac{\sin^2(\lambda N/2)}{2\pi N \sin^2(\lambda/2)}.$$

This is the so-called *Fejer kernel* of Fourier analysis and it is well known to have the *delta function property* as $N \to \infty$. That is, for all "well-behaved" functions $g(\lambda)$,

$$\lim_{N\to\infty} \int_{-\pi}^{\pi} |H_N(\mu - \lambda)|^2 g(\mu)\, d\mu = g(\lambda).$$

Except for the fact that all side lobes are positive, $|H_N(\lambda)|^2$ looks very much like the window function of Fig. 6.6. Consequently, if the spectral density function is continuous, say, then for sufficiently large N,

$$E|z_\nu^{(N)}|^2 \cong f(\lambda_\nu). \tag{8.5}$$

The smoother $f(\lambda)$ is in the vicinity of λ_ν the better is this approximation for moderate values of N. If $f(\lambda)$ were constant over $(-\pi, \pi)$, equality would hold in (8.5) for all N. For this reason, $f(\lambda_\nu)$ could be called the white noise variance of $z_\nu^{(N)}$.

Expression (8.5) suggests that the statistic

$$I_{N,\nu} = |z_\nu^{(N)}|^2 = \frac{1}{2\pi N}\left|\sum_{t=1}^{N} X(t)e^{-i\lambda_\nu t}\right|^2 \tag{8.6}$$

would be a reasonable estimator for the spectral density function at λ_ν. This estimator is the *periodogram* which has played such a significant role in time series analysis. We now investigate some of the more important statistical properties of the periodogram.

The Distribution of the Periodogram and Other of Its Properties

In theory, the periodogram can be defined for all frequencies λ, $-\pi < \lambda \leq \pi$, by the expression

$$I_N(\lambda) = |z^{(N)}(\lambda)|^2,$$

where

$$z^{(N)}(\lambda) = \frac{1}{(2\pi N)^{1/2}} \sum_{t=1}^{N} X(t)e^{-i\lambda t}. \tag{8.7}$$

In practice, this finite Fourier transform can only be calculated at a finite set of frequencies. This presents no problem, since, in fact, *function* (8.7) *is completely determined by its values at the frequencies* $\lambda_\nu = 2\pi\nu/N$, $-[(N-1)/2] \leq \nu \leq [N/2]$. This follows from a discrete, frequency domain version of the sampling theorem of Section 3.2 which we present at this point:

Frequency Domain Sampling Theorem *Let $a(t)$ be a function which is zero except for the arguments $t = 1, 2, \ldots, R$ and let $A(\lambda)$ be its Fourier transform*

$$A(\lambda) = \sum_{t=1}^{R} a(t)e^{i\lambda t}, \qquad -\pi < \lambda \leq \pi.$$

Then,

$$A(\lambda) = \sum_{\nu=-[(R-1)/2]}^{[R/2]} B_\nu^{(R)}(\lambda) A(2\pi\nu/R), \tag{8.8}$$

where, setting $\lambda_\nu = 2\pi\nu/R$,

$$B_\nu^{(R)}(\lambda) = \frac{\sin(R(\lambda - \lambda_\nu)/2)}{R\sin((\lambda - \lambda_\nu)/2)} \exp\left(-i\frac{R+1}{2}(\lambda - \lambda_\nu)\right).$$

Moreover, if $A(\lambda)$ *is any function for which representation* (8.8) *holds, then its Fourier coefficients* $a(t)$ *are necessarily zero for* $t \le 0$ *and* $t \ge R + 1$.

The proof of this theorem closely parallels the proof of the sampling theorem in Section 3.2 and is left to the reader.

With $R = N$ and $a(t) = X(t)/(2\pi N)^{1/2}$, we obtain

$$z^{(N)}(\lambda) = \sum_{\nu=-[(N-1)/2]}^{[N/2]} B_\nu^{(N)}(\lambda) z_\nu^{(N)}.$$

where $z_\nu^{(N)}$ is defined at the frequencies $\lambda_\nu = 2\pi\nu/N$ by (8.1). This is the result we wanted. It implies that the periodogram is also determined by the $z_\nu^{(N)}$'s. Consequently the statistical properties of $I_N(\lambda)$ depend only on the joint distribution of these random variables and we will first obtain the relevant properties of this distribution.

A simple but useful observation, based on definition (8.1), is that

$$z_{-\nu}^{(N)} = \bar{z}_\nu^{(N)} \tag{8.9}$$

for $|\nu| \le [(N-1)/2]$. This result will be needed shortly.

We now sketch an argument for the following important property: *If the spectral density function* $f(\lambda)$ *is continuous, then the random variables* $z_\nu^{(N)}$, $-[(N-1)/2] \le \nu \le [N/2]$ *are asymptotically independent as* $N \to \infty$. Since they are zero-mean, normal random variables, it suffices to show that the covariances tend to zero, i.e.,

$$\lim_{N\to\infty} E z_\mu^{(N)} \bar{z}_\nu^{(N)} = 0 \qquad \text{for} \quad \mu \ne \nu.$$

From (8.2) we obtain

$$E z_\mu^{(N)} \bar{z}_\nu^{(N)} = \int_{-\pi}^{\pi}\int_{-\pi}^{\pi} H_N\left(\alpha - \frac{2\pi\mu}{N}\right)\overline{H_N\left(\beta - \frac{2\pi\nu}{N}\right)} EZ(d\alpha)\overline{Z(d\beta)}$$

$$= \int_{-\pi}^{\pi} H_N\left(\alpha - \frac{2\pi\mu}{N}\right)\overline{H_N\left(\alpha - \frac{2\pi\nu}{N}\right)} f(\alpha)\, d\alpha. \tag{8.10}$$

Now, except for a factor of absolute value 1, the function $H_N(\alpha - (2\pi\mu/N))\cdot \overline{H_N(\alpha - (2\pi\nu/N))}$ is the product of two window functions of the type pictured in Fig. 6.6 centered at frequencies $\lambda_\mu = 2\pi\mu/N$ and $\lambda_\nu = 2\pi\nu/N$, respectively. However, since both frequencies tend to zero with N, these peaks become narrower and closer together. It follows from the continuity of $f(\lambda)$ that (8.10) will have the same limit as does

$$f\left(\frac{\lambda_\mu + \lambda_\nu}{2}\right)\int_{-\pi}^{\pi} H_N\left(\alpha - \frac{2\pi\mu}{N}\right)\overline{H_N\left(\alpha - \frac{2\pi\nu}{N}\right)}\,d\alpha.$$

By using (8.3), we obtain

$$\begin{aligned}
&\int_{-\pi}^{\pi} H_N\left(\alpha - \frac{2\pi\mu}{N}\right)\overline{H_N\left(\alpha - \frac{2\pi\nu}{N}\right)}\,d\alpha \\
&\quad = \frac{1}{2\pi N}\sum_{t=1}^{N}\sum_{s=1}^{N}\exp\left(-i\frac{2\pi\mu}{N}t\right)\exp\left(i\frac{2\pi\nu}{N}s\right)\int_{-\pi}^{\pi}\exp(i\alpha(t-s))\,d\alpha \\
&\quad = \frac{1}{N}\sum_{t=1}^{N}\exp\left(i\frac{2\pi(\nu-\mu)t}{N}\right),
\end{aligned}$$

since

$$\int_{-\pi}^{\pi} e^{i\alpha k}\,d\alpha = \begin{cases} 2\pi, & k = 0, \\ 0, & k \neq 0. \end{cases}$$

The final expression is 0 when $\nu \neq \mu$ and 1 when $\nu = \mu$. Consequently, we obtain

$$\lim_{N\to\infty} E z_\mu^{(N)}\bar{z}_\nu^{(N)} = 0 \qquad \text{for} \quad \mu \neq \nu.$$

Moreover, if $f(\lambda)$ is reasonably smooth, the approximations

$$E z_\mu^{(N)}\bar{z}_\nu^{(N)} \cong \begin{cases} f(\lambda_\nu), & \mu = \nu, \\ 0, & \mu \neq \nu, \end{cases} \tag{8.11}$$

are quite good for sufficiently large N. The first statement is the same as (8.5).

The distribution theory for the $z_\nu^{(N)}$'s is an asymptotic theory in the sense that it is based on replacing the approximate equality in (8.11) *by equality and proceeding as though these random variables were independent with the indicated variances. The results derived using this convention will be asymptotically correct and will hold to a good degree of approximation if N is not too small.*

Hereafter we will fix N and delete the superscript on the $z_\nu^{(N)}$'s. Thus, according to the convention of the last paragraph,

$$z_\nu \approx N(0, f(\lambda_\nu)),$$

and the z_ν's, $-[(N-1)/2] \le \nu \le [N/2]$, are independent. Now, represent z_ν in Cartesian form:

$$z_\nu = a_\nu - ib_\nu .$$

As in the representation of the co- and quad-spectral densities in Section 5.5, the purpose of the minus sign in this expression is to obtain the Fourier representations

$$a_\nu = \frac{1}{(2\pi N)^{1/2}} \sum_{t=1}^{N} X(t) \cos \lambda_\nu t \qquad \text{and} \qquad b_\nu = \frac{1}{(2\pi N)^{1/2}} \sum_{t=1}^{N} X(t) \sin \lambda_\nu t$$

from (8.7). Then, since

$$a_\nu = (z_\nu + \bar{z}_\nu)/2 \tag{8.12}$$

$$b_\nu = (z_\nu - \bar{z}_\nu)/2i, \tag{8.13}$$

it follows that a_ν and b_ν are real-valued, zero-mean, normal random variables. Moreover, the vector pairs (a_ν, b_ν) are independent since the z_ν's are.

One additional important property is that *a_ν is independent of b_ν for each ν.* To see this, first note that for $\nu \neq 0$ and $\nu \neq N/2$,

$$Ez_\nu^2 = Ez_\nu \bar{z}_{-\nu} = 0.$$

This follows from (8.9) and the independence of the z_ν's. Similarly, $E\bar{z}_\nu^2 = 0$ for $\nu \neq 0, N/2$. Consequently, by (8.11) and (8.12),

$$Ea_\nu b_\nu = -(1/4i)(Ez_\nu^2 - E\bar{z}_\nu^2) = 0.$$

For $\nu = 0$ and $\nu = N/2$ (which can occur only if N is even) this follows immediately from the fact that

$$a_0 = z_0, \qquad b_0 = 0, \qquad a_{N/2} = z_{N/2}, \qquad b_{N/2} = 0.$$

The same technique is used to compute the variances of a_ν and b_ν: From (8.11) and (8.12) we have

$$Ea_\nu^2 = \begin{cases} \frac{1}{4}(Ez_\nu^2 + 2Ez_\nu \bar{z}_\nu + E\bar{z}_\nu^2), & \nu \neq 0, N/2, \\ E|z_\nu|^2, & \nu = 0, N/2, \end{cases}$$

$$= \begin{cases} \frac{1}{2}f(\lambda_\nu), & \nu \neq 0, N/2, \\ f(0), & \nu = 0, \\ f(\pi), & \nu = N/2. \end{cases} \tag{8.14}$$

Similarly,

$$Eb_N^2 = \begin{cases} \frac{1}{2}f(\lambda_\nu), & \nu \neq 0, N/2, \\ 0, & \nu = 0, N/2. \end{cases} \tag{8.15}$$

We now have all of the information needed to compute the asymptotic distribution of the periodogram at the frequencies λ_ν, $-[(N-1)/2] \leq \nu \leq [N/2]$. From the definition of the periodogram,

$$I_{N,\nu} = |z_\nu|^2 = a_\nu^2 + b_\nu^2.$$

However, for $\nu \neq 0$, $a_\nu/(\frac{1}{2}f(\lambda_\nu))^{1/2}$ and $b_\nu/(\frac{1}{2}f(\lambda_\nu))^{1/2}$ are independent $N(0, 1)$ random variables. Consequently,

$$\chi^2_{2,\nu} = (a_\nu^2 + b_\nu^2)/\tfrac{1}{2}f(\lambda_\nu)$$

has the chi-square distribution with two degrees of freedom. Thus,

$$I_{N,\nu} = \tfrac{1}{2}f(\lambda_\nu)\chi^2_{2,\nu} \qquad \text{for} \qquad \nu \neq 0, N/2.$$

For $\nu = 0$, $N/2$, by the same argument,

$$I_{N,0} = a_0^2 = f(0)\chi^2_{1,0} \qquad \text{and} \qquad I_{N,N/2} = a^2_{N/2} = f(\pi)\chi^2_{1,N/2},$$

where $\chi^2_{1,\nu}$ is a random variable possessing the chi-square distribution with one degree of freedom.

In summary, the periodogram ordinates $I_{N,\nu} = I_N(\lambda_\nu)$ are asymptotically independent and have the asymptotic distributions of the indicated multiples of chi-square random variables.

This result and the expressions

$$E(\chi_r^2) = r, \qquad \operatorname{Var}(\chi_r^2) = 2r$$

for the mean and variance of a chi-square random variable with r degrees of freedom allow us to obtain *the asymptotic mean and variance of the periodogram*:

$$\begin{aligned} E(I_{N,\nu}) &= f(\lambda_\nu), \\ \operatorname{Var}(I_{N,\nu}) &= \tfrac{1}{4}f^2(\lambda_\nu)\operatorname{Var}(\chi_2^2) \\ &= f^2(\lambda_\nu), \qquad \nu \neq 0, N/2, \end{aligned}$$

and

$$\operatorname{Var}(I_{N,\nu}) = \begin{cases} 2f^2(0), & \nu = 0, \\ 2f^2(\pi), & \nu = N/2. \end{cases}$$

A more precise result is true. If λ is any frequency in $(-\pi, \pi]$ and ν_N is a sequence of integers such that as $N \to \infty$,

$$\lambda_{\nu_N} = 2\pi\nu_N/N \to \lambda$$

at the "right" rate [see, e.g., Brillinger (1970)], then the mean and variance of I_{N,ν_N} and $I_N(\lambda)$ have the same limits and these limits are obtained from the above expressions by replacing λ_ν by λ. That is,

$$\lim_{N\to\infty} E(I_N(\lambda)) = f(\lambda) \tag{8.16}$$

and

$$\lim_{N\to\infty} \operatorname{Var}\big(I_N(\lambda)\big) = \begin{cases} f^2(\lambda), & \lambda \neq 0, \pi, \\ 2f^2(0), & \lambda = 0, \\ 2f^2(\pi), & \lambda = \pi. \end{cases} \tag{8.17}$$

The following results from statistics will be needed in the following discussions. The *mean-square error* of an estimator $\hat{\vartheta}_n$ of a parameter ϑ is $E(\hat{\vartheta}_n - \vartheta)^2$. Consistency was defined in Section 2.10 where it was also stated that in order for an estimator to be consistent it is sufficient that $\lim_{n\to\infty} E(\hat{\vartheta}_n - \vartheta)^2 = 0$. It can be argued that this condition is also necessary. Now, an elementary application of the properties of expectation yields

$$E(\hat{\vartheta}_n - \vartheta)^2 = (E\hat{\vartheta}_n - \vartheta)^2 + \operatorname{Var}(\hat{\vartheta}_n).$$

The expression $E\hat{\vartheta}_n - \vartheta$ is called the bias of the estimator. The estimator is said to be *unbiased* if $E\hat{\vartheta}_n = \vartheta$ for all n and *asymptotically unbiased* if $\lim_{n\to\infty} E\hat{\vartheta}_n = \vartheta$. It follows that the estimator is consistent if and only if it is unbiased or asymptotically unbiased and $\lim_{n\to\infty} \operatorname{Var}(\hat{\vartheta}_n) = 0$.

Expressions (8.16) and (8.17) reveal the basic mean-square properties of the periodogram as an estimator of the spectral density function. It is asymptotically unbiased by (8.16). Thus, the asymptotic mean-square error is the same as the asymptotic variance. However, unless $f(\lambda) = 0$ the asymptotic variance is not zero. *Thus, in general, the periodogram is not a consistent estimator of* $f(\lambda)$. This (and the asymptotic independence of the $I_{N,\nu}$'s) largely accounts for the "erratic and unfaithful" behavior of $I_N(\lambda)$ as an estimator for $f(\lambda)$ mentioned in Section 8.1. In the next section we will consider estimators for $f(\lambda)$ which avoid this shortcoming while retaining the desirable property of asymptotic unbiasedness. Thus, for these estimators the mean-square error will go to zero. The asymptotic distributions of these estimators will be derived from the distribution theory we have developed for the periodogram.

8.3 DISTRIBUTION THEORY FOR UNIVARIATE SPECTRAL ESTIMATORS

Because of the Fourier series representation

$$f(\lambda) = \frac{1}{2\pi} \sum_{k=-\infty}^{\infty} e^{-i\lambda k} C(k)$$

for the spectral density function [see expression (3.23)], a "natural" estimator for the spectral density can be obtained by replacing $C(k)$ in this expression by an estimate based on the observations $X(1), X(2), \ldots, X(N)$. From the discussion of the ergodic theorems in Sections 2.10 and 3.3, a suitable choice would be

$$\hat{C}(k) = \begin{cases} \dfrac{1}{N} \displaystyle\sum_{t=1}^{N-|k|} X(t + |k|)X(t), & |k| \le N - 1, \\ 0, & |k| > N - 1. \end{cases} \tag{8.18}$$

However, by a straightforward change of variables, it can be shown that

$$\frac{1}{2\pi} \sum_{k=-\infty}^{\infty} e^{-i\lambda k} \hat{C}(k) = I_N(\lambda), \tag{8.19}$$

the periodogram. Thus, without further modification, this estimation scheme still leads to an inconsistent estimate of the spectral density. An analysis of the difficulty indicates the following: When the argument $|k|$ is large, i.e., near $N - 1$, the random variables $\hat{C}(k)$ are averages of a relatively small number of the products $X(t + |k|)X(t)$. Thus, the stabilizing influence of the averaging operation has not had a chance to take effect and these random variables retain about the same degree of variability no matter how large N is. This accounts for the fact that the periodogram has variance which never approaches zero with increasing N.

This analysis also indicates a possible means for eliminating the inconsistency of the periodogram without destroying its asymptotic unbiasedness. For an integer $M < N$, let $w_M(k)$, $k = 0, \pm 1, \ldots$ be a sequence of weights with the properties

(i) $0 \le w_M(k) \le w_M(0) = 1$,

(ii) $w_M(-k) = w_M(k)$ for all k,

(iii) $w_M(k) = 0$ for $|k| > M$.

The specific weight sequences that have been used in practice will be discussed later. At present, only these general properties will be needed.

Form the estimator

$$\hat{f}(\lambda) = \frac{1}{2\pi} \sum_{k=-\infty}^{\infty} e^{-i\lambda k} w_M(k) \hat{C}(k). \tag{8.20}$$

This is called a *weighted covariance estimator*. This type of estimator was introduced by Tukey (1949) and was the first kind adapted extensively to digital computer calculations. Grenander and Rosenblatt (1957) provided the distribution theory for this class of estimators under rather general assumptions about the underlying stochastic process. The weight function $w_M(k)$ is called a *lag window* and the integer M is the *lag number*.

Note that by property (iii) of the lag window, the products $w_M(k)\hat{C}(k)$ are zero for $|k| > M$. Thus, the more variable autocovariance estimates are omitted from (8.20) and by keeping M relatively small as N tends to infinity, it is reasonable to expect that a consistent sequence of estimators can be constructed. Moreover, by letting M tend to infinity it should be possible to design the weight sequence so as to retain the asymptotic unbiasedness of the estimator. The confirmation of these expectations will be indicated presently.

First, we derive an equivalent form for the estimators (8.20). The *spectral window* $W_M(\lambda)$, corresponding to the lag window $w_M(k)$, is its Fourier transform

$$W_M(\lambda) = \frac{1}{2\pi} \sum_{k=-\infty}^{\infty} e^{-i\lambda k} w_M(k). \tag{8.21}$$

From the properties of the lag window, it is easily seen that $W_M(\lambda)$ is real-valued and has the properties

(i) $W_M(-\lambda) = W_M(\lambda)$,

(ii) $\int_{-\pi}^{\pi} W_M(\lambda)\, d\lambda = 1$.

Note that (8.20) is the Fourier transform for the product $w_M(k)\hat{C}(k)$. It follows from the discussion of Fourier series in Example 1.2, that the Fourier transform of this product is the convolution of the Fourier transforms. Consequently, (8.20) is equivalent to

$$\hat{f}(\lambda) = \int_{-\pi}^{\pi} W_M(\lambda - \mu) I_N(\mu)\, d\mu. \tag{8.22}$$

An estimator with the same asymptotic distribution as (8.20) can be obtained by replacing the integral (8.22) by its *Riemann approximating sum*

$$\hat{f}(\lambda) = \frac{2\pi}{N} \sum_{\nu=-[(N-1)/2]}^{[N/2]} W_M(\lambda - \lambda_\nu) I_{N,\nu}.$$

Since,

$$\frac{2\pi}{N} \sum_{\nu=-[(N-1)/2]}^{[N/2]} W_M(\lambda_\nu) \cong \int_{-\pi}^{\pi} W_M(\lambda)\, d\lambda = 1$$

(Riemann approximation, again), this is asymptotically equivalent to an estimator of the form

$$\hat{f}(\lambda) = \sum_{\nu=-[(N-1)/2]}^{[N/2]} K(\lambda - \lambda_\nu) I_{N,\nu}, \tag{8.23}$$

where $K(\lambda)$ is a symmetric, periodic, real-valued weight function for which

$$\sum_{\nu=-[(N-1)/2]}^{[N/2]} K(\lambda_\nu) = 1.$$

Estimators of this form are called *smoothed periodogram estimators.*

This is the general form of the estimator first suggested by Daniell (1946). Since the fast Fourier transform algorithm was introduced by Cooley and Tukey (1965), the calculation of the periodogram has become extremely fast and for a period of time this type of spectral estimator largely replaced the weighted covariance estimator in applications of spectral analysis. As we will indicate in Chapter 9, the fast Fourier transform can also be used to calculate weighted covariance estimators and these estimators are experiencing a "comeback" in the rapidly changing spectral estimation scene.

The distribution theory for smoothed periodogram estimators is rather easily derived from the theory generated for the periodogram in the last section. Because of the asymptotic equivalence of the weighted covariance and smoothed periodogram estimators, we will also be able to translate the results for the smoothed periodogram to the weighted covariance estimator. We first derive the asymptotic mean, variance, and distribution of a simple smoothed periodogram estimator.

The Daniell Estimator and Its Moments

The most basic smoothed periodogram estimator is the uniformly weighted average suggested by Daniell. If the average is taken over n neighboring frequencies λ_ν, this estimator at frequency $\lambda_k = 2\pi k/N$ is

$$\hat{f}(\lambda_k) = \frac{1}{n} \sum_{\nu=k-[(n-1)/2]}^{k+[n/2]} I_{N,\nu}.$$

For simplicity, in this discussion we will assume that $\lambda_k - \lambda_\nu \neq 0, \pi$ for all ν. The asymptotic mean and variance of the estimator are easily calculated from the asymptotic theory of the last section:

$$\begin{aligned} E(\hat{f}(\lambda_k)) &= \frac{1}{n} \sum_{\nu=k-[(n-1)/2]}^{k+[n/2]} E(I_{N,\nu}) \\ &= \frac{1}{n} \sum_{\nu=-[(n-1)/2]}^{[n/2]} f(\lambda_k - \lambda_\nu). \end{aligned}$$

Similarly, because of the asymptotic independence of the $I_{N,\nu}$'s,

$$\begin{aligned} \operatorname{Var}(\hat{f}(\lambda_k)) &= \frac{1}{n^2} \sum_{\nu=k-[(n-1)/2]}^{k+[n/2]} \operatorname{Var}(I_{N,\nu}) \\ &= \frac{1}{n^2} \sum_{\nu=-[(n-1)/2]}^{[n/2]} f^2(\lambda_k - \lambda_\nu). \end{aligned}$$

Now, if n is sufficiently small and N sufficiently large that $f(\lambda)$ is effectively constant over every frequency interval of length $2\pi n/N$ (the length of the intervals over which $\lambda_k - \lambda_v$ varies for $-[(n-1)/2] \le v \le [n/2]$), then

$$E(\hat{f}(\lambda_k)) \cong f(\lambda_k) \tag{8.24}$$

and

$$\operatorname{Var}(\hat{f}(\lambda_k)) \cong f^2(\lambda_k)/n. \tag{8.25}$$

Moreover, the same type of computation can be used to calculate the asymptotic covariance between estimators at frequencies λ_k and λ_l yielding

$$\operatorname{Cov}(\hat{f}(\lambda_k), \hat{f}(\lambda_l)) \cong \begin{cases} \dfrac{n - |l-k|}{n^2} f(\lambda_k) f(\lambda_l), & |l-k| < n, \\ 0, & |l-k| \ge n. \end{cases} \tag{8.26}$$

As in the case of the periodogram, this is an asymptotically unbiased estimator of the spectral density. However, there is an essential difference between this estimator and the periodogram. Note that if n is allowed to tend to infinity with N in such a way that the length $2\pi n/N$ of the smoothing interval of frequencies tends to zero, then the variance of $\hat{f}(\lambda_k)$ tends to zero. That is, the estimator is both asymptotically unbiased and consistent. Also note that the estimators have zero covariance, hence are independent, for frequency spacings $2\pi n/N$ or larger. Thus, the width of the rectangular smoothing band determines the spacing between independent estimators. This *bandwidth* can be viewed as a measure of the resolution of the estimator. Moreover, the covariance between estimators at every pair of distinct frequencies tends to zero if $2\pi n/N \to 0$. Thus, the estimators are asymptotically independent. As we will see presently, the same simple argument can be used to obtain comparable properties for the general smoothed periodogram estimator. First, however, the asymptotic distribution of the Daniell estimator will be obtained.

Distribution of the Daniell Estimator

Suppose that N and n are large enough to make the asymptotic theory valid to a good approximation and that n/N is small enough for $f(\lambda)$ to be effectively constant over frequency intervals of length $2\pi n/N$. Then, as we saw in the last section, fixing k, the quantities $I_{N,v}/\frac{1}{2}f(\lambda_k)$ are (essentially) independent, chi-square random variables for $k - [(n-1)/2] \le v \le k + [n/2]$, each with two degrees of freedom. (We will take $\lambda_k \ne 0, \pi$ and assume that n/N is sufficiently small that $v \ne 0, [N/2]$.) Consequently, since the sum of independent chi-square variables is again chi-square and the degrees of freedom of the sum is the sum of the degrees of freedom [Tucker (1962, p. 75)] the random variable

$$\sum_{v=k-[(n-1)/2]}^{k+[n/2]} I_{N,v}/\tfrac{1}{2}f(\lambda_k) = \chi^2_{2n}$$

has the chi-square distribution with $2n$ degrees of freedom. However, then

$$\hat{f}(\lambda_k) = \frac{1}{n}\sum_{v=k-[(n-1)/2]}^{k+[n/2]} I_{N,v} = \frac{f(\lambda_k)}{2n}\chi^2_{2n}. \tag{8.27}$$

In other words, $2n\hat{f}(\lambda_k)/f(\lambda_k)$ *has (asymptotically) the chi-square distribution with* $2n$ *degrees of freedom.* This important result will be carried over, with an additional approximation, to the general smoothed periodogram.

Asymptotic Mean and Variance of the General Smoothed Periodogram Estimator and the Weighted Covariance Estimator

The asymptotic mean, covariance, and variance of the estimator (8.23) at frequencies $\lambda_v = 2\pi v/N$ can be written down immediately from the properties of the periodogram;

$$E\hat{f}(\lambda_k) = \sum_{v=-[(N-1)/2]}^{[N/2]} K(\lambda_k - \lambda_v)f(\lambda_v), \tag{8.28}$$

$$\begin{aligned}\operatorname{Cov}(\hat{f}(\lambda_k), \hat{f}(\lambda_l)) = {} & K(\lambda_k)K(\lambda_l)f^2(0) \\ & + \sum_{v=-[(N-1)/2]}^{[N/2]} K(\lambda_k - \lambda_v)K(\lambda_l - \lambda_v)f^2(\lambda_v) \\ & (+K(\lambda_k - \pi)K(\lambda_l - \pi)f^2(\pi)),\end{aligned} \tag{8.29}$$

and

$$\begin{aligned}\operatorname{Var}(\hat{f}(\lambda_k)) = {} & K^2(\lambda_k)f^2(0) \\ & + \sum_{v=-[(N-1)/2]}^{[N/2]} K^2(\lambda_k - \lambda_v)f^2(\lambda_v) \\ & (+K^2(\lambda_k - \pi)f^2(\pi)).\end{aligned} \tag{8.30}$$

The terms in parentheses are included only when N is even. As in the case of the Daniell estimator, all interesting smoothing functions will depend on a parameter which, when increased, will cause the function to become more and more sharply peaked at $\lambda = 0$. More precisely, for increasing parameter values, the functions $K(\lambda)$ will have the delta function property defined in Section 8.2. Since most of the important smoothing functions are (asymptotically) related to spectral windows of weighted covariance estimators by the expression

$$K(\lambda) \cong (2\pi/N)W_M(\lambda),$$

it is convenient to return to this form of the smoothing function. This will allow us to use the lag parameter M to govern the delta function behavior of the smoothing function and, at the same time, obtain the properties of the weighted covariance estimators.

With this substitution in (8.28)–(8.30) we can view the various sums as Riemann approximations to the corresponding integrals to obtain the asymptotic expressions

$$E\hat{f}(\lambda_k) \cong \frac{2\pi}{N} \sum_{v=-[(N-1)/2]}^{[N/2]} W_M(\lambda_k - \lambda_v) f(\lambda_v)$$

$$\cong \int_{-\pi}^{\pi} W_M(\lambda_k - \lambda) f(\lambda)\, d\lambda, \tag{8.31}$$

$$\operatorname{Cov}(\hat{f}(\lambda_k), \hat{f}(\lambda_l)) \cong \frac{2\pi}{N} \int_{-\pi}^{\pi} W_M(\lambda_k - \lambda) W_M(\lambda_l - \lambda) f^2(\lambda)\, d\lambda \tag{8.32}$$

if either $\lambda_k \neq 0, \pi$ or $\lambda_l \neq 0, \pi$, and

$$\operatorname{Var}(\hat{f}(\lambda_k)) \cong \frac{2\pi}{N} \int_{-\pi}^{\pi} W_M{}^2(\lambda_k - \lambda) f^2(\lambda)\, d\lambda \tag{8.33}$$

if $\lambda_k \neq 0, \pi$. For the time being, we will deal only with estimates at frequencies other than 0 and π. Then, the terms involving $f^2(0)$ and $f^2(\pi)$ in (8.29) and (8.30) are negligible in comparison to the others for large M and N and have been deleted in (8.32) and (8.33).

Now, if $W_M(\lambda)$ is concentrated in a peak about $\lambda = 0$ and $f(\lambda)$ is essentially constant over every frequency interval of length comparable to the width of the peak, then (8.31) and (8.33) reduce to

$$E\hat{f}(\lambda_k) \cong f(\lambda_k) \tag{8.34}$$

and

$$\operatorname{Var}(\hat{f}(\lambda_k)) \cong f^2(\lambda_k) \frac{2\pi}{N} \int_{-\pi}^{\pi} W_M{}^2(\lambda)\, d\lambda. \tag{8.35}$$

These less precise expressions will be needed to establish the commonly used approximation to the distribution of $\hat{f}(\lambda_k)$ to be given shortly.

One further refinement in this approximation is possible for most lag window-spectral window pairs used in practice. This refinement is based on the fact that the lag window sequence is almost always formed by scaling a real-valued (piecewise) continuous function $w(v)$ as follows. Let $w(v)$ be zero for $|v| > 1$. Then, the lag window is given by

$$w_M(k) = w(k/M).$$

If $w(v)$ is even and $0 \leq w(v) \leq w(0) = 1$ for $|v| < 1$ then this lag window has properties (i)–(iii) given earlier. By Parseval's relation and (8.21),

$$\begin{aligned} 2\pi \int_{-\pi}^{\pi} W_M{}^2(\lambda)\, d\lambda &= \sum_{k=-M}^{M} w^2(k/M) \\ &= M \sum_{k=-M}^{M} w^2(k/M)(1/M) \\ &\cong M \int_{-1}^{1} w^2(v)\, dv. \end{aligned}$$

It follows that (8.35) can be rewritten in the form

$$\operatorname{Var} \hat{f}(\lambda_k) \cong f^2(\lambda_k) c_w M/N, \tag{8.36}$$

where $c_w = \int_{-1}^{1} w^2(v)\, dv = \int_{-\infty}^{\infty} w^2(v)\, dv$. Thus, the variance depends on the specific window only through this integral.

We show in Section A8.1 that the spectral windows corresponding to these lag windows have the delta function property as $M \to \infty$. This establishes the asymptotic unbiasedness of the estimator. Moreover, if the rate at which $M \to \infty$ is controlled in such a way that $M/N \to 0$, then, by (8.36), the variance goes to zero. Thus, the estimator is consistent.

Asymptotic Distribution of the Smoothed Periodogram and Weighted Covariance Estimators

From the results of the last section, if $f(\lambda)$ is effectively constant over the width of the main lobe of $W_M(\lambda)$, then

$$\begin{aligned} \hat{f}(\lambda_k) &= \frac{2\pi}{N} \sum_{v=-[(N-1)/2]}^{[N/2]} W_M(\lambda_k - \lambda_v) I_{N,v} \\ &\cong \frac{f(\lambda_k)\pi}{N} \sum_{v=-[(N-1)/2]}^{[N/2]} W_M(\lambda_k - \lambda_v) U_{N,v}, \end{aligned}$$

where, for the values of v for which $\lambda_k - \lambda_v$ falls in this main lobe, the quantities

$$U_{N,v} = I_{N,v} / \tfrac{1}{2} f(\lambda_k)$$

are (asymptotically) independent chi-square random variables with two degrees of freedom. Thus, the approximate distribution of the spectral estimator for large M and N is that of a linear combination of independent chi-square random variables. Unfortunately, this distribution is quite complicated and calculations are never made with it. Perhaps this is true because a useful "folklore" approximation, sometimes called *Satterthwaite's approximation*, is available which is correct asymptotically and very simple to use. This approximation was introduced to time series analysis by Tukey (1949). Effectively, the window $W_M(\lambda)$ is replaced by a rectangular window as used in the Daniell estimator. In this case the distribution of $\hat{f}(\lambda_k)$ is that of a constant multiple of a chi-square random variable,

$$\hat{f}(\lambda_k) \approx c\chi_r^2.$$

Then, the constant c and degrees of freedom r are adjusted so that the first two moments of the given estimator and the estimator with rectangular spectral window agree;

$$E\hat{f}(\lambda_k) = E(c\chi_r^2), \qquad \operatorname{Var}(\hat{f}(\lambda_k)) = \operatorname{Var}(c\chi_r^2).$$

However, recall that

$$E(c\chi_r^2) = cr \qquad \text{and} \qquad \operatorname{Var}(c\chi_r^2) = 2c^2 r.$$

Thus, r and c can be solved for to yield

$$c = \operatorname{Var}(c\chi_r^2)/2E(c\chi_r^2) = \operatorname{Var}(\hat{f}(\lambda_k))/2E\hat{f}(\lambda_k)$$

and

$$r = 2[E\hat{f}(\lambda_k)]^2/\operatorname{Var}(\hat{f}(\lambda_k)). \tag{8.37}$$

The parameter (8.37) is known as the *equivalent degrees of freedom* (EDF) of the spectral estimator and is one of the parameters used extensively in the design of spectral analyses as we will see in the next chapter. If we use the asymptotic expressions (8.34) and (8.35), an alternate form for the equivalent degrees of freedom can be derived;

$$r = N\Big/\int_{-\pi}^{\pi} W_M^{\,2}(\lambda)\, d\lambda. \tag{8.38}$$

Then, the asymptotic relation $(2\pi/N)W_M(\lambda) \cong K(\lambda)$ leads to the smoothed periodogram equivalent degrees of freedom

$$r = 2\Big/\sum_{\nu=-[(N-1)/2]}^{[N/2]} K^2(\lambda_\nu). \tag{8.39}$$

By means of expression (8.36), weighted covariance estimators with lag windows of the form $w_M(k) = w(k/M)$ are seen to have EDF of the form

$$r = 2N/c_w M. \tag{8.40}$$

Moreover, from the same expression, we obtain

$$c = f(\lambda_k)/r.$$

Dropping the subscript k, if λ is a frequency for which a spectral estimate is calculated, these results can be summarized as follows: *If M and N are reasonably large or if N is large and $K(\lambda)$ well peaked at $\lambda = 0$ and if $f(\lambda)$ is sufficiently smooth near $\lambda' \neq 0$, π, then the distribution of $r\hat{f}(\lambda')/f(\lambda')$ is approximately chi-square with r degrees of freedom—where r is the equivalent degrees of freedom of the estimator given by* (8.39) *if $\hat{f}(\lambda')$ is a smoothed periodogram estimator and by* (8.38) *or* (8.40) *if a weighted covariance estimator.*

Confidence Intervals for the Spectral Density and Log Spectral Density

The above distributional results can be used to calculate confidence intervals for $f(\lambda)$ and $\log f(\lambda)$ at frequencies $\lambda' \neq 0$, π for which estimators are obtained. See Tucker (1962) for a discussion of the pertinent theory. A $100(1-\alpha)\%$ confidence interval for $f(\lambda')$ is computed as follows: First, the equivalent degrees of freedom r must be determined from (8.39) or (8.40). From Table A9.1 of the chi-square distribution given in the appendix to Chapter 9, two numbers a and b can be determined which satisfy the equations

$$P(\chi_r^2 \leq a) = \alpha/2, \qquad P(\chi_r^2 \leq b) = 1 - (\alpha/2).$$

It follows that

$$P(a \leq \chi_r^2 \leq b) = 1 - \alpha.$$

Thus, by the above result,

$$\begin{aligned} 1 - \alpha &\cong P\left(a \leq \frac{r\hat{f}(\lambda')}{f(\lambda')} \leq b\right) \\ &= P\left(\frac{r\hat{f}(\lambda')}{b} \leq f(\lambda') \leq \frac{r\hat{f}(\lambda')}{a}\right). \end{aligned}$$

It follows that

$$\frac{r\hat{f}(\lambda')}{b} \leq f(\lambda') \leq \frac{r\hat{f}(\lambda')}{a} \tag{8.41}$$

is a $100(1-\alpha)\%$ confidence interval for $f(\lambda')$. This is the equal tail probability confidence interval. For the description of the so-called best unbiased confidence intervals with a reference to the pertinent tables, see Hannan

(1970, p. 252). When the equivalent degrees of freedom is not too small, the two intervals are nearly the same.

The length of the above confidence interval varies with the magnitude of $\hat{f}(\lambda')$. Aside from the compression in scale realized by taking the logarithm, the most convincing reason for considering the log spectral density in practice is that a confidence interval of constant width (over frequency) can be realized for this parameter. This interval is obtained by taking the logarithm of all three terms of (8.41),

$$\log(r/b) + \log \hat{f}(\lambda') \le \log f(\lambda') \le \log(r/a) + \log \hat{f}(\lambda'). \tag{8.42}$$

This is a $100(1-\alpha)\%$ confidence interval for $\log f(\lambda')$ of length

$$\log(r/a) - \log(r/b) = \log b/a.$$

Of course, plotting the confidence limits for $f(\lambda)$ on a log scale has the same effect. As an illustration of the way these confidence intervals are displayed graphically, see Fig. 8.1.

Example 8.1 *A Spectral Density Confidence Interval Calculation*

Suppose that a spectrum analysis is performed on a time series with $N = 1000$ data points using a weighted covariance estimator with the lag window

$$w_M(k) = \begin{cases} 1, & \text{for } |k| \le M, \\ 0, & \text{otherwise,} \end{cases}$$

and $M = 100$ lags. [This is the Bartlett (1) window of Table 8.1.] Now suppose that the estimated spectral density at $\lambda = \lambda'$ rad/unit time is computed to be $\hat{f}(\lambda') = 30$. [The units of spectral density are squared amplitude per unit frequency. Thus, for example, if the time series were measured in centimeters and time in seconds, the units of $f(\lambda')$ and $\hat{f}(\lambda')$ would be centimeters squared per radian per second.] We will find a 95% confidence interval for $f(\lambda')$.

Either by direct calculation or from Table 8.1 we find the equivalent degrees of freedom for the Bartlett (1) window to be

$$r = N/M = 1000/100 = 10.$$

Since $\alpha = 0.05$, we must find the values of a and b from Table A9.1 which satisfy the equations

$$P(\chi^2_{10} \le a) = 0.05/2 = 0.025 \qquad \text{and} \qquad P(\chi^2_{10} \le b) = 1 - (0.05/2) = 0.975.$$

These values are seen to be

$$a = 3.247, \qquad b = 20.483.$$

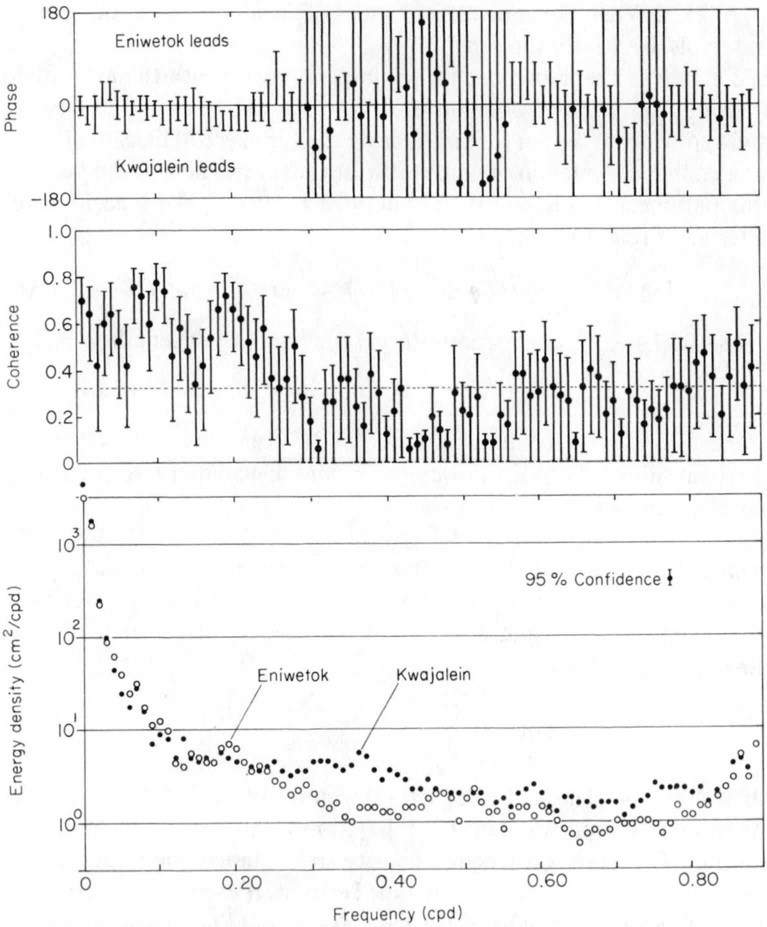

Fig. 8.1 *Graphs of estimated spectral densities, coherence, and phase from the Groves and Hannan study* (1968) (*Section* 5.6) *with indicated* 95% *confidence limits*. Source: Groves and Hannan (1968); copyright by American Geophysical Union.

Thus, the lower and upper limits of the 95% confidence interval for $f(\lambda')$ are

$$r\hat{f}(\lambda')/b = (10)(30)/20.483 = 14.65,$$
$$r\hat{f}(\lambda')/a = (10)(30)/3.247 = 92.3.$$

We can now assert that with 95% confidence, the true value of the spectral density at $\lambda = \lambda'$ lies in the interval

$$14.65 \leq f(\lambda') \leq 92.3.$$

Suppose we wish to test the hypothesis that $f(\lambda') = 50$ against the alternative $f(\lambda') \neq 50$ at significance level, $\alpha = 0.05$. Since 50 is in the computed confidence interval we would accept this hypothesis [see Tucker (1962, p. 211)]. The extreme length of the confidence interval indicates that the power of this test is rather poor. In Chapter 9 we will show how to design time series experiments to meet prespecified power and confidence interval length criteria.

Bandwidth of Spectral Estimators

One final parameter of importance for comparing spectral estimators is bandwidth. If the smoothing function is that of the Daniell estimator;

$$K(\lambda_v) = \begin{cases} 1/n, & -[(n-1)/2] \leq v \leq [n/2], \\ 0, & \text{otherwise}, \end{cases} \tag{8.43}$$

where $\lambda_v = 2\pi v/N$, then the width of the base of this rectangular window in frequency units,

$$\beta = 2\pi n/N, \tag{8.44}$$

is the natural definition of bandwidth. One possible definition of bandwidth for the general spectral estimators (8.20) and (8.23) again uses the device of replacing the spectral window of the general estimator by a rectangular one for which the first two moments of this estimator and the resulting Daniell estimator agree. The bandwidth will then be the length of the base of the fitted rectangle. To determine this length, note that the fitting criterion has the effect of equating the degrees of freedom of the Daniell estimator and the equivalent degrees of freedom of the general estimator;

$$r = 2n.$$

However, by relationship (8.44), it follows that the *equivalent bandwidth* (EBW) of the general estimator will be

$$\beta = \pi r/N. \tag{8.45}$$

This leads to the expression

$$\beta = 2\pi \Big/ N \sum_{v=-[(N-1)/2]}^{[N/2]} K^2(\lambda_v) \tag{8.46}$$

for the equivalent bandwidth of the general smoothed periodogram estimator because of (8.39) and to

$$\beta = 2\pi/c_w M \tag{8.47}$$

for the weighted covariance estimators for which $w_M(k) = w(k/M)$ by (8.40).

An interpretation of bandwidth as a descriptive parameter for spectral estimators will be given in Chapter 9. For the present, we will simply list the equivalent bandwidths of the more commonly used spectral estimators in Table 8.1.

Some Windows Used in Practice

The choice of the particular lag window to use in a weighted covariance estimator or the smoothing function for a smoothed periodogram estimator has received a great deal of attention in the time series literature. We will consider this question briefly in Chapter 9 when point estimates of time series parameters are discussed. For the present, we list in Table 8.1 several of the windows which have been used in spectral estimation studies along with their equivalent degrees of freedom and equivalent bandwidths.

Example 8.2 *Calculation of Approximate EDF and EBW of the Window Proposed by Jones*

Jones (1971) suggested the following procedure for generating lag and spectral windows which are very nearly shaped like normal density functions. The lag window weights are proportional to binomial coefficients and are generated recursively according to the following scheme:

$$w_0 = 1,$$

$$w_k = \frac{L+1-k}{L+k} w_{k-1}, \qquad k = 1, 2, \ldots .$$

The integer L is determined as follows: When L is large, w_k becomes very small with increasing values of k and computer underflow may be encountered. To avoid this, the iteration is stopped when w_k becomes smaller than some prespecified value ε. The index k at which the truncation takes place is equated with the lag parameter M. Using the normal approximation for the binomial coefficients [see, e.g., Feller (1968, p. 183)], it can be shown that

$$w_k \cong \exp(-k^2/L). \tag{8.48}$$

This leads to the approximate value for L as a function of ε and M;

$$L \cong [M^2/-\ln \varepsilon].$$

Substituting the (noninteger) value $L = M^2/(-\ln \varepsilon)$ into (8.48), we obtain the approximate expression for the lag window;

$$w(v) = \begin{cases} \exp(-(-\ln \varepsilon)v^2), & |v| \le 1, \\ 0, & \text{otherwise,} \end{cases}$$

Table 8.1

A Periodogram Smoothing Function and Several Covariance Weighting Functions Which Have Found Use in Practice along with Pertinent Parameters

Proposer	Window description	c_w	EDF	EBW
Daniell	$K(\lambda_\nu) = 1/n,\ -[(n-1)/2] \le \nu \le [n/2]$	—	$2n$	$2\pi n/N$
Bartlett(1)	$w(v) = I_{[-1,1]}(v)$[a]	2.00	N/M	π/M
Bartlett(2)	$w(v) = 1 - \lvert v \rvert$	0.67	$3N/M$	$3\pi/M$
Tukey	$w(v) = 1 - 2a + 2a \cos \pi v$	$2(1 - 4a + 6a^2)$	$(N/M)(1 - 4a + 6a^2)$	$(\pi/M)(1 - 4a + 6a^2)$
(i) $a = 0.23$(hamming)	$w(v) = 0.54 + 0.46 \cos \pi v$	0.80	$2.50N/M$	$2.50\pi/M$
(ii) $a = 0.25$(hanning)	$w(v) = \frac{1}{2}(1 + \cos \pi v)$	0.75	$2.67N/M$	$2.67\pi/M$
Parzen(1)	$w(v) = 1 - v^2$	1.07	$1.87N/M$	$1.87\pi/M$
Parzen(2)	$w(v) = \begin{cases} 1 - 6v^2 + 6\lvert v \rvert^3, & \lvert v \rvert \le \frac{1}{2}, \\ 2(1 - \lvert v \rvert)^3, & \frac{1}{2} \le \lvert v \rvert \le 1 \end{cases}$	0.54	$3.7N/M$	$3.7\pi/M$
Jones	$w(v) \simeq \exp(\ln \varepsilon) v^2,\ \varepsilon = 0.001$[b]	0.48	$4.18N/M$	$4.18\pi/M$

[a] $I_A(v)$ denotes the set characteristic function of A.

[b] This window is discussed in Example 8.2.

and

$$w_k \cong w(k/M).$$

It follows that the equivalent degrees of freedom and equivalent bandwidth can be determined from (8.40) and (8.47) after c_w has been computed for this window. However,

$$c_w = \int_{-\infty}^{\infty} w^2(v)\,dv \cong \int_{-\infty}^{\infty} \exp(-2(-\ln \varepsilon)v^2)\,dv.$$

Thus, evaluating this normal integral, we obtain

$$c_w \cong \left(\frac{\pi}{2(-\ln \varepsilon)}\right)^{1/2}.$$

When $\varepsilon = 0.001$, for example, $-\ln \varepsilon \cong 6.9$ and we obtain

$$c_w \cong 0.478.$$

This value is rounded to two places in Table 8.1.

8.4 DISTRIBUTION THEORY FOR MULTIVARIATE SPECTRAL ESTIMATORS WITH APPLICATIONS TO STATISTICAL INFERENCE

Suppose that

$$\mathbf{X}(t) = \begin{pmatrix} X_1(t) \\ \vdots \\ X_p(t) \end{pmatrix}, \qquad t = 0, \pm 1, \ldots,$$

is a multivariate, zero-mean, Gaussian process with continuous spectrum and $(p \times p)$ spectral density matrix

$$\mathbf{f}(\lambda) = [f_{l,m}(\lambda)].$$

The multidimensional weighted covariance estimate of $\mathbf{f}(\lambda)$ is based on the following covariance estimators:

$$\hat{C}_{l,m}(k) = \begin{cases} \dfrac{1}{N}\displaystyle\sum_{t=1}^{N-k} X_l(t+k)X_m(t), & k = 0, 1, \ldots, N-1, \\ \hat{C}_{m,l}(-k), & k = -1, -2, \ldots, -N+1, \\ 0, & |k| \geq N. \end{cases} \tag{8.49}$$

Let

$$\hat{\mathbf{C}}(k) = [\hat{C}_{l,m}(k)].$$

Then, the *weighted covariance estimators* can be represented in matrix form by the expression

$$\hat{\mathbf{f}}(\lambda) = \frac{1}{2\pi} \sum_{k=-\infty}^{\infty} e^{-i\lambda k} w_M(k) \hat{\mathbf{C}}(k), \tag{8.50}$$

with the usual understanding that the indicated operations are performed coordinatewise. The lag window and spectral window are the same as in the univariate case. With the same convention, the vector *finite Fourier transform* is

$$\mathbf{z}(\lambda) = \frac{1}{(2\pi N)^{1/2}} \sum_{t=1}^{N} \mathbf{X}(t) e^{-i\lambda t} \tag{8.51}$$

and the *periodogram matrix* is

$$\mathbf{I}_N(\lambda) = \mathbf{z}(\lambda)\mathbf{z}(\lambda)^*. \tag{8.52}$$

Then, the matrix analog of (8.22) is

$$\hat{\mathbf{f}}(\lambda) = \int_{-\pi}^{\pi} W_M(\lambda - \mu) \mathbf{I}_N(\mu)\, d\mu. \tag{8.53}$$

The *multivariate smoothed periodogram estimator is*

$$\hat{\mathbf{f}}(\lambda) = \sum_{\nu=-[(N-1)/2]}^{[N/2]} K(\lambda - \lambda_\nu) \mathbf{I}_N(\lambda_\nu), \tag{8.54}$$

where $\lambda_\nu = 2\pi\nu/N$. The argument outlined for the univariate case can be applied here as well to show that the distributions of both estimators (8.50) and (8.54) are asymptotically equivalent to the distribution of the Daniell estimator

$$\hat{\mathbf{f}}(\lambda_k) = \frac{1}{n} \sum_{\nu=-[(n-1)/2]}^{[n/2]} \mathbf{I}_N(\lambda_k - \lambda_\nu), \tag{8.55}$$

where $n = r/2$ and r is the equivalent degrees of freedom given by (8.38)–(8.40). Now, if the elements of $\hat{\mathbf{f}}(\lambda)$ are effectively constant over the equivalent bandwidth of the spectral window, it follows from (8.55) that $n\hat{\mathbf{f}}(\lambda_k)$ is asymptotically the sum of n independent, identically distributed random matrices of the form

$$\mathbf{I}_N(\lambda_\nu) = \mathbf{z}(\lambda_\nu)\mathbf{z}(\lambda_\nu)^*,$$

where $\mathbf{z}(\lambda_\nu)$ has the multivariate complex normal distribution with mean $\mathbf{0}$ and covariance matrix $\mathbf{f}(\lambda_k)$. (See Section 1.4.) When $\lambda_k \neq 0, \pi$, this sum has the complex p-dimensional Wishart distribution, denoted $W_p{}^C(n, \mathbf{f}(\lambda_k))$, which was introduced into time series analysis by Goodman (1957). The reader is referred to Brillinger (1970) and Hannan (1970, p. 295) for properties of this distribution. Now, *the asymptotic theory for estimators* (8.50) *and* (8.54) *is based on the assumption that* $n\hat{\mathbf{f}}(\lambda_k)$ *has exactly the* $W_p^C(n, \mathbf{f}(\lambda_k))$ *distribution for* $\lambda_k \neq 0, \pi$, where $n = \frac{1}{2}$(EDF). The distribution theory for estimators at $\lambda_k = 0, \pi$ is somewhat more delicate. In particular, the distribution for zero frequency depends on whether or not a dc correction is made (Hannan, 1970, p. 251). However, as before, we will be primarily concerned with the properties of estimators at frequencies $\lambda \neq 0, \pi$ and results for 0 and π will only occasionally be given.

We will be interested in the distributions of the estimators of the various spectral parameters introduced in Chapter 5. Each parameter was seen to be a function of the components of the spectral density matrix $\mathbf{f}(\lambda)$. Symbolically, if $\mathbf{l}(\lambda)$ is a vector of univariate spectral parameters, then

$$\mathbf{l}(\lambda) = \mathbf{h}(\mathbf{f}(\lambda)), \tag{8.56}$$

where $\mathbf{h}(\mathbf{x})$ is a "well-behaved" function of a matrix variable. The standard procedure for estimating $\mathbf{l}(\lambda)$ is to replace $\mathbf{f}(\lambda)$ by one of the estimators (8.50) or (8.54)—or, in certain cases to be discussed later, by an estimator with the same asymptotic distribution. Thus, the statistical properties of

$$\hat{\mathbf{l}}(\lambda) = \mathbf{h}(\hat{\mathbf{f}}(\lambda))$$

will depend on the statistical properties of $\hat{\mathbf{f}}(\lambda)$ and on the function $\mathbf{h}(\mathbf{x})$. In particular, the asymptotic distribution of $\hat{\mathbf{l}}(\lambda)$ can be obtained by standard (but complicated) transformation of variables procedures applied to the asymptotic distribution of $\hat{\mathbf{f}}(\lambda)$ given above.

Goodman (1957) accomplished the arduous task of obtaining the asymptotic marginal and joint distributions of the estimators of the bivariate parameters introduced in Chapter 5 as well as several others not so widely used in practice. The asymptotic distributions of estimates of partial and multiple coherence were computed by Goodman (1963) and by Khatri (1964) and the distributions of the regression parameter estimates are given by Hannan (1970) and Brillinger (1970). The principal use made of the distribution theory is to construct confidence intervals for the various spectral parameters. In the remainder of this section we will present the standard confidence intervals for these parameters and illustrate their uses. Some of the more interesting hypothesis tests will also be given. For details of the theory the reader is referred to the excellent summary by Hannan (1970).

Confidence Intervals for Coherence

The coefficient of coherence for two components $X_j(t)$, $X_k(t)$ of a multivariate time series was defined in expression (5.49). The corresponding estimator, which we will call the *sample coherence*, is obtained as described above;

$$\hat{\rho}_{j,k}(\lambda) = |\hat{f}_{j,k}(\lambda)|/(\hat{f}_{j,j}(\lambda)\hat{f}_{k,k}(\lambda))^{1/2}. \tag{8.57}$$

The distribution of (8.57), based on the complex Wishart approximation for the distribution of $\hat{\mathbf{f}}(\lambda)$, has been extensively tabulated by Amos and Koopmans (1963) and by Alexander and Vok (1963). Moreover, Enochson and Goodman (1965) have shown that for $n > 20$ and $0.4 \leq \rho^2 \leq 0.95$, the random variable

$$\varphi = \tanh^{-1}(\hat{\rho})$$

is approximately normally distributed with mean and variance

$$E(\varphi) = \tanh^{-1}(\rho) + \big(1/2(n-1)\big),$$
$$\operatorname{Var}(\varphi) = 1/2(n-1).$$

In these expressions, $\rho = \rho_{j,k}(\lambda)$, $\hat{\rho} = \hat{\rho}_{j,k}(\lambda)$, $\tanh^{-1}$ denotes the inverse hyperbolic tangent and $2n$ is the equivalent degrees of freedom of the estimator $\hat{\mathbf{f}}(\lambda)$. This permits a straightforward calculation of a $100(1-\alpha)\%$ confidence interval for ρ: If $u_{\alpha/2}$ is the upper $\alpha/2$ cutoff point for the standard normal $\big(N(0, 1)\big)$ distribution, then

$$P\big(-u_{\alpha/2} \leq (\varphi - E(\varphi))/(\operatorname{Var}(\varphi))^{1/2} \leq u_{\alpha/2}\big) = 1 - \alpha.$$

Thus, after some algebra, we obtain the $100(1-\alpha)\%$ confidence interval $\underline{\rho} \leq \rho \leq \bar{\rho}$, where

$$\begin{aligned}\underline{\rho} &= \tanh\{\tanh^{-1}(\hat{\rho}) - (u_{\alpha/2})(2(n-1))^{-1/2} - (2(n-1))^{-1}\},\\ \bar{\rho} &= \tanh\{\tanh^{-1}(\hat{\rho}) + (u_{\alpha/2})(2(n-1))^{-1/2} - (2(n-1))^{-1}\}.\end{aligned} \tag{8.58}$$

These limits can be computed using standard tables of the normal distribution and the hyperbolic and inverse hyperbolic functions [see, e.g., Abramowitz and Stegun (1964)].

For selected values of $n = \frac{1}{2}$EDF, convenient graphs of confidence limits for ρ are given by Amos and Koopmans (1963). These graphs are reproduced as Figs. A9.1 and A9.2. To obtain an 80 or 90% confidence interval, draw a horizontal line on the appropriate graph from the observed value of $\hat{\rho}$ until it crosses the curves labeled with the given value of n. The values of ρ on the abscissa corresponding to these crossing points are the lower and upper limits of the confidence interval. If the upper curve is not intersected by the horizontal line, the lower limit is 0.

Additional curves for other values of n or for a different confidence coefficient can easily be constructed from the Amos and Koopmans (1963) tables, or the confidence intervals can be calculated from the approximation (8.58). As an illustration of the way these intervals can be displayed graphically, 95% confidence intervals for coherence of the Groves and Hannan sea level data are plotted in Fig. 8.1.

Cutoff points for tests of the hypothesis $\rho = 0$ vs $\rho > 0$ can be obtained from Table A9.4 of the F distribution because of the fact that for $\lambda \neq 0, \pi$, if $\rho = 0$, then

$$(n-1)\hat{\rho}^2/(1-\hat{\rho}^2) = F_{2,2(n-1)}$$

has the F distribution with 2 and $2(n-1)$ degrees of freedom. Power curves for $\alpha = 0.10$ and 0.05 and selected values of n are given in Figs. A9.3 and A9.4. The cutoff values for these tests are given in Table A9.6. Power and cutoff values for unlisted degrees of freedom can be obtained by interpolating the graphical and tabulated values linearly in the reciprocals of the degrees of freedom.

Example 8.3 *Computation of a Confidence Interval for Coherence*

Suppose that the data of Example 8.1 is now assumed to be from a bivariate time series and we estimate the coherence at $\lambda = \lambda'$ to be

$$\hat{\rho}_{j,k}(\lambda') = 0.80,$$

using the same weighted covariance estimator as before. Since this estimator has 10 equivalent degrees of freedom we obtain $n = 5$. Then, a 90% confidence interval for $\rho_{j,k}(\lambda')$ can be obtained from Fig. A9.2. A horizontal line at $\hat{\rho} = 0.80$ intersects the two curves labeled $n = 5$ at 0.34 and 0.91. Thus, with 90% confidence,

$$0.34 \leq \rho_{j,k}(\lambda') \leq 0.91.$$

Confidence Intervals for the Gain of $\tilde{L}$

We defined the gain function of the filter which transforms $X_j(t)$ into the best mean-square approximation to $X_k(t)$ in expression (5.56). The natural estimator for this parameter is

$$\hat{\beta}_{j,k}(\lambda) = |\hat{f}_{j,k}(\lambda)|/\hat{f}_{j,j}(\lambda). \tag{8.59}$$

Goodman (1957) obtained a $100(1-\alpha)\%$ (two-dimensional) confidence interval for the transfer function $\tilde{B}_{j,k}(\lambda)$ of this filter. Because of the inequality, $||x| - |y|| \leq |x - y|$, a $100(1-\alpha)\%$ confidence interval for the gain $\beta_{j,k}(\lambda) = |\tilde{B}_{j,k}(\lambda)|$ is the set of values of the parameter satisfying the inequality

$$|\hat{\beta}_{j,k}(\lambda) - \beta_{j,k}(\lambda)| \leq \begin{cases} \left[\dfrac{\hat{f}_{k,k}(\lambda)(1-\hat{\rho}_{j,k}(\lambda)^2)}{(n-1)\hat{f}_{j,j}(\lambda)} F_{2,2_n-2}(\alpha)\right]^{1/2}, & \lambda \neq 0, \pi, \\ \left[\dfrac{\hat{f}_{k,k}(\lambda)(1-\hat{\rho}_{j,k}(\lambda)^2)}{v\hat{f}_{j,j}(\lambda)}\right]^{1/2} t_v(\alpha), & \lambda = 0, \pi, \end{cases} \tag{8.60}$$

where $v = n - 2$ if $\lambda = 0$ and $n - 1$ if $\lambda = \pi$.

The quantities $F_{2,\,2n-2}(\alpha)$ and $t_\nu(\alpha)$ are the upper α cutoff points of the F and t distributions with the indicated degrees of freedom;

$$P(F_{2,\,2n-2} \le F_{2,\,2n-2}(\alpha)) = 1 - \alpha,$$
$$P(t_\nu \le t_\nu(\alpha)) = 1 - \alpha.$$

These cutoff values can be obtained from Tables A9.3 and A9.4. Cutoff values for unlisted degrees of freedom can be obtained from both tables by interpolating linearly in the reciprocals of the degree of freedom.

This confidence interval is especially useful for establishing bands on the gain function of an in-service linear filter with output contaminated by noise as described in Example 5.4. When no noise is present, $\rho_{j,k}(\lambda) = 1$ and thus $\hat{\rho}_{j,k}(\lambda) = 1$ with probability 1. In this case the measured (sample) gain is identical to the true gain which is reflected in the fact that the right-hand side of (8.60) is zero.

Confidence Intervals for the Phase Angle

The estimator for phase is

$$\hat{\vartheta}_{j,\,k}(\lambda) = -\operatorname{Arctan} \hat{q}_{j,\,k}(\lambda)/\hat{c}_{j,\,k}(\lambda), \tag{8.61}$$

where, $\hat{c}_{j,\,k}(\lambda) = \operatorname{Re} \hat{f}_{j,\,k}(\lambda)$ and $\hat{q}_{j,\,k}(\lambda) = -\operatorname{Im} \hat{f}_{j,\,k}(\lambda)$ are the sample co- and quad-spectral densities [see expression (5.52)]. A $100(1-\alpha)\%$ confidence interval for $\vartheta_{j,\,k}(\lambda)$ is the set of all values of the parameter satisfying the inequality

$$|\sin(\hat{\vartheta}_{j,k}(\lambda) - \vartheta_{j,k}(\lambda))| \le \left\{\frac{1 - \hat{\rho}_{j,k}(\lambda)^2}{\hat{\rho}_{j,k}(\lambda)^2(2n-2)}\right\}^{1/2} t_{2n-2}\left(\frac{\alpha}{2}\right), \qquad \lambda \ne 0, \pi, \tag{8.62}$$

where $t_{2n-2}(\alpha/2)$ is the upper $\alpha/2$ cutoff point of the t distribution with $2n-2$ degrees of freedom [Hannan (1970, p. 257)].

Think of the frequency interval $-\pi < \vartheta \le \pi$ as being formed into a circle by joining the endpoints $-\pi$ and π as in Fig. 8.2. From tables of the arcsine function [e.g., Abramowitz and Stegun (1964)] one can solve for the angle ϑ^* for which $\sin \vartheta^*$ equals the right-hand side of (8.62). Then, the confidence interval is

$$\hat{\vartheta}_{j,\,k}(\lambda) - \vartheta^* \le \vartheta_{j,\,k}(\lambda) \le \hat{\vartheta}_{j,\,k}(\lambda) + \vartheta^*.$$

This interval may overlap the point $\pm\pi$ in which case an "interval" consisting of two segments as pictured in Fig. 8.2 would be obtained when the circle is straightened into the usual linear scale for $\vartheta_{j,\,k}(\lambda)$.

Inequality (8.62) also determines a second interval of length $2\vartheta^*$ centered at $\hat{\vartheta}_{j,\,k}(\lambda) + \pi$, because of the trigonometric relation

$$\sin(\vartheta + \pi) = -\sin \vartheta.$$

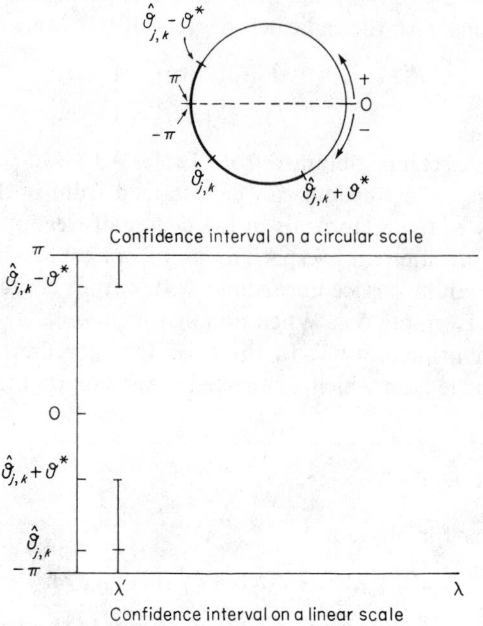

Fig. 8.2 *Illustration of a confidence interval for* $\vartheta_{j,k}(\lambda')$ *on a circular scale and on the corresponding linear scale.*

This interval is due to a possible (statistical) error in the sign of $\hat{c}_{j,k}(\lambda)$ or $\hat{q}_{j,k}(\lambda)$ which would cause the displacement of the estimate of $\vartheta_{j,k}(\lambda)$ by the amount π. When the coefficient of coherence is large, thus the right-hand side of (8.62) is small, an error in sign is improbable and the interval described above will have very nearly the quoted confidence coefficient. When $\rho_{j,k}(\lambda)$ is small, this confidence coefficient is less credible for the given interval. However, it is common practice to use only the interval centered at $\hat{\vartheta}_{j,k}(\lambda)$ until the right-hand side of (8.62) reaches or exceeds unity. At this point, the two intervals join and the "no information" interval $-\pi < \hat{\vartheta}_{j,k}(\lambda) \leq \pi$ results. This phenomenon is illustrated by the confidence intervals for phase graphed in Fig. 8.1. Note how the lengths of the confidence intervals for phase vary with the magnitude of the coherence.

Example 8.4 *Calculation of a Confidence Interval for the Phase Angle*

Suppose that the data of Examples 8.1 and 8.3 yield an estimated phase angle of

$$\hat{\vartheta}_{j,k}(\lambda') = 3.15 \quad \text{rad},$$

and that a 95% confidence interval for $\vartheta_{j,k}(\lambda')$ is desired. Recall that $n = 5$ and $\hat{\rho}_{j,k}(\lambda') = 0.80$. Entering Table A9.3 of Student's t-distribution with $2n - 2 = 8$ degrees of freedom, we obtain (since $\alpha = 0.05$),

$$t_8(0.025) = 2.306.$$

This allows us to compute the right-hand side of (8.62);

$$\{(1 - (0.8)^2)/(0.8)^2(8)\}^{1/2}(2.306) = 0.612.$$

From the arcsine tables of Abramowitz and Stegun (1964) we find

$$\vartheta^* = \arcsin(0.612) = 0.66.$$

Thus, to two decimal places, the 95% confidence interval is

$$3.15 - 0.66 \leq \vartheta_{j,k}(\lambda') \leq 3.15 + 0.66$$

or

$$2.49 \leq \vartheta_{j,k}(\lambda') \leq 3.81 \quad \text{rad.}$$

Now, the angles in this interval satisfying the inequality $\pi < \vartheta \leq 3.81$ can be represented in the interval $(-\pi, \pi]$ by replacing each ϑ by $\vartheta - 2\pi$. Thus, using the approximation $\pi \cong 3.14$, we obtain the final 95% confidence interval,

$$\begin{aligned} -3.14 \leq \vartheta_{j,k}(\lambda') &\leq -2.47 \quad \text{rad,} \\ 2.49 \leq \vartheta_{j,k}(\lambda') &\leq 3.14 \quad \text{rad.} \end{aligned}$$

Joint Confidence Intervals for Gain and Phase

The Goodman (1957) confidence interval for the transfer function of $\tilde{L}$ can be converted into a joint $100(1 - \alpha)$% confidence interval of the following form for the gain and phase functions:

$$\begin{aligned} \hat{\beta}_{j,k}(\lambda) - \hat{s}(\lambda) &\leq \beta_{j,k}(\lambda) \leq \hat{\beta}_{j,k}(\lambda) + \hat{s}(\lambda), \\ \hat{\vartheta}_{j,k}(\lambda) - \Delta\hat{\vartheta}(\lambda) &\leq \vartheta_{j,k}(\lambda) \leq \hat{\vartheta}_{j,k}(\lambda) + \Delta\hat{\vartheta}(\lambda), \end{aligned}$$

$\lambda \neq 0, \pi$, where

$$\hat{s}^2(\lambda) = \frac{1}{n-1} F_{2,2n-2}(\alpha)[1 - \hat{\rho}^2_{j,k}(\lambda)] \frac{\hat{f}_{j,j}(\lambda)}{\hat{f}_{k,k}(\lambda)}$$

and

$$\Delta\hat{\vartheta}(\lambda) = \arcsin[\hat{s}(\lambda)/\hat{\beta}_{j,k}(\lambda)].$$

Note that the confidence interval for the gain is the same as before.

These confidence intervals are useful for giving joint bounds on the gain and phase of a linear system for which the observed output is contaminated by noise as in Example 5.4. It is easily seen that the lengths of both confidence intervals depend on the signal-to-noise ratio defined therein. The computation of these intervals can be accomplished by means of the computer program BMDX68 (Dixon, 1969).

Confidence Interval for Multiple Coherence

The multiple coherence $R^2_{j\cdot\mathbf{m}}(\lambda)$ was shown in Section 5.6 to be an important parameter both in multivariate spectral correlation analysis and regression time series analysis. Recall that $\mathbf{m} = (m_1, \ldots, m_q)$ represents the indices of the components of the random vector $\mathbf{X}(t) = (X_1(t), \ldots, X_p(t))$ upon which $X_j(t)$ is being regressed. The number of elements q in the index set will be one of the parameters of the distribution of $\hat{R}^2_{j\cdot\mathbf{m}}(\lambda)$. Let $\hat{\mathbf{f}}^X(\lambda)$ represent one of the estimators of the spectral density matrix $\mathbf{f}^X(\lambda)$ given earlier in this section. Then, the estimator for multiple coherence is

$$\hat{R}^2_{j\cdot\mathbf{m}}(\lambda) = \hat{\mathbf{f}}^X_{j,\mathbf{m}}(\lambda)\hat{\mathbf{f}}_{\mathbf{m}}{}^X(\lambda)^{-1}\hat{\mathbf{f}}^X_{j,\mathbf{m}}(\lambda)/\hat{f}^X_{j,j}(\lambda),$$

where $\hat{\mathbf{f}}^X_{j,\mathbf{m}}(\lambda)$ is the $1 \times q$ vector with elements $\hat{f}^X_{j,m_r}(\lambda)$, $r = 1, 2, \ldots, q$, and $\hat{\mathbf{f}}_{\mathbf{m}}{}^X(\lambda)$ is the $q \times q$ matrix of elements $\hat{f}^X_{m_r,m_s}(\lambda)$; $r, s = 1, 2, \ldots, q$. The distribution of this estimator depends on $n = \frac{1}{2}(\text{EDF})$, q, and the "true" multiple coherence $R^2 = R^2_{j\cdot\mathbf{m}}(\lambda)$. The distribution for $\lambda = 0$ is the same as for $0 < \lambda < \pi$ except that q is replaced by $\frac{1}{2}q$ and n by $\frac{1}{2}(n-1)$. (This assumes the use of the standard mean correction wherein the sample mean is subtracted from each series.) For $\lambda = \pi$, the same result holds with q replaced by $\frac{1}{2}q$ and n by $\frac{1}{2}n$. Tables of the cumulative distribution function are given by Alexander and Vok (1963). For our purposes, a convenient table of the upper and lower end points of 95% and 99% confidence intervals for R^2 is supplied by Groves and Hannan (1968). This table is reproduced as Table A9.5 in the appendix to Chapter 9.

Example 8.5 *Calculation of a Confidence Interval for Multiple Coherence*

Suppose that a 95% confidence interval for $R^2_{1\cdot\mathbf{m}}(\lambda')$ at $\lambda' \neq 0, \pi$ is desired, where $X_1(t)$ is regressed on $q = 4$ other component time series. Suppose also that the estimator at λ' is based on 30 degrees of freedom and has the computed value 0.64. Using the notation $\hat{R} = \hat{R}_{1,\mathbf{m}}(\lambda')$, we can enter the table with $\alpha = 0.05, q = 4, n - q = 15 - 4 = 11 \cong 10$ and $\hat{R} = 0.8$ to obtain the confidence interval

$$0.44 \le R^2_{1\cdot\mathbf{m}}(\lambda') \le 0.86.$$

Rather crude forms of interpolation, such as plotting the confidence limits from the table on graph paper against the parameter of interest, should provide sufficiently accurate confidence limits for other than the given values of q, n, and R. To obtain better accuracy, the more comprehensive tables of Alexander and Vok (1963) will be required.

For the purpose of testing the hypothesis $R^2_{j\cdot\mathbf{m}}(\lambda) = 0$ against the alternate $R^2_{j\cdot\mathbf{m}}(\lambda) > 0$, it is important to note that when the parameter has the value 0, the statistic

$$\frac{n-q}{q}\left(\frac{\hat{R}^2_{j\cdot\mathbf{m}}(\lambda)}{1-\hat{R}^2_{j\cdot\mathbf{m}}(\lambda)}\right) = F_{2q,\,2n-2q}$$

has the F distribution with $2q$ and $2n - 2q$ degrees of freedom for $\lambda \neq 0, \pi$. This result makes it possible to establish the cutoff value for a test based on statistic $\hat{R}^2_{j\cdot\mathbf{m}}(\lambda)$. For, if $d = F_{2q,\,2n-2q}(\alpha)$ is the upper α cutoff value of the F-distribution, the appropriate decision rule for a level α test is to reject the hypothesis if

$$\hat{R}^2_{j\cdot\mathbf{m}}(\lambda) \geq qd/(n + q(d+1)).$$

The value of d can be obtained from Table A9.4. However, the power of this test must be computed from tables such as those of Alexander and Vok (1963).

Confidence Intervals for Partial Coherence

The estimator of partial coherence can be obtained as described above from expression (5.77). Equivalently, the sample residual spectral matrix can be computed by inserting estimated spectral densities in (5.74), then calculating the ordinary coherence from the appropriate elements to obtain the sample partial coherence. This is the computational procedure used, for example, in the multivariate spectral analysis computer program BMDX68 (Dixon, 1969). It is plausible from this method of computing the sample partial coherence that its probability distribution should be the same as that of the ordinary coherence with a reduction in the degrees of freedom to account for the data used to estimate the q regression coefficients. This is, in fact, the case and for $\lambda \neq 0$ the asymptotic distribution of $\hat{\rho}_{j,k\cdot\mathbf{m}}(\lambda)$ is the same as that of the ordinary coherence with the parameter n replaced by $n - q + 1$. For $\lambda = 0$, n is replaced by $n - q$ [see Hannan (1970, p. 262)]. Thus, confidence intervals for the partial coherence can be obtained from Figs. A9.1 and A9.2.

Moreover, the cutoff value for a test of zero partial coherence can be obtained from the fact that if $\rho_{j,k\cdot\mathbf{m}}(\lambda) = 0$, then,

$$(n-q)\hat{\rho}^2_{j,k\cdot\mathbf{m}}(\lambda)/(1 - \hat{\rho}^2_{j,k\cdot\mathbf{m}}(\lambda))$$

has the F-distribution with 2 and $2(n-q)$ degrees of freedom. The power of this test for selected degrees of freedom can be obtained from Figs. A9.3 and A9.4 for significance levels 0.10 and 0.05. Cutoff values for the degrees of freedom given in the graphs are listed in Table A9.6.

Some Distributional Results for Multivariate Spectral Regression Parameters

The regression model introduced in Section 5.6 is

$$\mathbf{Y}(t) = \mathbf{L}(\mathbf{X}(t)) + \boldsymbol{\eta}(t),$$

where $\mathbf{X}(t)$ is a $p \times 1$ vector stochastic process and $\mathbf{Y}(t)$ and $\boldsymbol{\eta}(t)$ are $q \times 1$ processes, $\boldsymbol{\eta}(t)$ uncorrelated with $\mathbf{X}(t)$. In addition, the processes are now assumed to be zero-mean Gaussian processes with continuous spectra. The important questions to be dealt with on the basis of this model concern the strength and form of the linear dependence of $\mathbf{Y}(t)$ on $\mathbf{X}(t)$. These questions can be suitably answered through hypothesis tests and confidence regions for the $(q \times p)$-dimensional transfer function $\mathbf{B}(\lambda)$ of $\mathbf{L}$ and for the generalized coherence $\boldsymbol{\rho}^2(\lambda)$ defined by expression (5.83). We give only a few details concerning the distribution theory required to carry out these procedures. For complete details both for this model and for the case in which $\mathbf{X}(t)$ is assumed to be nonstochastic, see Brillinger (1970).

Joint $100(1-\alpha)\%$ confidence intervals for the gain $\beta_{j,k}(\lambda)$ and phase $\vartheta_{j,k}(\lambda)$ of each component of $\mathbf{B}(\lambda)$ at $\lambda \neq 0, \pi$ are as follows:

$$\hat{\beta}_{j,k}(\lambda) - \hat{s}_{j,k}(\lambda) \le \beta_{j,k}(\lambda) \le \hat{\beta}_{j,k}(\lambda) + \hat{s}_{j,k}(\lambda),$$

$$\hat{\vartheta}_{j,k}(\lambda) - \hat{\psi}_{j,k}(\lambda) \le \vartheta_{j,k}(\lambda) \le \vartheta_{j,k}(\lambda) + \hat{\psi}_{j,k}(\lambda),$$

where

$$\hat{s}^2_{j,k}(\lambda) = \frac{p}{n-p} \hat{f}^Y_{j,j}(\lambda)(1 - \hat{R}^2_{j\cdot\mathbf{p}}(\lambda))(\hat{\mathbf{f}}^X(\lambda)^{-1})_{k,k} F_{2p,2n-2p}(\alpha)$$

and

$$\hat{\psi}_{j,k}(\lambda) = \arcsin\left(\hat{s}_{j,k}(\lambda)/\hat{\beta}_{j,k}(\lambda)\right).$$

The random variables $\hat{R}^2_{j\cdot\mathbf{p}}(\lambda)$ and $(\hat{\mathbf{f}}^X(\lambda)^{-1})_{k,k}$ are the multiple coherence of $Y_j(t)$ on all components of $\mathbf{X}(t)$ and the element in the kth row and column of $\hat{\mathbf{f}}^X(\lambda)^{-1}$, respectively. Again, $F_{2p,2n-2p}(\alpha)$ is the upper α cutoff point of the F-distribution with $2p$ and $2n-2p$ degrees of freedom, where $2n$ is the equivalent degrees of freedom of the estimators of $\hat{\mathbf{f}}^X(\lambda)$, $\hat{\mathbf{f}}^Y(\lambda)$, and $\hat{\mathbf{f}}^{Y,X}(\lambda)$ [see Bendat and Piersol (1966, p. 234)]. The estimators of $\mathbf{f}^X(\lambda)$ and $\mathbf{f}^Y(\lambda)$ are given by (8.50) or (8.54) with $\mathbf{X}(t)$ and $\mathbf{Y}(t)$ entered in (8.49) or (8.51). The estimator for $\mathbf{f}^{Y,X}(\lambda)$ is obtained by replacing $\hat{\mathbf{C}}(k)$ by the $q \times p$ matrix function

$$\hat{\mathbf{C}}^{Y,X}(k) = \begin{cases} \dfrac{1}{N} \displaystyle\sum_{t=1}^{N-k} \mathbf{Y}(t+k)\mathbf{X}(t)', & k = 0, 1, \ldots, N-1, \\ \dfrac{1}{N} \displaystyle\sum_{t=1}^{N-|k|} \mathbf{Y}(t)\mathbf{X}(t+|k|)', & k = -1, -2, \ldots, -N+1, \\ \mathbf{0}, & |k| \geq N, \end{cases}$$

in the weighted covariance case (8.50) and by replacing $\mathbf{I}_N(\lambda_v)$ by

$$\mathbf{I}_N^{Y,X}(\lambda_v) = \mathbf{z}^Y(\lambda_v)\mathbf{z}^X(\lambda_v)^*$$

in (8.54) where $\mathbf{z}^X(\lambda)$ and $\mathbf{z}^Y(\lambda)$ are the multivariate finite Fourier transforms of $\mathbf{X}(t)$ and $\mathbf{Y}(t)$, respectively, defined by expression (8.51).

In the case of a single output series ($q = 1$), the matrix coherence $\boldsymbol{\rho}^2(\lambda)$ was seen in Section 5.6 to reduce to the multiple coherence of $Y(t)$ on $\mathbf{X}(t)$. Thus, confidence intervals for $\boldsymbol{\rho}^2(\lambda)$ can be obtained from the above results for multiple coherence. Note that the number of time series being regressed upon is p rather than q in this case. Moreover, the hypothesis of no linear regression, $\boldsymbol{\rho}^2(\lambda) = \mathbf{0}$, or equivalently, $\mathbf{B}(\lambda) = \mathbf{0}$, can be tested by means of the test for zero multiple coherence given above.

These results hold equally well if the input is one-dimensional ($p = 1$), for then $\boldsymbol{\rho}^2(\lambda)$ reduces to the multiple coherence of $X(t)$ on $\mathbf{Y}(t)$. Thus, confidence intervals and tests for $\boldsymbol{\rho}^2(\lambda)$ can again be obtained from those for multiple coherence. When both p and q are greater than one it is necessary to obtain the distributions of functions of the eigenvalues of $\boldsymbol{\rho}^2(\lambda)$ in order to obtain comparable information about this parameter. We will not pursue this topic in this book.

APPENDIX TO CHAPTER 8

A8.1 Delta Function Property of the Spectral Window Corresponding to Lag Windows $w_M(k) = w(k/M)$

Let $w(v)$ satisfy the conditions stated in the text which imply properties (i)–(iii) for $w_M(k)$. First note that the Fourier transform

$$W(\lambda) = \frac{1}{2\pi} \int_{-\infty}^{\infty} w(v) e^{-iv\lambda}\, dv$$

is bounded and continuous (Example 1.4). Moreover, it is easily shown to be absolutely integrable with

$$\int_{-\infty}^{\infty} W(\lambda)\, d\lambda = 1.$$

Since $w_M(k)$ is essentially a sampled version of $w(v)$, by an argument similar to the one given in Section 3.2, it can be shown that the Fourier transforms of the two functions are related by the equation

$$W_M(\lambda) = M \sum_{k=-\infty}^{\infty} W(M\lambda + 2\pi Mk), \qquad -\pi < \lambda \leq \pi.$$

Now, if $g(\lambda)$ is any bounded, continuous periodic function, we have

$$\int_{-\pi}^{\pi} W_M(\mu) g(\lambda - \mu)\, d\mu = \sum_{k=-\infty}^{\infty} \int_{-\pi}^{\pi} MW(M(\mu + 2\pi k)) g(\lambda - \mu)\, d\mu$$

$$= \sum_{k=-\infty}^{\infty} \int_{-M\pi}^{M\pi} W(v + 2\pi Mk) g(\lambda - (v/M))\, dv,$$

by the change of variables $v = M\mu$. Since $W(\lambda)$ is absolutely integrable and $g(\lambda)$ is bounded, we can choose a number A sufficiently large such that $\gamma = \max_\lambda |g(\lambda)| \int_{|v|>A} |W(v)|\, dv$ is arbitrarily small. Then if M is chosen to exceed A/π, the absolute difference between

$$\int_{-\pi}^{\pi} W_M(\mu) g(\lambda - \mu)\, d\mu \quad \text{and} \quad \int_{-A}^{A} W(v) g(\lambda - (v/M))\, dv$$

will be no larger than γ. However, since $g(\lambda)$ is continuous, M can be chosen large enough so that $|g(\lambda - (v/M)) - g(\lambda)| < \gamma / \int_{-A}^{A} |W(v)|\, dv$ for $|v| \leq A$. Then, the absolute difference between $\int_{-A}^{A} W(v) g(\lambda - (v/M))\, dv$ and $g(\lambda) \int_{-A}^{A} W(v)\, dv$ will not exceed γ. Finally, this last expression and $g(\lambda)$ do not differ by more than γ in absolute value and we conclude that for M sufficiently large,

$$\left| \int_{-\pi}^{\pi} W_M(\mu) g(\lambda - \mu)\, d\mu - g(\lambda) \right| < 3\gamma.$$

Thus,

$$\lim_{M \to \infty} \int_{-\pi}^{\pi} W_M(\mu) g(\lambda - \mu)\, d\mu = g(\lambda)$$

as was to be shown.

A8.2 Why a Spectral Line Appears as a Peak in the Estimated Spectrum

By extending the definition of spectral density to the discrete component of the spectrum by means of the Dirac delta function of Section 6.4, we can use some of the expressions derived in this chapter to indicate why spectral lines show up as peaks in the estimated spectrum. Recall from expression

(2.6) that the discrete spectral distribution of a discrete-time weakly stationary process can be represented in terms of the spectral function as

$$F_d(A) = \sum_{\lambda_j \in A} p(\lambda_j).$$

Then, if $\delta_p(\lambda)$ is the periodic Dirac delta function introduced in Section 6.4, we can define a "density function" for this spectral distribution by the expression

$$f_d(\lambda) = \sum_{j=-\infty}^{\infty} p(\lambda_j)\, \delta_p(\lambda - \lambda_j).$$

Note that, indeed,

$$F_d(A) = \int_A f_d(\lambda)\, d\lambda.$$

Now, assuming for the moment that the spectrum is pure discrete, we obtain from (8.7) and an easy extension of (8.10) that

$$EI_N(\lambda) = \int_{-\pi}^{\pi} |H_N(\alpha - \lambda)|^2 f_d(\alpha)\, d\alpha.$$

Substituting the above expression for $f_d(\lambda)$ into this integral and (formally) evaluating the resulting sum, this becomes

$$EI_N(\lambda) = \sum_{j=-\infty}^{\infty} p(\lambda_j) |H_N(\lambda - \lambda_j)|^2.$$

For a spectral estimator of the form (8.22), with N reasonably large in comparison to M, the approximation $|H_N(\lambda)|^2 \cong \delta_p(\lambda)$ is quite good. Thus, taking the expectation of this estimator, we obtain

$$E\hat{f}(\lambda) = \int_{-\pi}^{\pi} W_M(\lambda - \mu) EI_N(\mu)\, d\mu$$

$$\cong \sum_{j=-\infty}^{\infty} p(\lambda_j) W_M(\lambda - \lambda_j).$$

A graph of this function consists of a series of peaks centered at the points of discrete spectral power. With high probability, this will also be the appearance of $\hat{f}(\lambda)$. If a continuous spectral component is also present, then these peaks will be superimposed on the estimate of the spectral density. See Fig. 6.12 for an excellent example of this phenomenon.

Note that the shape of the peaks will be roughly the shape of the spectral window. Thus, the peak at λ_j will have height proportional to $p(\lambda_j)/\beta$, where β is the bandwidth of the estimator, and width approximately equal to β. In the ideal situation in which β is small in comparison to the distances between peaks the spectrum will exhibit tall, narrow, isolated peaks as shown in Fig. 6.12.

CHAPTER

9

Sampling Properties of Spectral Estimates, Experimental Design, and Spectral Computations

9.1 INTRODUCTION

In this final chapter we will treat some of the more practical aspects of spectral analysis. First we will introduce exact expressions for the mean, variance, and covariance of spectral estimators in order to discuss the dependence of the bias and variability of these estimators on such parameters as the number of data points and the equivalent bandwidth. This will also allow us to discuss criteria for the selection of spectral windows. A special bias problem sometimes encountered when estimating coherence will also be considered.

In some instances the time series analyst is pleasantly surprised by being consulted about the measurement of a spectrum before the data is gathered. This makes it possible to select the appropriate sample size and, in the case of a continuous-time series, the sampling rate, to produce spectral estimates with preselected attributes. We will consider the guidelines for experimental design which have been most useful in practice.

Next, methods for computing spectral estimators will be discussed. The invention of the digital computer made applications of spectral analysis to real problems possible beginning in the 1950s. Since then, the spread of these techniques to more complex problems has been tied closely to advances in computer technology. A major breakthrough in the calculation of spectra was made possible by the introduction of the fast Fourier transform algorithm into time series analysis by Cooley and Tukey (1965). We will discuss this algorithm and its implications for the computation of spectra in this chapter.

Finally various aspects of data modification and analysis such as removing "wild" values and dealing with missing observations will be considered.

9.2 PROPERTIES OF SPECTRAL ESTIMATORS AND THE SELECTION OF SPECTRAL WINDOWS

The distribution theory of the last chapter is an asymptotic theory and the derived distributions provide good approximations to the true distributions only if M and N are reasonably large. Under the assumption that the underlying process is Gaussian, expressions for the mean, variance, and covariance of spectral estimators can be derived which are valid for all M and N. By studying these expressions we will be able to better understand the properties of spectral estimators as point estimators. This understanding will provide us with practical criteria for selecting spectral windows, thus lag windows, for weighted covariance estimators and smoothing functions for smoothed periodogram estimators.

We again consider spectral estimation for *univariate* processes first. Thus, $X(t)$, $t = 0, \pm 1, \ldots$, will represent a real-valued, zero-mean, Gaussian process with continuous spectrum and continuous spectral density function $f(\lambda)$. For the purposes of this and succeeding sections it is convenient to base our discussion on the modified process

$$b_t X(t), \qquad t = 0, \pm 1, \ldots, \tag{9.1}$$

where b_0, $b_{\pm 1}$, ... is a sequence of real numbers called a *data window*. By selecting particular values for the b_t's, a variety of situations can be simulated. For example, the usual sample $X(1)$, $X(2)$, ..., $X(N)$ can be viewed as arising from (9.1) through the use of the data window

$$b_t = \begin{cases} 1, & t = 1, 2, \ldots, N, \\ 0, & \text{otherwise.} \end{cases} \tag{9.2}$$

A sample with missing observations corresponds to taking $b_t = 1$ when the value of $X(t)$ is observed and $b_t = 0$ when it is missing. We will also be interested in data windows for which the b_t's tend gradually from one near $t = N/2$ to zero near $t = 1$ and $t = N$. Such windows are called *tapers* or *faders* and, as we will see, they have important uses for correcting bias in smoothed periodogram estimates.

At present, *we allow the values of* b_t *to be arbitrary for* $1 \leq t \leq N$ *and take* $b_t = 0$ *for* t *outside of this range.* In this section, $B(\lambda)$ will denote the

Fourier transform of this sequence

$$B(\lambda) = \frac{1}{2\pi} \sum_{t=-\infty}^{\infty} e^{-i\lambda t} b_t.$$

The weighted covariance and smoothed periodogram estimators appropriate to the windowed data (9.1) are obtained as follows. A modified autocovariance estimator is defined to be

$$\hat{C}(k) = \begin{cases} \dfrac{1}{\sum_{t=-\infty}^{\infty} b_t^2} \sum_{t=-\infty}^{\infty} b_{t+k} X(t+k) b_t X(t), & k \geq 0, \\ \hat{C}(-k), & k < 0. \end{cases} \tag{9.3}$$

Observe that this reduces to our previous definition of the sample autocovariance, (8.18), when the data window is given by (9.2). The weighted covariance estimator is then defined as before;

$$\hat{f}(\lambda) = \frac{1}{2\pi} \sum_{k=-\infty}^{\infty} e^{-i\lambda k} w_M(k) \hat{C}(k).$$

The smoothed periodogram estimator is based on the following modification of the finite Fourier transform:

$$z(\lambda) = \left(2\pi \sum_{t=-\infty}^{\infty} b_t^2\right)^{-1/2} \sum_{t=-\infty}^{\infty} e^{-i\lambda t} b_t X(t). \tag{9.4}$$

We will consider the following smoothed periodogram estimator of $f(\lambda)$ in this section: Let $I(\lambda) = |z(\lambda)|^2$ and define

$$\hat{f}(\lambda) = \int_{-\pi}^{\pi} W(\lambda - \mu) I(\mu)\, d\mu, \tag{9.5}$$

where $W(\lambda)$ is a real-valued, symmetric weight function for which

$$\int_{-\pi}^{\pi} W(\lambda)\, d\lambda = 1.$$

This is actually equivalent to the more familiar form (8.23) for a smoothed periodogram estimator because of the frequency domain sampling theorem (Section 8.2). However, this form will better suit our needs here.

The weight function $W(\lambda)$ is the Fourier transform of a square-summable weight sequence w_k in the time domain;

$$W(\lambda) = \frac{1}{2\pi} \sum_{k=-\infty}^{\infty} e^{-i\lambda k} w_k.$$

The weights are necessarily real-valued, symmetric, and have $w_0 = 1$. In all cases of interest $|w_k| \leq w_0$. Consequently, except for the fact that w_k need not be zero for $|k| > M$, this sequence has all of the properties of a lag window for a weighted covariance estimator (Section 8.3).

It is now easy to show that both the weighted covariance and smoothed periodogram estimators can be written as the sum

$$\hat{f}(\lambda) = \sum_{s=-\infty}^{\infty} \sum_{t=-\infty}^{\infty} a_{s,t} X(s) X(t) e^{-i\lambda(s-t)}, \tag{9.6}$$

where

$$a_{s,t} = b_t w_{s-t} b_s \Big/ 2\pi \sum_{t=-\infty}^{\infty} b_t^2 \tag{9.7}$$

and $w_k = w_M(k)$ in the weighted covariance case. Our treatment will follow, in most essentials, the paper of Jones (1971). First, as is shown in the Appendix, if $A(\lambda, \mu)$ is the two-dimensional Fourier transform of $a_{s,t}$,

$$A(\lambda, \mu) = \sum_{s=-\infty}^{\infty} \sum_{t=-\infty}^{\infty} a_{s,t} e^{-i(\lambda s - \mu t)}, \tag{9.8}$$

then

$$E(\hat{f}(\lambda)) = \int_{-\pi}^{\pi} A(\lambda - \alpha, \lambda - \alpha) f(\alpha)\, d\alpha. \tag{9.9}$$

It is also shown that $A(\lambda, \lambda)$ can be put in the form

$$A(\lambda, \lambda) = \int_{-\pi}^{\pi} |B(\alpha)|^2 W(\lambda - \alpha)\, d\alpha \Big/ \int_{-\pi}^{\pi} |B(\alpha)|^2\, d\alpha. \tag{9.10}$$

Moreover,

$$\begin{aligned} \operatorname{Cov}(\hat{f}(\lambda), \hat{f}(\mu)) = &\int_{-\pi}^{\pi} \int_{-\pi}^{\pi} A(\lambda - \alpha, \lambda - \beta) \overline{A(\mu - \alpha, \mu - \beta)} f(\alpha) f(\beta)\, d\alpha\, d\beta \\ &+ \int_{-\pi}^{\pi} \int_{-\pi}^{\pi} A(\lambda - \alpha, \lambda - \beta) \overline{A(\mu + \beta, \mu + \alpha)} f(\alpha) f(\beta)\, d\alpha\, d\beta. \end{aligned} \tag{9.11}$$

Thus,

$$\begin{aligned} \operatorname{Var}(\hat{f}(\lambda)) = &\int_{-\pi}^{\pi} \int_{-\pi}^{\pi} A(\lambda - \alpha, \lambda - \beta) \overline{A(\lambda + \beta, \lambda + \alpha)} f(\alpha) f(\beta)\, d\alpha\, d\beta \\ &+ \int_{-\pi}^{\pi} \int_{-\pi}^{\pi} A(\lambda - \alpha, \lambda - \beta) \overline{A(\lambda + \beta, \lambda + \alpha)} f(\alpha) f(\beta)\, d\alpha\, d\beta. \end{aligned} \tag{9.12}$$

These are exact expressions based on the Gaussian assumption. Unfortunately, the ones for the variance and covariance are too complicated to be very useful and, in fact, all useful expressions are approximations based on various simplifying assumptions. We now develop some of these expressions and indicate the assumptions on which they are based.

Alternate Expressions for Variance and Covariance, Variance Leakage

It is easy to check that $\overline{A(\lambda, \mu)} = A(-\lambda, -\mu)$. Moreover, in virtually all cases of interest, most of the "mass" of $A(\mu, \lambda)$ will be concentrated in the vicinity of $(\mu, \lambda) = (0, 0)$. Consequently, the product $A(\lambda - \alpha, \lambda - \beta) \times \overline{A(\mu + \beta, \mu + \alpha)} = A(\lambda - \alpha, \lambda - \beta)A(-\mu - \beta, -\mu - \alpha)$ in (9.11) will be nearly zero for all α, β. Thus the integral will essentially vanish unless λ is close to $-\mu$. The corresponding integral in (9.12) nearly vanishes unless $\lambda \cong 0$ or π, in which cases it equals the first integral. *Consequently, our first approximation will consist of dropping the second terms in* (9.11) *and* (9.12) *with the understanding that we will always be interested in the covariance of estimators at frequencies of the same sign and the expression for the variance of* $\hat{f}(\lambda)$ *is valid for* $\lambda \neq 0, \pi$. *This expression is to be multiplied by* 2 *at* $\lambda = 0, \pi$.

The next approximation is derived from assuming that for each λ, $f(\alpha)$ and $f(\beta)$ are nearly constant, thus equal to $f(\lambda)$, for all values of α and β for which $A(\lambda - \alpha, \lambda - \beta)$ is appreciably different from zero. This can be made to hold to a good degree of approximation by making the bandwidth of the spectral window sufficiently small. The adjustment of bandwidth is necessary to control bias as we will see presently. Thus, with these two assumptions and the observation that $A(\lambda, \mu)$ is periodic of period 2π and even in each variable, (9.11) and (9.12) become

$$\begin{aligned} \operatorname{Cov}(\hat{f}(\lambda), \hat{f}(\mu)) &\cong f(\lambda)f(\mu)\int_{-\pi}^{\pi}\int_{-\pi}^{\pi} A(\lambda - \alpha, \lambda - \beta)\overline{A(\mu - \alpha, \mu - \beta)}\, d\alpha\, d\beta. \\ &= f(\lambda)f(\mu)\int_{-\pi}^{\pi}\int_{-\pi}^{\pi} A(\alpha, \beta)\overline{A(\mu - \lambda + \alpha, \mu - \lambda + \beta)}\, d\alpha\, d\beta, \end{aligned} \tag{9.13}$$

$$\begin{aligned} \operatorname{Var}(\hat{f}(\lambda)) &\cong f^2(\lambda)\int_{-\pi}^{\pi}\int_{-\pi}^{\pi} |A(\lambda - \alpha, \lambda - \beta)|^2\, d\alpha\, d\beta \\ &= f^2(\lambda)\int_{-\pi}^{\pi}\int_{-\pi}^{\pi} |A(\alpha, \beta)|^2\, d\alpha\, d\beta. \end{aligned} \tag{9.14}$$

Jones (1971) calls these expressions the *white noise variance* and *covariance*, since if the spectral densities were those of white noise processes, they would be valid without anything being assumed about the bandwidth of the window. He also notes a situation of practical importance in which (9.14) is not an especially good approximation to (9.12). When $a_{s,t}$ depends on s and t only through the difference $s - t$, which occurs only when the data window is of the form (9.2), it is easily shown that $A(\lambda, \mu)$ is concentrated on the diagonal $\lambda = \mu$, $-\pi < \mu \leq \pi$. Thus, the variance of $\hat{f}(\lambda)$ is an average of the values of $f^2(\alpha)$ for frequencies α near λ by (9.12). If, as is the case for tapered or missing-value data windows, $a_{s,t}$ is not a "difference kernel," then it is possible for $A(\lambda - \alpha, \lambda - \beta)$ to have appreciable off-diagonal mass which can bring power into (9.12) from frequencies at some distance from λ. Jones calls this *variance leakage*. This phenomenon appears to be a problem primarily in the case of missing data when a large proportion of the sampled values are missing and when the missing values are rather regularly spaced (Jones, 1972). Otherwise, under the above stated condition, (9.13) and (9.14) are adequate approximations.

Convenient computing expressions for (9.13) and (9.14) can be obtained by substituting (9.8) into the expressions and evaluating the integrals. We find that

$$H(\gamma) = \int_{-\pi}^{\pi} \int_{-\pi}^{\pi} A(\alpha, \beta)\overline{A(\alpha + \gamma, \beta + \gamma)}\, d\alpha\, d\beta$$

$$= 4\pi^2 \sum_{s=-\infty}^{\infty} \sum_{t=-\infty}^{\infty} a_{s,t}^2 e^{i\gamma(s-t)}.$$

Now, with the substitution of (9.7) for $a_{s,t}$, a change of indices yields

$$H(\gamma) = \sum_{k=-\infty}^{\infty} d_k w_k^2 \cos \gamma k,$$

where

$$d_k = \sum_{t=-\infty}^{\infty} b_{t+k}^2 b_t^2 \Big/ \left(\sum_{t=-\infty}^{\infty} b_t^2\right)^2. \tag{9.15}$$

Thus, (9.13) and (9.14) become

$$\operatorname{Cov}(\hat{f}(\lambda), \hat{f}(\mu)) \cong f(\lambda) f(\mu) \sum_{k=-\infty}^{\infty} d_k w_k^2 \cos(\lambda - \mu)k \tag{9.16}$$

and

$$\operatorname{Var}(\hat{f}(\lambda)) \cong f^2(\lambda) \sum_{k=-\infty}^{\infty} d_k w_k^2. \tag{9.17}$$

If we denote the correlation coefficient of $\hat{f}(\mu)$ and $\hat{f}(\mu - \lambda)$ by $r(\lambda)$, it follows that

$$r(\lambda) \cong \sum_{k=-\infty}^{\infty} d_k w_k^2 \cos k\lambda \Big/ \sum_{k=-\infty}^{\infty} d_k w_k^2. \tag{9.18}$$

This is an easily computable function which makes it possible to calculate the correlation between spectral estimators for any frequency separation λ to the accuracy of the approximation leading to (9.13) and (9.14). Moreover, the variance of the spectral estimator is especially simple to evaluate by means of (9.17) to within the unknown factor $f^2(\lambda)$.

One further approximation is possible when the data window can be represented in the form

$$b_t = b(t/N),$$

where $b(v)$ is a bounded continuous function which is zero outside of the interval (0, 1]. This representation holds for most of the tapers considered in practice. Then, if N is large and w_k is (essentially) zero for indices $|k| > M$ for M relatively small in comparison to N, we can write

$$d_k = \sum_{t=-\infty}^{\infty} b^2\left(\frac{t+k}{N}\right) b^2\left(\frac{t}{N}\right) \cdot \frac{1}{N} \bigg/ N\left(\sum_{t=-\infty}^{\infty} b^2\left(\frac{t}{N}\right) \cdot \frac{1}{N}\right)^2$$

$$\cong \int_0^1 b^4(v)\, dv \bigg/ N\left(\int_0^1 b^2(v)\, dv\right)^2$$

for $|k| \leq M$. Thus, if we define

$$\kappa_b = \int_0^1 b^4(v)\, dv \bigg/ \left(\int_0^1 b^2(v)\, dv\right)^2,$$

it follows from (9.16) and (9.17) that

$$\operatorname{Cov}(\hat{f}(\lambda), \hat{f}(\mu)) \cong \frac{f(\lambda) f(\mu)}{N} \kappa_b \sum_{k=-\infty}^{\infty} w_k^{\,2} \cos(\lambda - \mu)k,$$

and

$$\operatorname{Var}(\hat{f}(\lambda)) \cong \frac{f^2(\lambda)}{N} \kappa_b \sum_{k=-\infty}^{\infty} w_k^{\,2}. \tag{9.19}$$

This leads to an approximation for the correlation coefficient of $\hat{f}(\lambda)$ and $\hat{f}(\mu)$ which does not depend on the data window. This indicates, in some measure, the crudeness of the approximation. However, it does lead to expressions for the variance which can be used to modify the asymptotic expressions given in the last chapter for equivalent bandwidth and equivalent degrees of freedom when a taper is employed in the data. Thus, for example, if $W(\lambda)$ is the weight function of a smoothed periodogram estimator, the Parseval relation yields

$$\sum_{k=-\infty}^{\infty} w_k^{\,2} = 2\pi \int_{-\pi}^{\pi} W^2(\lambda)\, d\lambda$$

and (9.19) becomes

$$\operatorname{Var}(\hat{f}(\lambda)) \cong \frac{2\pi f^2(\lambda)\kappa_b}{N} \int_{-\pi}^{\pi} W^2(\lambda)\, d\lambda. \tag{9.20}$$

This differs from expression (8.35) by the factor κ_b.

Note that the Schwarz inequality yields

$$\left(\int_0^1 b^2(v)\, dv\right)^2 \leq \left(\int_0^1 1\, dv\right)\left(\int_0^1 b^4(v)\, dv\right),$$

where equality holds only if $b^2(v) = 1$, $0 < v \leq 1$. However, since $b(v)$ is continuous, this means that $b(v) = 1$, $0 < v \leq 1$. It follows that $\kappa_b \geq 1$ and the inequality is strict except for the data window (9.2) corresponding to normal sampling. Thus, tapering increases the variance of the usual estimator.

In addition, when the weights w_k are the lag window of a weighted covariance estimator of the form

$$w_k = w(k/M),$$

where $w(v)$ satisfies the conditions given in Section 8.3, then the analog of the variance approximation (8.36) for tapered data is

$$\operatorname{Var}(\hat{f}(\lambda)) \cong f^2(\lambda)\kappa_b c_w M/N. \tag{9.21}$$

The following convention for correcting the EBW and EDF values in Table 8.1 can be obtained from these results: *If* (9.20) *and* (9.21) *are used to calculate the equivalent bandwidth and equivalent degrees of freedom through the definitions given in Section* 8.3, *the values of these quantities given in Table* 8.1 *can be corrected for the effects of tapering by dividing by* κ_b. For a more careful derivation which leads to the same result, the reader is referred to Brillinger (1970) and Hannan (1970, p. 265). Finally, a useful expression which relates the asymptotic variance to the equivalent bandwidth β is

$$\operatorname{Var}(\hat{f}(\lambda)) \cong 2\pi f^2(\lambda)/\beta N. \tag{9.22}$$

Example 9.1 *Two Important Tapers and a General Method for Constructing Tapers*

Two useful tapers are obtained from the relation $b_t = b(t/N)$ for the following functions:

Trapezoidal Taper: For $0 \leq a \leq \frac{1}{2}$,

$$b_1(v) = \begin{cases} v/a, & 0 < v \leq a, \\ 1, & a \leq v \leq 1 - a, \\ (1 - v)/a, & 1 - a \leq v \leq 1, \\ 0, & \text{otherwise.} \end{cases}$$

Cosine Taper: For $0 \le a \le \frac{1}{2}$,

$$b_2(v) = \begin{cases} \frac{1}{2}(1 - \cos(\pi v/a)), & 0 < v \le a, \\ 1, & a \le v \le 1 - a, \\ \frac{1}{2}(1 - \cos(\pi(1 - v)/a)), & 1 - a \le v \le 1, \\ 0, & \text{otherwise.} \end{cases}$$

More generally, if $h(y)$ is any monotone continuous function on [0, 1] with $h(0) = 0$ and $h(1) = 1$, we can construct a tapering function $b(v)$ by letting

$$b(v) = \begin{cases} h(v/a), & 0 < v \le a, \\ 1, & a \le v \le 1 - a, \\ h((1 - v)/a), & 1 - a \le v \le 1, \\ 0, & \text{otherwise,} \end{cases}$$

for $0 \le a \le \frac{1}{2}$. The characteristic κ_b is then easily computed as follows: For any integer n,

$$\begin{aligned} \int_0^1 b^n(v)\, dv &= \int_a^{1-a} b^n(v)\, dv + 2\int_0^a b^n(v)\, dv \\ &= 1 - 2a + 2\int_0^a h^n(v/a)\, dv \\ &= 1 - 2a\left(1 - \int_0^1 h^n(y)\, dy\right). \end{aligned}$$

Then,

$$\kappa_b = \left[1 - 2a\left(1 - \int_0^1 h^4(y)\, dy\right)\right] \Big/ \left[1 - 2a\left(1 - \int_0^1 h^2(y)\, dy\right)\right]^2.$$

For the trapezoidal taper, $h(y) = y$ and it follows that $\int_0^1 h^n(y)\, dy = 1/(n + 1)$. Thus,

$$\kappa_{b_1} = (1 - \tfrac{8}{5}a)/(1 - \tfrac{4}{3}a)^2.$$

As a ranges from 0 to $\frac{1}{2}$, κ_{b_1} varies monotonely from 1 to 1.8. Thus, the equivalent degrees of freedom and equivalent bandwidth are multiplied by a factor between 1 and 0.556. Nearly half the equivalent degrees of freedom are lost for the triangular taper corresponding to $a = \frac{1}{2}$. However, normally a would be taken to be between 1/10 and $\frac{1}{4}$. Thus a less extreme loss of stability would be incurred with this taper.

We consider next the important topic of bias of spectral estimates. The loss of degrees of freedom for tapered series will be shown to have compensating advantages from the viewpoint of bias.

Bias of Estimates of the Spectral Density

There are two principal sources of bias exhibited by estimators of the spectral density. The first is the inability of the estimators to distinguish fine structure in the spectrum under certain conditions. Thus, this type of bias is associated with the concept of *resolution*. The second type of bias is a distortion introduced by the *side lobes* of the spectral window when the underlying spectral density has one or more large peaks. We will discuss resolution first and indicate the role of bandwidth as a measure of resolution.

Effect of Varying Bandwidth on Resolution

We will first look at bandwidth in the case of the Daniell estimator of the last chapter when N is large. It was seen that if λ is one of the frequencies $2\pi k/N$, then the (asymptotic) expectation of this estimator is

$$E(\hat{f}(\lambda)) = \frac{1}{n} \sum_{\nu=-[(n-1)/2]}^{[n/2]} f(\lambda - \lambda_\nu).$$

From the natural definition of bandwidth

$$\beta = 2\pi n/N,$$

this sum is the Riemann approximation to the expression

$$E(\hat{f}(\lambda)) = \frac{1}{\beta} \int_{-\beta/2}^{\beta/2} f(\lambda - \mu)\, d\mu. \tag{9.23}$$

Thus, the expectation of $\hat{f}(\lambda)$ is the average of the true spectral density over an interval of frequencies of width equal to the bandwidth of the estimator centered at λ.

The bias of the estimator is $E\hat{f}(\lambda) - f(\lambda)$. Since we have assumed $f(\lambda)$ to be continuous, a basic theorem of the integral calculus yields

$$\lim_{\beta \to 0} \frac{1}{\beta} \int_{-\beta/2}^{\beta/2} f(\lambda - \mu)\, d\mu = f(\lambda).$$

What happens is that as β becomes smaller and smaller, $f(\mu)$ behaves more and more like a constant with value $f(\lambda)$ for frequencies near λ and, effectively, $f(\lambda - \mu)$ is factored out of (9.23) as this constant. It follows from this observation that the bias of the estimator can be made as small as desired by taking the bandwidth small enough. That is, the resolution of the estimator can be improved by decreasing the bandwidth. Lack of resolution is due to the smudging of the spectrum over the frequency interval of length β.

An alternate interpretation of the bandwidth for the Daniell estimator is that it is the spacing between (asymptotically) independent estimates of the spectrum. This follows easily from properties of the periodogram given in Section 8.2. Another is that two sharp peaks in the spectrum will not be distinguished as separate features unless the bandwidth of the estimator is smaller than the spacing $\Delta\lambda$ between the peaks. This is illustrated in Fig. 9.1. The bandwidth at which the peaks start to become resolved is $\beta = \Delta\lambda$.

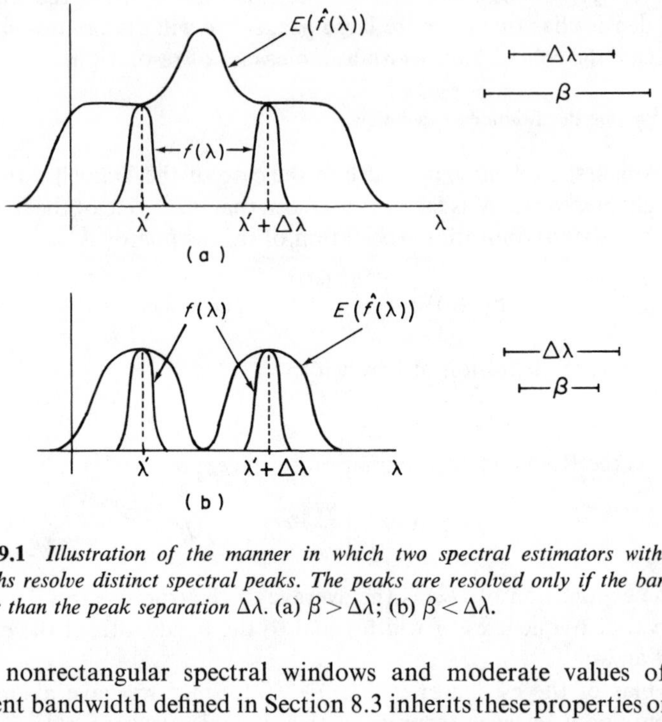

Fig. 9.1 *Illustration of the manner in which two spectral estimators with different bandwidths resolve distinct spectral peaks. The peaks are resolved only if the bandwidth* β *is smaller than the peak separation* $\Delta\lambda$. (a) $\beta > \Delta\lambda$; (b) $\beta < \Delta\lambda$.

For nonrectangular spectral windows and moderate values of N the equivalent bandwidth defined in Section 8.3 inherits these properties only in an approximate sense. The same is true of the other definitions of equivalent bandwidth which have been proposed [see, e.g., Grenander and Rosenblatt (1957) and Parzen (1961)]. However, it is useful to think of the equivalent bandwidth of a spectral estimator or spectral window as being the approximate spacing between independent estimators and as having the above described connections with resolution.

Hereafter, for convenience we will drop the modifier "equivalent" when discussing bandwidth and degrees of freedom. As is seen from Table 8.1, the bandwidths of weighted covariance estimators with lag windows truncated at $\pm M$ depend inversely on M. Thus, resolution increases as M increases. Moreover, once the window has been specified, the bandwidth of the estimator is completely determined by the value of M. Thus, in designing experiments in which a weighted covariance estimator with given

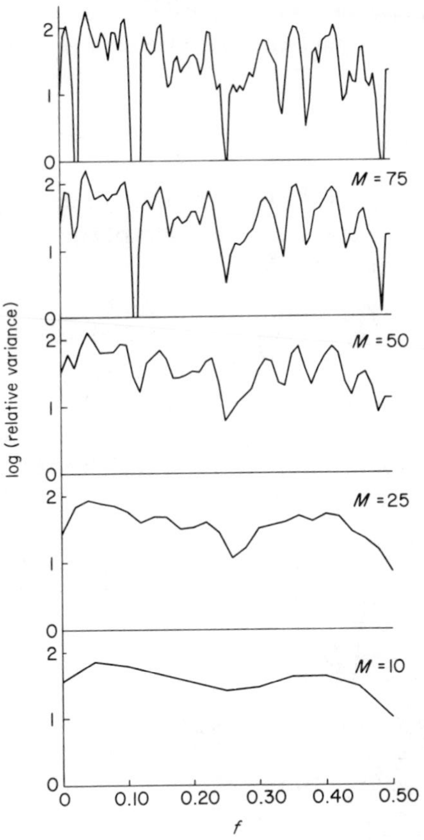

Fig. 9.2 *Graphs of weighted covariance estimates of a geological time series with $N = 121$ for varying numbers of lags M.* Source: Anderson and Koopmans (1963).

lag window is to be used, the number of lags can be determined on the basis of meeting some criterion for resolution. An illustration of the variation of resolution with M for a weighted covariance estimator is given in Fig. 9.2.

Note from expression (9.22) for $\mathrm{Var}[\hat{f}(\lambda)]$ that by decreasing the bandwidth of the spectral estimator to gain resolution, the variance increases unless compensated for by an increase in sample size. If the sample size cannot be increased, stability and resolution must be traded off to achieve a reasonable compromise. This reciprocal relationship between resolution and stability is called the *Grenander uncertainty principle* after U. Grenander, one of the early investigators in the field of statistical spectrum analysis. This principle is so named because of its similarity to the Heisenberg uncertainty principle of quantum mechanics.

Side Lobe Distortion (Window Leakage)

We return to expressions (9.9) and (9.10) for the expectation of $\hat{f}(\lambda)$. The expectation is a weighted average of the values of $f(\mu)$ with respect to a weight function or spectral window;

$$E(\hat{f}(\lambda)) = \int_{-\pi}^{\pi} A(\lambda - \alpha) f(\alpha)\, d\alpha,$$

where we have used the abbreviated notation $A(\lambda) = A(\lambda, \lambda)$ for the spectral window. This window will have a main lobe centered at the frequency of interest and side lobes as pictured in Fig. 9.3. This figure illustrates the manner

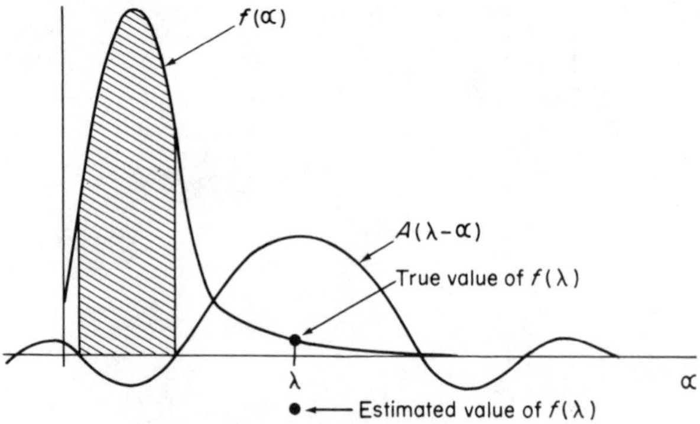

Fig. 9.3 *Illustration of the distortion introduced into an estimate of the spectral density at frequency λ by the side lobes of the spectral window and a peak in the spectrum. The cross-hatched power is averaged into the estimate with negative weight.*

in which the side lobes can transmit power from peaks in the spectrum into estimates at frequencies some distance away. The side lobes permit features of the spectrum outside of the bandwidth of the estimator to influence the value of $E(\hat{f}(\lambda))$, thus, with high probability, the value of $\hat{f}(\lambda)$. In a situation such as that pictured in Fig. 9.3 it is quite possible for both $E(\hat{f}(\lambda))$ and $\hat{f}(\lambda)$ to be negative.

Not all spectral windows have negative side lobes as pictured in Fig. 9.3. Some lag windows, such as the two Parzen windows given in Table 8.1, are *nonnegative definite* and, thus, have nonnegative Fourier transforms. (See Section A6.2 for a justification of this statement.) These windows always lead to positive spectral estimates—but they do not eliminate the bias due to side lobe distortion. Now, a spectral peak centered on a side lobe will lead to an overestimate of $f(\lambda)$ rather than an underestimate. In either case, this form of bias can cause real difficulties in the interpretation of spectral estimates and steps should always be taken to minimize it. The easiest way to do this is to decrease the bandwidth of the spectral estimator until it is substantially smaller than the width of the narrowest spectral peak. This makes it impossible for the kind of problem illustrated in Fig. 9.3 to occur.

Unfortunately, because of the trade-off between bandwidth and stability, if the sample size is fixed it is not always possible to make the bandwidth as small as desired. Then, an attempt can be made to remove the peaks before estimation by a prewhitening technique (see Chapter 6) and a spectral window with minimal side lobes can be used. We now return to expression (9.10) and a discussion of the influence of the data window on the bias of the estimator.

Reduction of Bias by Data Tapering

From (9.10) it is seen that spectral window $A(\lambda)$ of an estimator based on tapered data is related to the weight function $W(\lambda)$ by the expression

$$A(\lambda) = \int_{-\pi}^{\pi} D(\alpha) W(\lambda - \alpha)\, d\alpha,$$

where

$$D(\lambda) = |B(\lambda)|^2 \bigg/ \int_{-\pi}^{\pi} |B(\mu)|^2 \, d\mu.$$

Recall that $B(\lambda)$ is the Fourier transform of the data window b_t. We first consider the weight function (9.2) for ordinary sampling. The following discussion parallels that of the modification of transfer functions by truncation given in Section 6.4. The role of the ideal transfer function is played by the spectral window $W(\lambda)$ which is transformed into the spectral window $A(\lambda)$ by means of the averaging kernel $D(\lambda)$. This function depends on the rectangular data window and, in fact, $D(\lambda)$ is a normalized version of the square of the function $C_K(\lambda)$ pictured in Fig. 6.6, with $2K + 1 = N$. Thus, except for the fact that the side lobes of $D(\lambda)$ are all positive, they still contain a sizeable portion of the total mass of the window.

This fact is of little consequence for weighted covariance estimators where M is generally taken rather small in comparison to N and for which the spectral window $W(\lambda)$ is continuous and smooth. The reason for this is that the bandwidth of $D(\lambda)$ is of order $1/N$ while that of $W(\lambda)$ is of order $1/M$. Thus, $W(\lambda)$ will appear to be nearly constant over every frequency interval of length equal to the bandwidth of $D(\lambda)$ and it will follow that

$$A(\lambda) \cong W(\lambda).$$

Thus, the bias of the estimator will not depend to any extent on the data window.

This is not the case with the Daniell smoothed periodogram estimator, however. For this estimator, the weight function $W(\lambda)$ is rectangular and the side lobes of $D(\lambda)$ are transferred to the edges of this rectangle much as in Fig. 6.7. In the present context, $W(\lambda)$ is represented by the function $B(\lambda)$ of the graph and $A(\lambda)$ is comparable to $B_K(\lambda)$. The technique for reducing the side lobes recommended by Cooley and Tukey (1965) is to taper the data. This is one of the methods we discussed in Section 6.4 for improving the fidelity of digital filters. The effect of using the Hanning taper, which is simply a cosine taper with $a = \frac{1}{2}$, is comparable to that pictured in Fig. 6.9.

It is evident that a substantial reduction in side lobe distortion can be achieved by tapering with only a modest loss in resolution. When it is suspected that the spectral density contains one or more peaks of substantial power it is highly advisable to use a data taper with the Daniell estimator and any other smoothed periodogram estimator with sharp window features. Additional details concerning data tapering are given by Cooley *et al.* (1967).

Comparisons of untapered Daniell estimates and weighted covariance estimates demonstrating the effects of window leakage for some actual data are given by Edge and Liu (1970).

The Selection of a Spectral Window

Tapering can be viewed as a technique for modifying spectral windows to improve one of the properties of spectral estimators. The selection and modification of spectral windows is a topic that has received a great deal of attention in spectral analysis beginning with the pioneering work of Tukey (1948 approx.) and continuing to the present. In order to make a selection from among the collection of possible windows it is clearly necessary to apply some goodness criterion or figure of merit for windows. It became apparent rather early in the history of the subject that no single figure of merit could suffice for all possible underlying spectra. Moreover, no single window can be best with respect to all of the possible criteria. Consequently, this subject has progressed through the introduction of windows by individual investigators who support their candidates on the basis of the properties they feel are most important. Table 8.1 contains the more commonly used windows and their proposers. The most complete discussion of the criteria underlying the selection of windows in vogue before the mid 1960s can be found in the papers of Parzen (1967). Since then, this field has remained alive through the development of new computational techniques which make the use of more esoteric windows possible and through new theoretical developments. See, for example, the recent introduction of spline spectral windows in the paper of Cogburn and Davis (1972).

In practice it has been found that when the sample size N is large enough to achieve adequate resolution with good stability most of the windows of Table 8.1 yield comparable estimators when properly matched for resolution and stability. The larger side lobes of the two Bartlett windows make them somewhat less desirable than the others. Also, the Daniell estimator should be used only with tapered data. The greatest difficulties occur when N is fixed at a value too small to achieve proper resolution and stability. Unfortunately, this is often the situation with time series arising in such fields as economics. In this case although the selection of the spectral window is more important, it cannot solve all of the problems likely to be encountered. Various filtering and prewhitening operations will probably prove to be more critical than the selection of a particular spectral window.

A further limitation on the selectability of spectral windows for the average time series analyst is that most standard computer routines make only one window available. For example, the biomedical weighted covariance program BMD02T (Dixon, 1970) uses the Hanning window. Even the more recent programs which calculate smoothed periodogram estimators generally use the Daniell estimator with a single tapering option. [See, for example, the program BMDX92 in Dixon (1969) which uses a cosine taper.] Consequently, unless one is disposed to constructing his own computer programs, the selection of windows is a moot point in practice.

Coherence Bias

A rather subtle form of bias occasionally affects the sample coherence $\hat{\rho}_{j,k}(\lambda)$. If the phase angle $\vartheta_{j,k}(\lambda)$ is a rapidly varying function of λ in a neighborhood of the frequency at which the coherence is to be estimated, the estimate can be biased downward to such an extent that a strong coherence will be masked. Intuitively, the expectation of the sample coherence at λ' can be thought of as being the absolute value of an average over frequency of the complex coherence $\rho_{j,k}(\lambda)e^{i\vartheta_{j,k}(\lambda)}$. If $\vartheta_{j,k}(\lambda) \cong \tau\lambda$ and $\rho_{j,k}(\lambda) \cong c$ (c near 1) over the interval $(\lambda' - (\beta/2), \lambda' + (\beta/2))$, where β is the bandwidth of the estimator, then the average over frequency will behave like the integral

$$\int_{\lambda'-(\beta/2)}^{\lambda'+(\beta/2)} ce^{i\tau\lambda}\, d\lambda = c\beta e^{i\tau\lambda'}\left(\frac{1}{\beta\tau}\int_{-\beta\tau/2}^{\beta\tau/2} e^{iu}\, du\right).$$

Now, the parenthetical expression will be close to zero if $\beta\tau$ is large [see expression (1.11)].

One way to avoid this form of bias is to make the filter bandwidth small. If the sample size is limited and this cannot be accomplished, it is often possible to estimate τ and *realign* the two series by displacing the time variable of $X_j(t)$ by the estimate $\hat{\tau}$ of τ:

$$X_j'(t) = X_j(t - \hat{\tau}).$$

This is equivalent to filtering $X_j(t)$ with a linear filter with transfer function $e^{-i\hat{\tau}\lambda}$. If $\vartheta'_{j,k}(\lambda)$ denotes the phase shift of $X_j'(t)$ relative to $X_k(t)$, then

$$\vartheta'_{j,k}(\lambda') = \vartheta_{j,k}(\lambda') - \hat{\tau}\lambda' \cong (\tau - \hat{\tau})\lambda' \cong 0.$$

Since this transformation does not affect the coherence, the estimated coherence will now approximate the true value of this parameter more closely. Other methods for dealing with this problem have been proposed by Shapiro (1962), Tick (1967), Jones (1969), and Hannan and Thomson (1971).

We will consider next the problem of determining the bandwidth and degrees of freedom necessary to achieve the desired resolution and stability of a spectral estimator with a fixed but arbitrary spectral window. In general, we would use one of the windows listed in Table 8.1, but the results will apply equally well to any window for which the EBW and EDF are known.

9.3 EXPERIMENTAL DESIGN

It is almost always the case that physical time series are recorded in continuous-time (analog) form. Consequently, since we will only be concerned with digital estimation techniques, the first design problem is the determination of the sampling interval Δt for analog-to-digital conversion. Then, after some preprocessing to be discussed in Section 9.4, the data would be ready for spectral analysis by one of the standard computer programs. The investigator is required to furnish the bandwidth of the estimator and the number of data points (sample size) N. Once N and Δt are specified, the total length T of analog record required for the analysis is

$$T = N\,\Delta t \quad \text{time units.}$$

With this parameter, the measurement and analysis of the data can begin.

The selection of the design parameters Δt, N and bandwidth can be guided by the theory we have developed in this chapter and Chapter 8, but not completely determined by it. Other factors, such as physical or economic limitations on T and limited knowledge of the underlying spectrum, play a critical role, moving experimental design from the realm of science into the realm of "art." Since different conditions and restrictions exist for time series in different fields of investigation, the required "art" will not be exactly the same in any two fields. Only experience and a good understanding of the basic properties of the operations performed on the data can lead to the development of the necessary skills. In this section both the theoretical and the more generally applicable practical considerations of experimental design will be discussed. An example or two will be given to illustrate how one can occasionally take advantage of special information in designing experiments.

The Selection of Δt

From the discussion in Section 3.2, the principal objective in the selection of the sampling interval is to avoid aliasing. To do this, it would seem natural to make Δt several times smaller than the most conservative value required to make the Nyquist folding frequency $\lambda_N = \pi/\Delta t$ larger than the point beyond which the power in all component time series is essentially zero. There is a practical limitation on how small Δt can be taken, however. First, note that the expression for bandwidth, when $\Delta t \neq 1$, is

$$\beta = \pi r / N\, \Delta t, \tag{9.24}$$

where r is the degrees of freedom. This follows from (8.45) and the method for converting frequencies to the correct units when $\Delta t \neq 1$ discussed in Section 6.2. If bandwidth and degrees of freedom are held constant, thus maintaining the same resolution and stability, N is seen to vary inversely with Δt. Consequently, a decrease in Δt must be accompanied by an increase in sample size. Although this is becoming less of a problem as computer speed and memory capacity increase, there are and will continue to be upper bounds placed on N by computer programs, especially for time series with several components. For example, BMD02T (Dixon, 1970) requires $N \leq 1000$. Even the program BMDX92 (Dixon, 1969) based on the fast Fourier transform can handle at most 1000 data points with 14 data channels (time series components) and 2000 points with 6 channels. Other programs using disc or tape storage as "virtual" memories can accept larger sample sizes but at increased computing costs. Consequently, effective limitations on sample size, thus sampling interval, still exist.

The selection of the appropriate value of Δt would seem to depend critically on knowing the shapes of the spectra of the component time series for large values of $|\lambda|$. In practice this is not often the case. The electronic and/or mechanical system which senses, records, and amplifies the physical time series is usually very nearly a linear filter with a known (or computable) transfer function. Invariably, the gain function of this mechanism tends to zero as $|\lambda|$ becomes large. The "fall-off" is usually quite rapid, dropping several orders of magnitude over a relatively short interval of frequencies. Since we observe the output of this filter, and since it is unusual for physical time series to have large concentrations of power in high frequencies, the spectra of the observed series will inherit the "fall-off" characteristics of the sensing-recording-amplifying system. Consequently, if B is the frequency at which the gain function of the recorder is down from its peak value by two or three orders of magnitude and we calculate Δt to satisfy the inequality

$$2\pi/3B \leq \Delta t \leq \pi/B,$$

then the Nyquist frequency will fall in the range from B to $\frac{3}{2}B$. Using this criterion, the aliasing problem will not arise and a reasonable value of Δt from the viewpoint of sample size will usually be obtained.

Many recorder–amplifier systems have built-in, selectable, analog low-pass filters. In this case, if estimates of the spectrum are desired only for frequencies $|\lambda| \leq \lambda^*$, a filter can be selected with gain function reasonably flat at λ^* but down from its peak value two or three orders of magnitude at a point B larger than λ^*. If Δt is selected as in the last paragraph using this value of B, no aliasing will occur. Moreover, except for a constant scale factor, the spectrum will not have to be corrected for filter bias over the frequency range of interest. In situations where it is necessary to correct for filter bias, the method discussed in Section 4.4 can be used.

When nothing is known about the recording equipment or when the above discussion is otherwise inapplicable, a safe but rather expensive technique for avoiding aliasing is to perform complete spectrum analyses for two or three or possibly more values of Δt. When no appreciable difference in the shape of the estimated spectrum is observed in going from one value of Δt to a smaller value, it can be reasonably concluded that no aliasing is occurring.

The Selection of Bandwidth

The bandwidth is the most difficult of the design parameters to determine objectively. This is because bandwidth determines the resolution of the spectral estimator and the appropriate resolution to achieve a reasonably unbiased estimate of the spectrum depends on a knowledge of the fine structure of the spectrum. This is seldom available in advance of the analysis.

The bandwidth for a weighted covariance estimator with lag window of the form $w_M(k) = w(k/M)$ is (asymptotically)

$$\beta = 2\pi/c_w M\,\Delta t \quad \text{rad/unit time.} \tag{9.25}$$

(This is expression (8.47) divided by Δt to correct for the fact that $\Delta t \neq 1$.) We assume that Δt has been determined and that $w(v)$, thus c_w, is fixed. With the exception of the Daniell estimator, the smoothed periodogram estimators of interest have weight functions of the form $K(\lambda) \cong (2\pi/N)W_M(\lambda)$ where $W_M(\lambda)$ corresponds to a lag window of this type. Thus, their bandwidths are (asymptotically) the same as the bandwidths (9.25) of the corresponding weighted covariance estimators. It follows that the selection of bandwidth for both kinds of estimators will be equivalent to the selection of the lag parameter M. We will first determine M to achieve the desired resolution and then establish the value of N on the basis of one of the stability criteria to be discussed in the next subsection.

The bandwidth of the Daniell estimator with taper weights $b_t = b(t/N)$ is (asymptotically)

$$\beta = 2\pi n / N \kappa_b \, \Delta t, \tag{9.26}$$

where n is the number of periodogram values included in the average. For this type of estimator the stability requirement is applied first to fix the degrees of freedom $2n/\kappa_b$ and then N is determined by means of (9.26) to achieve the desired resolution. Consequently, the order of applying the resolution and stability criteria is reversed. However, with one exception, the criteria are the same for this estimator and those described above, and a discussion concentrating primarily on weighted covariance estimators can be easily adapted to the Daniell estimator.

In the unusual situation in which something of the fine structure of the spectrum is known, the bandwidth is relatively easy to specify. For example, if it is known that the spectrum contains two or more narrow peaks and the minimum separation of the peaks is known approximately, then all peaks will be resolved (with high probability) if the bandwidth is taken to be about $\frac{1}{2}$ of the minimum separation. (The qualification "with high probability" requires an estimator with relatively good stability. Otherwise the variability of the estimator can mask peaks even when the bandwidth is adequate to resolve them.) Thus, for example, if $\Delta\lambda$ represents the minimum separation between peaks, the assignment

$$\beta \cong \tfrac{1}{2} \Delta\lambda$$

leads to

$$M \cong 4\pi / c_w \, \Delta\lambda \, \Delta t$$

lags for the weighted covariance estimator.

Example 9.2 *A Spectrum for Which the Minimum Peak Separation Was Given*

The earth responds to an earthquake like a huge spherical bell to the blow of a hammer and in the process of "ringing" it displays a series of characteristic harmonics. These harmonics appear as peaks in long period seismograph records, the positions of which are determined by the physical constitution of the earth. From partial knowledge of this constitution, geophysicists have constructed models which predict the positions of these peaks. With the installation of the appropriate instrumentation just prior to the Chilean earthquake of 1963 it was possible, for the first time, to run a spectrum analysis on an actual seismogram to precisely determine the positions and strength of the harmonics. The specification of the number of lags was based on the minimum peak separations predicted by the models.

For spectra with relatively broad peaks, it is often important to determine the shapes and scope of the peaks rather precisely. To do this, the bandwidth can be so determined that the width of the narrowest peak is three, four, or more bandwidths. Often the spectra of several time series of similar origins will be obtained and by experimenting with the resolution on the first few of them, a good selection of bandwidth can be made for all.

A special property of weighted covariance estimators should be noted. Since we have assumed that lag windows $w_M(k)$ vanish for $|k| > M$, the weighted covariance estimator can be written in the form

$$\hat{f}(\lambda) = \frac{\Delta t}{2\pi} \sum_{k=-M}^{M} e^{-i\lambda k} w_M(k) \hat{C}(k).$$

Thus, by the frequency domain sampling theorem given in Section 8.2 with the assignment $R = 2M + 1$, it is seen that $\hat{f}(\lambda)$ is completely determined for all λ by its values at the frequencies

$$\lambda_v = 2\pi v/(2M + 1)\,\Delta t \cong \pi v/M\,\Delta t, \qquad -M \le v \le M.$$

Standard computer programs for calculating weighted covariance estimators produce estimates at the frequencies $\pi v/M\,\Delta t$ for $v = 0, 1, \ldots, M$. Estimates at all other values of λ can then be obtained from expression (8.8). Note that the spacing of the computed estimates is smaller than the bandwidth of all but a couple of the windows of Table 8.1. Consequently, by using the above bandwidth criteria for resolution we are also guaranteed of having estimators sufficiently closely spaced to adequately define the spectral peaks without interpolation. Moreover, this observation leads to a quick, "bandwidth-free" method for determining M: Choose M so as to have 3 or 4 estimates of the spectrum between peaks and 4 or more estimates within the narrowest peak to be defined.

It is sometimes important to estimate the lower portion of the spectrum beginning with some prescribed frequency λ^*. Since the first nonzero estimate occurs at $\lambda_1 = \pi/M\,\Delta t$, the criterion $\lambda_1 \le \lambda^*$ determines a lower bound for M;

$$M \ge \pi/\lambda^*\,\Delta t.$$

It would seem that this device would not be necessary if we simply start with the estimator for $\lambda = 0$ and use the interpolation formula from the frequency domain sampling theorem. However, the dc correction, which usually consists of subtracting the arithmetic mean from each term of the time series, biases the spectral estimate at zero frequency so that the first reliable estimate is at λ_1. Fishman (1969) gives a formula for correcting this bias and the bias at frequencies $0 \le \lambda \le \lambda_1$ based on the (usually unsafe) assumption that the discrete power at zero frequency is zero.

When little is known about the underlying spectrum or if the information one does have is suspect, the safest procedure is to run two or more analyses with different values of M. One stops when a larger value of M produces no improvement in resolution. The ideal is to be able to select N to give the desired stability for the value of M required to achieve the maximum resolution. Even when this is not possible, however, it is desirable to obtain three estimates of the spectrum; one with a small value of M relative to N, e.g., such that $N/M = 100$ or more; one with an intermediate value of M ($N/M \cong 20$); and one with a large value of M ($M \cong N$). This provides three different pictures of the spectrum progressing from an accurate (stable) view of the gross trend in spectral power with frequency to a relatively unstable but highly resolved view. With this procedure there is little chance of overlooking unsuspected features of the data such as narrow peaks and it is often possible to arrive at a reasonable value of M for use in future studies of comparable data. We now give an example in which sufficient information is available to enable us to select Δt and M.

Example 9.3 *A Design Example from Seismology*

A seismogram is obtained by means of a linear system with a gain function down 3 orders of magnitude from is peak response at 10 Hz. The seismogram is known to consist of two principal peaks due to different kinds of seismic waves in the range 2–7 Hz. The physics of the waves and the transmission medium allow us to predict that the peaks will be separated by at least 0.5 Hz. A spectrum analysis is to be designed to resolve the peaks. The available computer program uses a weighted covariance estimator with the Hanning window.

Presumably the output of the linear system does not have much power beyond 10 Hz, so the selection of the Nyquist frequency

$$\lambda_N = 2\pi(10) \quad \text{rad/sec}$$

would seem reasonable. Since $\Delta t = \pi/\lambda_N$ from expression (3.9) this yields the sampling interval

$$\Delta t = 0.05 \text{ sec},$$

or a sampling rate of 20 observations/sec. If we take the peak spacing to be $\Delta\lambda = 2\pi$ (0.5) rad/sec and require that the bandwidth satisfy the equation $\beta = \frac{1}{2}\Delta\lambda$, then, as argued above, the number of lags can be computed as

$$M \cong 4\pi/c_w \, \Delta\lambda \, \Delta t = 4\pi/(0.75)(2\pi(0.5))(0.05) = 106.6.$$

The value of c_w for the Hanning window was obtained from Table 8.1. Since M must be an integer, we would take the next larger integer value,

$$M = 107.$$

This number must be viewed as tentative and the analyst should be prepared to repeat the analysis with a more suitable value of M if the desired resolution of peaks is not realized. The results of the first analysis can be used to aid in the design of the second.

Selection of N

The last factor to control in the spectral estimation process is the variability of the estimator. In fact, what will actually be controlled is the relative variance $\mathrm{Var}(\hat{f}(\lambda))/f^2(\lambda)$, which, by converting (9.22) to the correct frequency units, can be written

$$\mathrm{Var}(\hat{f}(\lambda))/f^2(\lambda) \cong 2\pi/\beta N\, \Delta t.$$

(Note that previous expressions involving bandwidth can be converted to the case $\Delta t \neq 1$ by simply replacing each occurrence of β by $\beta\, \Delta t$.) With the bandwidth and Δt determined by the above considerations, the relative variance depends only on the sample size. Consequently, any criterion which specifies the relative variance, directly or indirectly, will determine the sample size, thus the data length $T = N\, \Delta t$.

Indirect control of the relative variance can be accomplished through specifying the maximum length of a confidence interval or the power of a hypothesis test for the spectrum itself or for one of the spectral parameters introduced in Chapter 5. Since the asymptotic distributions of the estimators of these parameters given in Chapter 8 depend directly on the equivalent degrees of freedom r or on the quantity $n = \frac{1}{2}r$, such specifications will lead to the determination of r. Then, the sample size can be obtained from the expression (8.45) for bandwidth;

$$N = \pi r/\beta\, \Delta t. \tag{9.27}$$

For smoothed periodogram and weighted covariance estimators associated with weight functions $w(v)$, this expression can be converted to the following useful formula by means of (9.25);

$$N = rc_w M/2. \tag{9.28}$$

We will illustrate the use of this formula shortly. An example of the use of (9.27) to calculate the sample size for the Daniell estimator will also be given.

Confidence Interval and Power Criteria for Determining Degrees of Freedom

We will illustrate the calculation of the degrees of freedom and sample size in a few important instances. As is the case in determining sample size from test power and confidence interval criteria in statistics, a "natural" power or confidence interval length is seldom provided as part of the design specifications. Thus, there will almost always be an opportunity to balance power or confidence interval length against whatever limitations on sample size or data length may exist in the situation at hand. As argued earlier, it is never possible to escape the reality of some practical or physical limitation on the length of data to be used in the analysis.

A restriction on data length is often imposed by the nature of the time series itself. The generating mechanism of virtually every real time series changes at some temporal rate. It is usually possible to divide time into epochs, within which the mechanism is essentially invariant. The lengths of the epochs will depend on how rapidly the mechanism is changing. Within epochs, the time series will be nearly stationary and each epoch will be characterized by its own spectrum. Thus, in order to estimate the spectrum for a given epoch, the data must be taken from the appropriate interval of time. This requires that the data length be smaller than the "scale of stationarity" for the time series. This restriction can often be relaxed to some degree by correcting the time series for nonstationarity through a transformation of variables, removal of trend, or other procedure. This becomes especially critical when the "scale of stationarity" is very small compared to the data lengths required for an adequate analysis, such as frequently occurs with data from economics and medicine, for example. The type of data correction depends critically on the type of nonstationarity encountered. Granger and Hatanaka (1964) discuss the methods most commonly used in business and economics. However, their methods can also be applied to many other kinds of time series.

At present, we will assume that the above mentioned limitations on data length can be disregarded and we will base our calculations purely on confidence interval length and test power considerations.

Prescribed Length Confidence Intervals for $\log f(\lambda)$

In Section 8.3 we showed that (asymptotic) $100(1-\alpha)\%$ confidence intervals for $f(\lambda)$ and $\log f(\lambda)$ are

$$r\hat{f}(\lambda)/b \leq f(\lambda) \leq r\hat{f}(\lambda)/a$$

and

$$\log(r/b) + \log \hat{f}(\lambda) \leq \log f(\lambda) \leq \log(r/a) + \log \hat{f}(\lambda),$$

where r is the EDF and the numbers a and b are determined by the equations

$$P(\chi_r^2 \leq a) = \alpha/2 \qquad \text{and} \qquad P(\chi_r^2 \leq b) = 1 - (\alpha/2)$$

[see expressions (8.41) and (8.42)]. The length of the interval for $\log f(\lambda)$ is $\log b/a$ and we will show how to design the experiment in order to guarantee, within the accuracy of the various approximations, that for a prescribed number L,

$$\log b/a \leq L. \tag{9.29}$$

Note that this criterion is equivalent to the specification that the ratio of the upper and lower confidence limits for $f(\lambda)$ not exceed $\exp(L)$.

Now, $\log b/a$ depends on the degrees of freedom r and, as would be expected, as r increases this quantity decreases. Consequently, the appropriate value of r for the design is the smallest integer for which (9.29) is satisfied. The calculation of r is facilitated by the use of Table A9.2 in which $\log b/a$ is tabulated as a function of r for confidence coefficients $1 - \alpha = 0.90, 0.95, 0.99$. Read down the column corresponding to the given confidence coefficient until the first value of $\log b/a$ smaller than (or equal to) L is found. Then read the value of r at the beginning of that row.

Example 9.4 *Computation of Data Length for the Data of Example 9.3*

Suppose that in order to properly define the peaks in the spectrum of Example 9.3 it is desirable to make the lengths of the confidence intervals no longer than 1.5 units of log power. Moreover, 95% confidence intervals are required.

From Table A9.2 we find $r = 15$ to be the smallest value for which $\log b/a \leq 1.5$. The sampling interval and bandwidth were computed to be $\Delta t = 0.05$ sec and $\beta = 0.5\pi$ rad/sec. Consequently, the sample size can be determined from (9.27) as

$$N = \pi r/\beta\, \Delta t = \pi(15)/(0.5\pi)(0.05) = 600.$$

This requires a data length of

$$T = N\, \Delta t = 600(0.05) = 30 \text{ sec.}$$

Prescribed Maximum Length Confidence Intervals for Coherence

It is argued by Amos and Koopmans (1963) that the longest confidence interval for coherence $\rho = \rho_{j,k}(\lambda)$ in Figs. A9.1 and A9.2 occurs at the largest value of $\hat{\rho}$ for which the lower confidence limit is zero. This can be used to determine the value of n for which the maximum lengths of the 80 or 90% confidence intervals do not exceed a presdesignated length L. This is accomplished as follows: Locate the value $\rho = L$ on the horizontal axis of the appropriate graph and draw a vertical line at that point. At the point where each upper curve intersects the vertical axis, draw a horizontal line to the point of intersection with the lower curve labeled with the same value of n. The appropriate value of n for the design is then the smallest n for which this point of intersection lies to the left of the vertical line. This calculation can also be carried out numerically for more densely spaced values of n and for other confidence coefficients by appealing to the tables of Amos and Koopmans (1963).

Example 9.5 *Calculation of Sample Size Based on the Maximum Confidence Interval Length Criterion for Coherence*

Consider the problem of determining the sample size required to obtain a 90% confidence interval for ρ of maximum length $L = 0.4$. A weighted covariance estimator with $M = 100$ lags, based on Jones' spectral window, is to be used in the analysis.

Following the above procedure on Fig. A9.2, we find that the smallest value of n for which graphs are given and for which the maximum confidence interval length is smaller than 0.4 is

$$n = 50.$$

Thus, the EDF is

$$r = 100.$$

The value of c_w for the Jones' window is found in Table 8.1 to be 0.48. Thus, from expression (9.28) we obtain the sample size

$$N = rc_w M/2 = (100)(0.48)(100)/2 = 2400 \quad \text{data points.}$$

Prescribed Power for the Hypothesis Test of Zero Coherence

It is frequently of interest to test whether or not the coherence between two time series at a given frequency is zero. If we require that the power of the test have a prescribed minimum value at a given positive value of $\rho = \rho_{j,k}(\lambda)$, a (minimum) value of n to achieve this goal can be obtained for tests with significance levels $\alpha = 0.05$ and 0.10 from Figs. A9.3 and A9.4. To illustrate this, and at the same time provide an example of a data length calculation for a Daniell estimator, consider the following computation.

Example 9.6 *Data Length Calculation for the Daniell Estimator Based on a Test for Zero Coherence*

A 5% test of the hypothesis $H: \rho = 0$ vs $A: \rho > 0$ at a prescribed frequency is required to have power at least 0.75 at $\rho = 0.30$. The test rejects H in favor of A if $\hat{\rho} > c$, where $\hat{\rho}$ is the standard estimator for coherence based on the Daniell spectral estimator. A trapezoidal taper with $a = 1/10$ is to be used on the data. In order to achieve satisfactorily resolved and unaliased spectral estimators, it has been determined that the bandwidth and sampling interval must be

$$\Delta t = 10^{-2} \quad \text{hr} \qquad \text{and} \qquad \beta = 2.5\pi \quad \text{rad/hr}.$$

Determine the data length required to meet the power criterion and the cutoff value c for the hypothesis test.

Solution: It is seen from Fig. A9.3 that the curve corresponding to the smallest value of n which rises above 0.75 at $\rho = 0.3$ has the index $n = 50$. From Table A9.6, we find that the cutoff value for the test with $n = 50$ is $c = 0.244$. That is, the hypothesis of zero coherence will be rejected if the sample coherence exceeds 0.244.

Now, the EDF for the Daniell estimator with trapezoidal tapering is

$$r = 2n/\kappa_{b_1},$$

where κ_{b_1} is obtained from the expression in Example 9.1 with $a = 1/10$;

$$\kappa_{b_1} = (1 - \tfrac{8}{5}(\tfrac{1}{10}))/(1 - \tfrac{4}{3}(\tfrac{1}{10}))^2 \cong 1.12.$$

This yields

$$r \cong 100/1.12 = 89.4.$$

From (9.27) we obtain the (approximate) sample size

$$N = \pi r/\beta\, \Delta t = \pi(89.4)/(2.5\pi)(10^{-2}) \cong 3580 \quad \text{data points.}$$

Thus, the required data length is approximately

$$T = (3580)(10^{-2}) = 35.80 \quad \text{hr.}$$

Criteria based on other time series parameters can be used to determine sample sizes and data lengths through comparable calculations. For example, the above calculations based on coherence can also be used for partial coherence with the appropriate adjustment in degrees of freedom (see Section 8.4). Also, Table A9.5 can be used to construct graphs similar to Figs. A9.1 and A9.2 from which prescribed maximum length confidence intervals for multiple coherence can be determined. Parameters such as phase and gain have confidence interval lengths which depend on estimators of other parameters and prescribed length confidence intervals cannot be constructed for them based on a single sample. That is, it is not possible to determine a sample size in advance which will guarantee a prescribed confidence level and maximum confidence interval length. It is possible to obtain such confidence intervals in two stages of sampling, however, based on techniques such as those given by Koopmans and Qualls (1971). In effect, the first stage of sampling is used to obtain high-probability bounds for the estimators involved in the endpoints of the confidence intervals. Then a second sample size can be determined which guarantees the prescribed length and confidence coefficient. In this way, prescribed length confidence intervals can also be obtained for the spectrum. A need for these intervals appears to occur rather seldom in practice. Moreover, the inconvenience of sampling time series in two stages restricts the appeal of these procedures.

A final word concerning our discussion of experimental design is in order. Because of the many approximations involved, these procedures must be used with caution. They provide guidelines and not necessarily incontrovertible answers. In the final analysis, the "proof" of a good design is that the results of the analysis make physical sense. When this is not the case, one must make an educated guess as to what went wrong and redesign the analysis and/or preprocessing of the data accordingly. Only experience and experimentation with a variety of ideas and methods will lead to good results in some instances.

It should be added that perhaps the most interesting use of spectrum analysis is as an exploratory tool for learning about a physical process. Thus, one welcomes unexpected features in the spectrum and other spectral parameters for the possible new information they convey. However, it is always necessary to make sure that the features are properties of the generating mechanism and have not simply been imposed on the data somewhere in the chain of processing and analysis operations. To do this, it is necessary to be familiar with the workings and idiosyncracies of each operation. A good discussion of electronic data acquisition systems and analog-to-digital converters is given by Enochson and Otnes (1968) and they will not be considered in this book. We have discussed the operation and potential pitfalls of digital filters in Chapter 6. In the next section we will discuss the actual procedures that have been used and are being used for computing spectra.

9.4 METHODS FOR COMPUTING SPECTRAL ESTIMATORS

Early smoothed (or unsmoothed) periodogram estimators and weighted covariance estimators were calculated by carrying out all of the operations exactly as indicated in the defining expressions. Thus, the finite Fourier transformation of a sample $X(1), \ldots, X(N)$ from a univariate time series required N operations each of multiplication, addition and the calculation of a complex exponential for each of N frequencies $\lambda_\nu = 2\pi\nu/N$, or an order of N^2 operations (see the discussion of the order of the number of operations for a computing method in Section 6.3). With the additional operations required to form the smoothed periodogram estimator, the total number of operations remains of order N^2. The strain this amount of computing placed on primitive computing equipment, aside from the problems of instability suffered with the periodogram, would be enough to explain the early interest in weighted covariance estimators.

For this type of estimator, the number of additions and multiplications required to calculate the covariances is of order NM and the Fourier transform of the covariances by the standard method requires an order M^2 additional operations. To obtain an asymptotically unbiased and consistent estimator, M must tend to infinity with N at such a rate that $M/N \to 0$. This would be the case if, for example, we assume that M is of order $\log N$. Moreover, the number of operations required to compute the weighted covariance estimators is then of order $N \log N$. This is a substantial improvement over the smoothed periodogram calculation and it accounts for the dominance of the weighted covariance method until 1965.

In that year, two publications appeared which initiated a swing back to smoothed periodogram estimators. In the first, Jones (1965) pointed out that with the computing equipment of that time, the actual computing times of the two methods were not too far apart for moderate sample sizes and by using the Daniell estimator, a nearly rectangular spectral window with the concomittant ideal properties described above and in the last chapter were available. More importantly, using the standard method of Fourier transformation, the smoothed periodogram method was faster than the weighted covariance method for spectral calculations involving multivariate time series of even relatively small dimension. The reason for this is that the smoothed periodogram requires only one Fourier transform per dimension, while a Fourier transform must be executed for each pair of coordinate series in the weighted covariance case.

The most crucial factor in reviving interest in the smoothed periodogram was the publication in that year of the fast Fourier transform algorithm by Cooley and Tukey (1965). This algorithm had appeared in print earlier [see Cooley *et al.* (1967) for its history], but it required modern computing technology and the current demands for Fourier computations in time series analysis for it to register the immense impact it has had. Virtually every method for computing spectra now uses the fast Fourier transform algorithm in some way.

We will give a brief description of how the fast Fourier transform algorithm works and how it achieves an improvement in the order of the number of operations required to compute a finite Fourier transform. The algorithm also achieves an improvement in computing accuracy. For a discussion of this and a more detailed description of the algorithm, how it is programmed, and its properties see Cooley *et al.* (1967). A precise error estimate is given by Gentleman and Sande (1966).

The Fast Fourier Transform Algorithm

Given numbers $x(1), x(2), \ldots, x(N)$, we have defined the finite Fourier transform to be

$$a(\lambda_\nu) = \sum_{t=1}^{N} x(t)e^{i\lambda_\nu t}$$

for the N frequencies $\lambda_\nu = 2\pi\nu/N$, $-[(N-1)/2] \leq \nu \leq [N/2]$. By extending $x(t)$ periodically outside of the range $1 \leq t \leq N$ and letting $a_\nu = a(\lambda_\nu)$ this transform can be expressed in a more convenient form for this discussion as

$$a_\nu = \sum_{t=0}^{N-1} x(t)e^{2\pi i\nu t/N}. \tag{9.30}$$

Moreover, since the transform is periodic of period N, the range of the frequency index can be taken to be $0 \leq \nu \leq N-1$.

Now, suppose that N can be expressed as the product of two factors, $N = r \cdot s$. The index t can be written in the form

$$t = rj + k$$

and, as j takes on all integer values from 0 to $s-1$ and k from 0 to $r-1$, t will assume all values from 0 to $N-1$. Then, (9.30) becomes

$$a_\nu = \sum_{j=0}^{s-1} \sum_{k=0}^{r-1} x(rj+k)e^{2\pi i\nu(rj+k)/N}. \tag{9.31}$$

However, since $r/N = 1/s$, we have

$$e^{2\pi i\nu(rj+k)/N} = e^{2\pi i\nu j/s}e^{2\pi i\nu k/N}.$$

Consequently, changing the order of summation in (9.31), we can write

$$a_\nu = \sum_{k=0}^{r-1} b_{k,\nu}\, e^{2\pi i\nu k/N}, \tag{9.32}$$

where

$$b_{k,\nu} = \sum_{j=0}^{s-1} x(rj+k)e^{2\pi i\nu j/s}. \tag{9.33}$$

Thus, the Fourier transform can be computed in two stages. The Fourier transforms (9.33) are calculated first and then they are transformed by (9.32) to obtain a_ν. The key observation to the efficiency of the algorithm is that, since (9.33) defines a function periodic of period s for each k, it follows that

$$b_{k,\nu} = b_{k,\nu+ls} \tag{9.34}$$

for every integer l. Thus, if transform (9.33) is carried out for each k and for $0 \leq \nu \leq s-1$, the values of $b_{k,\nu}$ for all ν, $0 \leq \nu \leq N-1$, are determined by (9.34). This reduces the number of operations (each consisting of a complex addition, multiplication, and exponential computation) in (9.33) from Ns to s^2 for each k. Thus, to compute $b_{k,\nu}$ for all values of k and ν, rs^2 operations are required. In addition, (9.32) requires r operations for each ν or Nr operations in all. The total number of operations to compute the finite Fourier transform is then

$$rs^2 + Nr = N(r+s).$$

This is to be compared with N^2 operations for a direct computation of (9.30).

Now, if

$$N = p_1, p_2 \cdots p_K$$

is the unique representation of N as a product of prime numbers, then the above process of decomposition can be carried out a number of times to obtain a sophisticated if somewhat complicated algorithm. The process is applied first with $r = p_1$ and $s = p_2 \cdots p_K$. Then, a further decomposition with $r = p_2$ and $s = p_3 \cdots p_K$ can be made, etc. Ultimately, the process reduces to a large number of short Fourier transforms involving a total of $N(p_1 + p_2 + \cdots + p_K)$ operations. The resulting algorithm can lead to a large time saving over the conventional method.

For example, if $N = 1500$, the prime factor decomposition is

$$N = 2 \cdot 2 \cdot 3 \cdot 5 \cdot 5 \cdot 5.$$

Then, $p_1 + \cdots + p_K = 22$ and

$$N^2/N(p_1 + \cdots + p_K) = 1500/22 \cong 70.$$

Thus, the "arithmetic" of the fast Fourier transform calculation is 70 times faster than that of the conventional method.

If N is a power of 2,

$$N = 2^p,$$

then the number of operations for the algorithm is $2pN$ which is of order $N \log N$. The algorithm for this configuration of sample size (the base 2 algorithm) is the easiest to program and many fast Fourier transform computer routines are of this type. By adding the appropriate number of zeros to the end of the data, every sample size can be put in this form. However, as N increases, the gaps between the appropriate powers of two become rather large, requiring an excessive amount of computing. Consequently, more sophisticated computer routines have been written to take advantage of the general form of the algorithm. Early programs appearing in the IBM support system SHARE are available to every IBM computer user. Spectral estimation programs based on the fast Fourier transform algorithm are also commonly available. For example, the time series spectrum estimation program BMDX92 (Dixon, 1969) uses the base 2 algorithm to compute Daniell estimates of spectra and cross-spectra for multivariate time series of several dimensions.

Weighted Covariance Estimators Revisited

As was indicated in Section 8.3, the following expression for the periodogram is an algebraic identity:

$$I_N(\lambda) = \frac{1}{2\pi} \sum_{k=-\pi+1}^{N-1} e^{-i\lambda k} \hat{C}(k), \tag{9.35}$$

where $\hat{C}(k)$ is the sample autocovariance function

$$\hat{C}(k) = \frac{1}{N} \sum_{t=1}^{N-|k|} X(t+|k|)X(t)$$

for $|k| \le N-1$. If instead of extending this function to be zero for $|k| \ge N$ we extend it to be periodic of period $2N-1$, then (9.35) will represent the finite Fourier transform of $\hat{C}(k)$ at frequencies $\lambda_\nu' = 2\pi\nu/(2N-1)$, $|\nu| \le N-1$. Thus, by inversion, we can exactly reconstruct $\hat{C}(k)$ for $|k| \le N-1$ from $I_N(\lambda)$ at these frequencies;

$$\hat{C}(k) = \frac{2\pi}{2N-1} \sum_{v=-N+1}^{N-1} e^{i\lambda_\nu' k} I_N(\lambda_\nu'). \tag{9.36}$$

This makes available an alternate method for computing covariances: First, the original data is Fourier transformed and the result "squared" to obtain $I_N(\lambda)$. Then the periodogram is Fourier transformed to obtain $\hat{C}(k)$. Note that a small problem occurs here. The finite Fourier transform, thus the method for calculating it, provides the same number of output values as input values. Thus, the transform of $X(1), \ldots, X(N)$ yields periodogram values at $\lambda_\nu = 2\pi\nu/N$ for $-[(N-1)/2] \le \nu \le [N/2]$. However, (9.36) requires the periodogram ordinates at the $2N - 1$ frequencies λ_ν'. The fast Fourier transform algorithm can be "tricked" into providing the periodogram at these frequencies by simply augmenting the original data set (or sets in the multivariate case) by $N - 1$ zeros at the beginning or end of the data, i.e., apply the algorithm to

$$X(1), X(2), \ldots, X(N), \underbrace{0, 0, \ldots, 0}_{N-1}.$$

Except for a scale factor, this set has the same finite Fourier transform as the original set and the transform will be given at the frequencies λ_ν'.

Now, using the order for the base 2 fast Fourier transform for comparison purposes, this method for calculating covariances takes an order of $N \log N$ operations. The number of operations required to compute all covariances by the standard method is of order N^2. Thus, the double transform method is clearly superior.

As we saw above, if a weighted covariance estimator is to be computed, only M covariances are required, where M is the number of lags. If M is of order $\log N$, it follows that the number of operations required to compute first the covariances and then the weighted covariance estimates by fast Fourier transform is of order $N \log N$ as is the direct method for computing weighted covariances estimates. The smoothed periodogram estimators of practical interest, those having smoothing functions with nonzero weights for n periodogram ordinates, also require an order $N \log N$ operations if n is taken to be of order $\log N$. [This produces asymptotically unbiased and consistent estimates, since the bandwidth is then of order $(\log N)/N$ and the variance is of order $1/\log N$.] Thus, it is not possible to choose among these three methods for computing spectra on the basis of an order argument.

Recently, evidence has been accumulating which favors the use of weighted covariance estimators computed by three applications of the fast Fourier transform [see, e.g., Parzen (1972)]. This method proves to be computationally faster than the others for many combinations of sample size and design parameters encountered in practice. Also, one has available the flexible family of lag windows which were developed during the period prior to 1965 in which these estimators were predominant.

Moreover, the newer developments in "window carpentry," such as the spline windows of Cogburn and Davis (1972) are producing lag windows which utilize all of the covariances. In doing so, much better rates of convergence of bias to zero and substantially better mean-square error properties are possible than could be produced by earlier windows. To obtain comparable properties using the equivalent smoothed periodogram estimator would require an order N^2 operations while the $N \log N$ order is retained by the weighted covariance estimators. Consequently, the fast Fourier transform version of the weighted covariance estimator is clearly favored. It is to be expected that packaged programs utilizing the fast Fourier transform to calculate weighted covariance estimates with the new types of lag windows will become commonly available in the near future.

Other Forms of Spectral Estimation

Other methods of computing spectral estimators have been considered in the course of the history of spectral analysis. Although they have not achieved the general popularity of the basic types of estimators discussed above, each has special features which have made it well suited to certain types of analysis. We will give a brief description of the more important methods here.

The Autoregression Estimator

An important recent method for estimating spectra is based on the finite autoregressive model of Section 7.3. This technique has been supported and used by Parzen (1972) and Gersh (1970), for example. The utility of this method is derived from two facts: First, virtually every time series encountered in practice (after appropriate preprocessing) can be approximated to any desired accuracy by a finite autoregressive model of sufficiently high degree. When the underlying spectrum is reasonably smooth, in fact the degree need not be too large. Second, there is a simple relationship between the coefficients of the autoregression and the covariances of the process. This relationship is embodied in the Yule–Walker equations derived in Section 7.3:

$$\sum_{k=1}^{p} b_k C(k-l) = -C(l), \qquad l = 1, 2, \ldots, p, \tag{9.37}$$

where the autoregressive process is

$$\sum_{k=0}^{p} b_k X(t-k) = \xi(t).$$

As before, $b_0 = 1$ and $\xi(t)$ is the innovation process defined in Section 7.3. Moreover, the innovation variance is

$$\sigma^2 = \sum_{k=0}^{p} b_k C(k). \tag{9.38}$$

Recall that the spectrum of the process is an elementary function of the coefficients;

$$f(\lambda) = (\sigma^2/2\pi)\left(1 \Big/ \left|\sum_{k=0}^{p} b_k e^{-i\lambda k}\right|^2\right). \tag{9.39}$$

Although we discuss only the case of a univariate process, the method can also be used for multivariate processes.

The following procedure is suggested by expressions (9.37)–(9.39). First, estimate the covariances of the given process. The fast Fourier transform algorithm can be used for this purpose. By some procedure, fit a value of p to the process. Next, determine estimates of the coefficients by solving (9.37) with the $C(k)$'s replaced by their estimates. Finally, enter these values and the estimated innovation variance into (9.39) to obtain the estimated spectrum.

The matrix of coefficients of (9.37) is of a special diagonal type known as a *Toeplitz matrix*. Especially efficient inversion programs are available which make it possible to solve this system of equations very quickly. A Fortran program for this purpose is given by Robinson (1967, p. 45), for example.

A simple method for fitting the degree of the autoregression is to plot the estimated innovation variances, obtained by replacing $C(k)$ and b_k by their estimates in (9.38), for a number of values of p. As a rule, the estimated variances will decrease with p but will "level off" at some point. Then estimate p to be the first value for which this "leveling off" occurs [see Whittle (1963, p. 37) for an illustration of this method]. Another procedure is given by Akaike (1970).

This method of spectral estimation works well and is computationally very efficient for time series with reasonably smooth spectra. However, the methods for estimating p and the b_k's are rather insensitive to sharp, local features of the spectrum. Consequently, it is not uncommon for rather prominent spectral peaks to be overlooked by this method. For this reason, it cannot be recommended for preliminary, exploratory analyses of data.

Complex Demodulation

This technique was discussed briefly in Section 2.10 for continuous-time series. The idea is to multiply the time series by a number of complex exponentials with different frequencies, forming complex-valued series

$$X_\mu(t) = e^{-i\mu t} X(t).$$

By the spectral representation of the process,

$$X_\mu(t) = \int e^{i(\lambda-\mu)t} Z(d\lambda) = \int e^{i\xi t} Z(\mu + d\xi).$$

Thus, the complex amplitude of $X_\mu(t)$ near zero frequency ($\xi = 0$) is the same as the complex amplitude of $X(t)$ near $\lambda = \mu$. Consequently, by subjecting $X_\mu(t)$ to a narrow-band, low-pass filter L one can very nearly isolate the spectral amplitude of $X(t)$ in a small interval around μ;

$$L(X_\mu(t)) \cong Z(d\mu) = |Z(d\mu)| e^{i\vartheta(\mu)}.$$

It follows that $|L(X_\mu(t))|^2 \cong |Z(d\mu)|^2$ is an estimate of $E|Z(d\mu)|^2 = f(\mu)\, d\mu$. By varying μ over the appropriate frequency range, an estimate of the spectrum can be obtained. Several possible low-pass filters are discussed in Chapters 4 and 6. Note that this method also yields an estimate of the phase $\vartheta(\mu)$ of $X(t)$ relative to the time origin of the data. This is a useful quantity in some time series applications.

Another useful application of complex demodulation is to the study of the variation of power with time of a nonstationary process. Call the length of the smallest time interval $[-r\,\Delta t,\ s\,\Delta t]$, containing, say, 95% of the squared weight $\sum_{k=-\infty}^{\infty} w_k^2$ of a digital convolution filter, the *operating length* of the filter. By taking the operating length of the filter L short relative to the "scale of stationarity" of the series, the estimates will follow, continuously, the changes in the spectral density and phase of the process at each frequency as they evolve in time from one epoch to another. For applications of complex demodulation to neurophysiology, see Walter (1969). Early applications of the technique were made by Tukey (1961).

The Faded Overlap Method

One of the earliest methods of "modern" spectral estimation, proposed by Bartlett (1948), was to subdivide the total data length into reasonably short pieces of equal length, estimate the spectrum for each piece, then average the resulting estimates frequency by frequency. There has been a continuing interest in a modified version of this idea. The modification consists of performing conventional spectral analyses on short, overlapping segments of the original data. Then, averages frequency by frequency can be formed or not as the application dictates.

One advantage of this method is that a considerable saving in time can be realized by analyzing several short data lengths rather than a single long data length. [See Cooley *et al.* (1967) for actual time comparisons using smoothed periodogram estimates with tapering]. The statistical properties of this type of estimator are also analyzed by Groves and Hannan (1968).

Another advantage of this method is that it provides another way to follow the variations of power of nonstationary series with time. Sequential plots of the spectra of the short segments provide useful visual displays of this variation. A plot of sequential spectra for a geological time series is given as an illustration in Fig. 9.4.

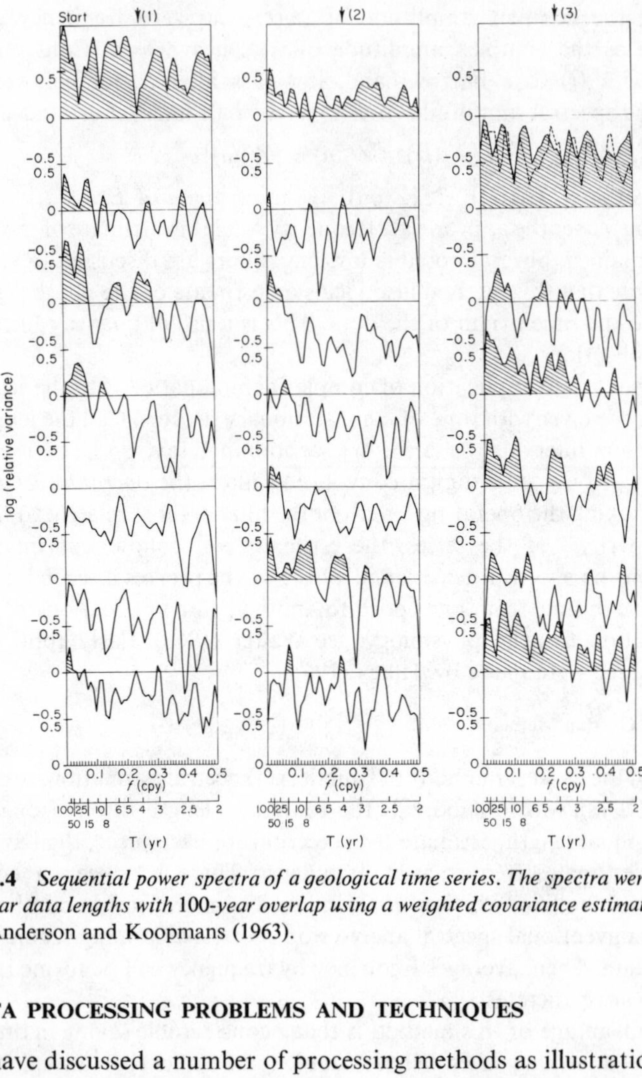

Fig. 9.4 *Sequential power spectra of a geological time series. The spectra were computed for* 200*-year data lengths with* 100*-year overlap using a weighted covariance estimator.* Source: Anderson and Koopmans (1963).

9.5 DATA PROCESSING PROBLEMS AND TECHNIQUES

We have discussed a number of processing methods as illustrations of the theory throughout the text. In this section, we will reference them briefly as well as present other problems that can arise and techniques for dealing with them.

Correcting Deviations from the Spectral Analysis Hypotheses

We based the statistical theory of spectral analysis on certain hypotheses: (i) that the underlying process is stationary and Gaussian, and (ii) that the process mean is zero and the spectrum is continuous. Some of the preprocessing operations ordinarily performed on data before a spectrum analysis is carried out are intended to bring the data into reasonable conformity with these hypotheses. For example, virtually every spectrum analysis program has a built-in routine for estimating and subtracting from each data point the line which best fits the data in the least-squares sense. There is usually an option for taking the slope equal to zero, which reduces the operation to the removal of the (constant) sample mean. In this way, the data is corrected for a nonzero mean and a possible linear trend, which can be either an indication of process nonstationarity or the effect of a "drift" in the recording equipment. In the first case, the linear term is retained as an important regression term and in the second it is simply discarded. The resulting residual will, in either event, conform more closely to the stationarity hypothesis.

It is particularly important that the sample mean be subtracted off. Recall that a nonzero mean m (the dc component) contributes a discrete component to the spectrum of power m^2 at $\lambda = 0$. To the spectral analysis program this will appear as a peak of height proportional to $Nm^2\,\Delta t$ over a frequency band of approximately one bandwidth, $\beta = \pi r/N\,\Delta t$, where r is the degrees of freedom of the estimator and N is the number of data points (see Section A8.2). Then, due to the inevitable side lobe distortion (filter leakage) of the spectral estimates, the estimates in the low end of the spectrum can be badly biased. For this same reason, discrete power peaks anywhere in the spectrum should be removed. See Sections 2.10 and 6.5 for suggested methods for removing peaks.

Large peaks in the continuous spectrum should be reduced as much as possible for the same reasons. The prewhitening method discussed in Section 6.6 will prove quite satisfactory in most instances.

There is evidence that the distributions of the more important spectral parameters are not overly sensitive to deviations from the Gaussian hypotheses [see, e.g., Benningus (1969)]. This is fortunate, since there are few general methods for correcting the process for such deviations. The methods used to correct data for nonnormality found in standard statistics texts would, in general, change the covariances, thus the spectrum of the process.

There are isolated techniques for removing various forms of nonstationarity, but again, except in a few cases, it is difficult to relate the spectrum of the corrected process to properties of the original data and its generating mechanism. For a good compendium of methods, see Granger and Hatanaka (1964).

One form of nonstationarity for which a great deal of theory exists is the case in which the process is a sum of a deterministic regression function of known form with a weakly stationary residual. Polynomial regression functions, of which the linear trend is a special case, are especially important. We gave one method for removing polynomial trends in Section 6.6. For details of the general regression theory, see Hannan (1970).

The Problem of Missing Data

Because of the vagaries of nature and of recording equipment it is often the case that one or more data points will be missing from a set of time series observations. When the number of missing points is small, the simple expedient of interpolating the missing values from neighboring data points in any reasonable way—e.g., linearly—will lead to a spectrum that differs very little from the one that would have been obtained if no data had been missing. Often, however, stretches of data will be missing at one or more places in the data because of such problems as intermittent recorder failure. Again, if sample values are available for a time series only on days during the business week, for example, the regularly spaced stretches of data corresponding to weekends will be missing. When the amount of missing data makes up a sizable part of the total data length, a substantial bias in the spectral estimates can result.

Intuitively, it can be seen that different parts of the spectrum are biased to different degrees. For example, if missing values are regularly spaced, estimates for periods small in comparison to this spacing will be relatively unaffected, since there will be relatively long stretches of uninterrupted data for the construction of these estimates. Also, long period spectral estimates will be comparatively unaffected. To see this, note that a filtering and decimation procedure (Section 6.4) could be applied to lower the Nyquist folding frequency without affecting the statistical properties of the low-frequency estimates. However, in this case the missing values would constitute a substantially smaller proportion of the total length, thus less of a problem. For example, if data were missing on weekends for daily observations, by decimating to weekly observations there would be no data missing if, say, the Thursday data were used. Thus, estimates for frequencies smaller than $\frac{1}{2}$ cycle per week would be unaffected by the missing data.

The actual effect of missing data on properties of the spectral estimates can be evaluated by a method proposed by Jones (1971). The expressions developed in Section 9.2 are specifically designed for this purpose. First, recall that a missing data window is a window b_t such that $b_t = 0$ for all t for which data is missing and $b_t = 1$ otherwise. Weighted covariance and smoothed periodogram estimates are then based on the modified covariances and Fourier transforms (9.3) and (9.4). The asymptotic variances and covariances of the estimates can then be conveniently calculated from (9.16) and (9.17). Some indication of bias can be obtained by looking at the modified spectral window (9.10).

Jones actually proposed a weighted covariance estimator based on unbiased estimates of the covariances. This produces spectral estimates with the same expected values as the corresponding estimators when no data is missing. The effects of the missing observations then show up as variance leakage, inflated variances, and decreased degrees of freedom. He demonstrated these effects by means of actual calculations for data with a large proportion (about 50%) of the observations missing. Surprisingly, reasonably good spectral estimates were still obtained.

Outliers (Wild Values) and Their Influence

Occasionally, numerical values will be found among observations on a time series which are obviously out of scale and, thus, not characteristic of the phenomenon under study. Sometimes the origin of these outliers can be pinpointed. Electrical "surges" in recorders or other electronic processing equipment are not uncommon, for example. Whatever the cause, it is essential that outliers be removed. The simplest procedure is to replace them by interpolated values from neighboring data. If there are many outliers and interpolation is not practical or possible, it would be reasonable to set them all equal to zero and proceed as though they were missing observations.

The reason it is essential to remove outliers is that they have large amplitudes, thus contain a substantial portion of the power in the data. They have a pronounced effect on the spectral estimates. Each outlier represents a "spike" in the time data which is converted into a periodic ripple in the frequency domain. This is a consequence of the correspondence between periodic components and spectral lines and the interchangeability of the Fourier transform domains. Since a periodic term in the time domain produces a line or "spike" in the frequency domain, a "spike" in the time domain produces a periodic term in the frequency domain. These periodic ripples give the appearance of a multipeaked spectrum when there are relatively few outliers. Consequently, interesting and provocative spectral peaks can be simply the result of not having corrected for outliers. When there are many outliers, the spectrum of interest is submerged in the outlier "noise" spectrum and a meaningful analysis is usually impossible.

Data Plots

From the discussion of outliers it is evident that it would be very useful to have a graph of the data. Such a plot is also useful for other purposes, such as deciding on the regression function to be used to correct for nonstationarity, obtaining indications of other forms of nonstationarity or possibly estimating the "scale of stationarity" for the data. Occasionally, obvious periodicities or strong cycles can be detected visually and some indications of the type of filtering needed to remove or reduce them can be obtained. These plots are usually well worth the expense of making them.

Other Graphs and Plots

Because of the vast amount of numerical information contained in the estimates of spectral parameters, graphs should be obtained for everything computed. Spectra, coherences, phase angles, gain and phase functions for filters should all be graphed against frequency. Spectral features which are difficult to see in lists of numerical output will show up at a glance in a graph. Fortunately, plotters are becoming increasingly available in even the smaller computer installations and graphs can be obtained by simply requesting them on the instruction cards for many standard spectrum programs. This is the case, for example, for the biomedical programs mentioned earlier in the text.

APPENDIX TO CHAPTER 9

A9.1 Sampling Properties of the Spectral Estimator (9.6)

The expressions for the mean, variance, and covariance of (9.6) depend on the spectral representation for the process and its properties given in Chapter 2. Substituting the representation for $X(s)$ and $X(t)$ in (9.6), with an interchange of integrals and sums we obtain

$$\hat{f}(\lambda) = \int_{-\pi}^{\pi} \int_{-\pi}^{\pi} A(\lambda - \alpha, \lambda - \beta) Z(d\alpha) \overline{Z(d\beta)}.$$

This leads to expression (9.9) for $E(\hat{f}(\lambda))$, since

$$E\big(Z(d\alpha)\overline{Z(d\beta)}\big) = \begin{cases} f(\alpha)\, d\alpha, & \text{if } \beta = \alpha, \\ 0, & \text{otherwise.} \end{cases}$$

Expressions (9.11) and (9.12) follow from this result and the Isserlis theorem for normal random variables by a computation very similar to the one given in Section A2.2. Here, the computation is applied to $E\hat{f}(\mu)\overline{\hat{f}(\lambda)}$. The form of $A(\lambda, \lambda)$ given in (9.10) can be justified as follows. From (9.7) and (9.8) we obtain

$$A(\lambda, \lambda) = \sum_{k=-\infty}^{\infty} c_k w_k e^{i\lambda k},$$

where

$$c_k = \sum_{t=-\infty}^{\infty} b_{t+k} b_t \Big/ 2\pi \sum_{t=-\infty}^{\infty} b_t^2.$$

Now, the Fourier transform of a product is the convolution of the Fourier transforms. Thus,

$$A(\lambda, \lambda) = \int_{-\pi}^{\pi} C(\alpha) W(\lambda - \alpha)\, d\alpha,$$

where $C(\lambda)$ and $W(\lambda)$ are the Fourier transforms of the c_k and w_k sequences.

However, the numerator of the expression for c_k is the discrete convolution s_k of the sequences b_t and $r_t = b_{-t}$;

$$s_k = \sum_{t=-\infty}^{\infty} r_{-k-t} b_t.$$

Since b_t is real-valued, it is easy to show that the Fourier transform of r_t is the complex conjugate of the transform of b_t. Thus, s_k has Fourier transform

$$S(\lambda) = |B(\lambda)|^2.$$

By the Parseval relation,

$$2\pi \sum_{t=-\infty}^{\infty} b_t^2 = \int_{-\pi}^{\pi} |B(\alpha)|^2 \, d\alpha.$$

It follows that the Fourier transform of $c_k = s_k/(2\pi \sum_{t=-\infty}^{\infty} b_t^2)$ is

$$C(\lambda) = |B(\lambda)|^2 \Big/ \int_{-\pi}^{\pi} |B(\alpha)|^2 \, d\alpha.$$

A9.2 Tables and Graphs for Confidence Intervals, Hypothesis Tests, and Experimental Design

List of Tables and Graphs

Index to Use of Tables and Graphs in Text

Table A9.1

Critical Values for the Chi-Square Distribution[a]

	γ					
r	0.005	0.025	0.05	0.95	0.975	0.995
1	-	0.001	0.004	3.841	5.024	7.879
2	0.010	0.051	0.103	5.991	7.378	10.597
3	0.072	0.216	0.352	7.815	9.348	12.838
4	0.207	0.484	0.711	9.488	11.143	14.860
5	0.412	0.831	1.145	11.071	12.833	16.750
6	0.676	1.237	1.635	12.592	14.449	18.548
7	0.989	1.690	2.167	14.067	16.013	20.278
8	1.344	2.180	2.733	15.507	17.535	21.955
9	1.735	2.700	3.325	16.919	19.023	23.589
10	2.156	3.247	3.940	18.307	20.483	25.188
11	2.603	3.816	4.575	19.675	21.920	26.757
12	3.074	4.404	5.226	21.026	23.337	28.299
13	3.565	5.009	5.892	22.362	24.736	29.819
14	4.075	5.629	6.571	23.685	26.119	31.319
15	4.601	6.262	7.261	24.996	27.488	32.801
16	5.142	6.908	7.962	26.296	28.845	34.267
17	5.697	7.564	8.672	27.587	30.191	35.718
18	6.265	8.231	9.390	28.869	31.526	37.156
19	6.844	8.907	10.117	30.144	32.852	38.582
20	7.434	9.591	10.851	31.410	34.170	39.997
21	8.034	10.283	11.591	32.671	35.479	41.401
22	8.643	10.982	12.338	33.924	36.781	42.796
23	9.260	11.689	13.091	35.172	38.076	44.181
24	9.886	12.401	13.848	36.415	39.364	45.559
25	10.520	13.120	14.611	37.652	40.646	46.928
26	11.160	13.844	15.379	38.885	41.923	48.290
27	11.808	14.573	16.151	40.113	43.194	49.645
28	12.461	15.308	16.928	41.337	44.461	50.993
29	13.121	16.047	17.708	42.557	45.722	52.336
30	13.787	16.791	18.493	43.773	46.979	53.672
31	14.458	17.539	19.281	44.985	48.232	55.003
32	15.134	18.291	20.072	46.194	49.480	56.328
33	15.815	19.047	20.867	47.400	50.725	57.648
34	16.501	19.806	21.664	48.602	51.966	58.964
35	17.192	20.569	22.465	49.802	53.203	60.275
36	17.887	21.336	23.269	50.998	54.437	61.581
37	18.586	22.106	24.075	52.192	55.668	62.883
38	19.289	22.878	24.884	53.384	56.896	64.181
39	19.996	23.654	25.695	54.572	58.120	65.476
40	20.707	24.433	26.509	55.758	59.342	66.766
41	21.421	25.215	27.326	56.942	60.561	68.053
42	22.138	25.999	28.144	58.124	61.777	69.336
43	22.859	26.785	28.965	59.304	62.990	70.616
44	23.584	27.575	29.787	60.481	64.201	71.893
45	24.311	28.366	30.612	61.656	65.410	73.166

[a] Pr(χ^2 r.v. with r degrees of freedom $\leq$ tabled value) $= \gamma$. Source: D. B. Owen (1962), *Handbook of Statistical Tables*, Addison-Wesley, Reading, Mass. Courtesy of U.S. Atomic Energy Commission.

Table A9.1 (*Cont.*)

	γ					
r	0.005	0.025	0.05	0.95	0.975	0.995
46	25.041	29.160	31.439	62.830	66.617	74.437
47	25.775	29.956	32.268	64.001	67.821	75.704
48	26.511	30.755	33.098	65.171	69.023	76.969
49	27.249	31.555	33.930	66.339	70.222	78.231
50	27.991	32.357	34.764	67.505	71.420	79.490
51	28.735	33.162	35.600	68.669	72.616	80.747
52	29.481	33.968	36.437	69.832	73.810	82.001
53	30.230	34.776	37.276	70.993	75.002	83.253
54	30.981	35.586	38.116	72.153	76.192	84.502
55	31.735	36.398	38.958	73.311	77.380	85.749
56	32.490	37.212	39.801	74.468	78.567	86.994
57	33.248	38.027	40.646	75.624	79.752	88.236
58	34.008	38.844	41.492	76.778	80.936	89.477
59	34.770	39.662	42.339	77.931	82.117	90.715
60	35.534	40.482	43.188	79.082	83.298	91.952
61	36.300	41.303	44.038	80.232	84.476	93.186
62	37.068	42.126	44.889	81.381	85.654	94.419
63	37.838	42.950	45.741	82.529	86.830	95.649
64	38.610	43.776	46.595	83.675	88.004	96.878
65	39.383	44.603	47.450	84.821	89.177	98.105
66	40.158	45.431	48.305	85.965	90.349	99.330
67	40.935	46.261	49.162	87.108	91.519	100.554
68	41.713	47.092	50.020	88.250	92.689	101.776
69	42.494	47.924	50.879	89.391	93.856	102.996
70	43.275	48.758	51.739	90.531	95.023	104.215
71	44.058	49.592	52.600	91.670	96.189	105.432
72	44.843	50.428	53.462	92.808	97.353	106.648
73	45.629	51.265	54.325	93.945	98.516	107.862
74	46.417	52.103	55.189	95.081	99.678	109.074
75	47.206	52.942	56.054	96.217	100.839	110.286
76	47.997	53.782	56.920	97.351	101.999	111.495
77	48.788	54.623	57.786	98.484	103.158	112.704
78	49.582	55.466	58.654	99.617	104.316	113.911
79	50.376	56.309	59.522	100.749	105.473	115.117
80	51.172	57.153	60.391	101.879	106.629	116.321
81	51.969	57.998	61.261	103.010	107.783	117.524
82	52.767	58.845	62.132	104.139	108.937	118.726
83	53.567	59.692	63.004	105.267	110.090	119.927
84	54.368	60.540	63.876	106.395	111.242	121.126
85	55.170	61.389	64.749	107.522	112.393	122.325
86	55.973	62.239	65.623	108.648	113.544	123.522
87	56.777	63.089	66.498	109.773	114.693	124.718
88	57.582	63.941	67.373	110.898	115.841	125.913
89	58.389	64.793	68.249	112.022	116.989	127.106
90	59.196	65.647	69.126	113.145	118.136	128.299

Table A9.1 (*Cont.*)

	γ					
r	0.005	0.025	0.05	0.95	0.975	0.995
91	60.005	66.501	70.003	114.268	119.282	129.491
92	60.815	67.356	70.882	115.390	120.427	130.681
93	61.625	68.211	71.760	116.511	121.571	131.871
94	62.437	69.068	72.640	117.632	122.715	133.059
95	63.250	69.925	73.520	118.752	123.858	134.247
96	64.063	70.783	74.401	119.871	125.000	135.433
97	64.878	71.642	75.282	120.990	126.141	136.619
98	65.694	72.501	76.164	122.108	127.282	137.803
99	66.510	73.361	77.046	123.225	128.422	138.987
100	67.328	74.222	77.929	124.342	129.561	140.169
102	68.965	75.946	79.697	126.574	131.838	142.532
104	70.606	77.672	81.468	128.804	134.111	144.891
106	72.251	79.401	83.240	131.031	136.382	147.247
108	73.899	81.133	85.015	133.257	138.651	149.599
110	75.550	82.867	86.792	135.480	140.917	151.948
112	77.204	84.604	88.570	137.701	143.180	154.294
114	78.862	86.342	90.351	139.921	145.441	156.637
116	80.522	88.084	92.134	142.138	147.700	158.977
118	82.185	89.827	93.918	144.354	149.957	161.314
120	83.852	91.573	95.705	146.567	152.211	163.648
122	85.520	93.320	97.493	148.779	154.464	165.980
124	87.192	95.070	99.283	150.989	156.714	168.308
126	88.866	96.822	101.074	153.198	158.962	170.634
128	90.543	98.576	102.867	155.405	161.209	172.957
130	92.222	100.331	104.662	157.610	163.453	175.278
132	93.904	102.089	106.459	159.814	165.696	177.597
134	95.588	103.848	108.257	162.016	167.936	179.913
136	97.275	105.609	110.056	164.216	170.175	182.226
138	98.964	107.372	111.857	166.415	172.412	184.538
140	100.655	109.137	113.659	168.613	174.648	186.847
142	102.348	110.903	115.463	170.809	176.882	189.154
144	104.044	112.671	117.268	173.004	179.114	191.458
146	105.741	114.441	119.075	175.198	181.344	193.761
148	107.441	116.212	120.883	177.390	183.573	196.062
150	109.142	117.985	122.692	179.581	185.800	198.360
200	152.241	162.728	168.279	233.994	241.058	255.264
250	196.161	208.098	214.392	287.882	295.689	311.346
300	240.663	253.912	260.878	341.395	349.874	366.844
400	330.903	346.482	354.641	447.632	457.305	476.606
500	422.303	439.936	449.147	553.127	563.852	585.207
600	514.529	534.019	544.180	658.094	669.769	692.982
700	607.380	628.577	639.613	762.661	775.211	800.131
800	700.725	723.513	735.362	866.911	880.275	906.786
900	794.475	818.756	831.370	970.904	985.032	1013.036
1000	888.564	914.257	927.594	1074.679	1089.531	1118.948

Table A9.2

Tabled Value Is the Length of the $100(1-\alpha)\%$ *Equal Tail Confidence Interval for Log Spectral Density Based on an Estimate with r Degrees of Freedom*[a]

	$1-\alpha$				$1-\alpha$		
r	0.90	0.95	0.99	r	0.90	0.95	0.99
2	4.07	4.39	6.96	90	0.492	0.587	0.773
3	3.10	3.77	5.19	100	0.467	0.557	0.733
4	2.59	3.14	4.27	120	0.427	0.508	0.669
5	2.26	2.73	3.71	140	0.394	0.470	0.618
				160	0.368	0.440	0.578
6	2.04	2.45	3.31				
7	1.87	2.25	3.02	180	0.348	0.414	0.545
8	1.73	2.09	2.80	200	0.329	0.393	0.517
9	1.63	1.95	2.61	220	0.134	0.375	0.492
10	1.54	1.84	2.46	240	0.300	0.360	0.472
				260	0.288	0.345	0.453
11	1.46	1.75	2.33				
12	1.40	1.67	2.22	280	0.278	0.333	0.437
13	1.34	1.59	2.12	300	0.269	0.321	0.422
14	1.28	1.53	2.04	400	0.232	0.278	0.365
15	1.24	1.48	1.96	500	0.209	0.248	0.326
				600	0.191	0.227	0.299
16	1.20	1.43	1.90				
17	1.15	1.39	1.83	700	0.176	0.209	0.276
18	1.12	1.35	1.78	800	0.165	0.196	0.259
19	1.10	1.30	1.73	900	0.156	0.185	0.243
20	1.06	1.27	1.68	1,000	0.147	0.176	0.230
				2,000	0.103	0.124	0.163
21	1.04	1.24	1.64	3,000	0.084	0.101	0.133
22	1.01	1.21	1.60	4,000	0.073	0.088	0.114
23	0.989	1.18	1.56	5,000	0.065	0.078	0.103
24	0.965	1.15	1.53	10,000	0.047	0.055	0.073
25	0.952	1.13	1.49				
26	0.927	1.11	1.46				
27	0.915	1.08	1.44				
28	0.889	1.06	1.41				
29	0.877	1.05	1.39				
30	0.864	1.03	1.36				
40	0.743	0.887	1.17				
50	0.663	0.792	1.04				
60	0.605	0.721	0.951				
70	0.560	0.667	0.880				
80	0.523	0.624	0.822				

[a] Source: Computed from Table 3.3, D. B. Owen (1962), *Handbook of Statistical Tables.* Addison-Wesley, Reading, Massachusetts. Courtesy of U.S. Atomic Energy Commission.

Table A9.3

Critical Values of Student's t-Distribution[a]

k \ γ	.75	.90	.95	.975	.99	.995	.9995
1	1.000	3.078	6.314	12.706	31.821	63.657	636.619
2	.816	1.886	2.920	4.303	6.965	9.925	31.598
3	.765	1.638	2.353	3.182	4.541	5.841	12.941
4	.741	1.533	2.132	2.776	3.747	4.604	8.610
5	.727	1.476	2 015	2.571	3.365	4.032	6.859
6	.718	1.440	1.943	2.447	3.143	3.707	5.959
7	.711	1.415	1.895	2.365	2.998	3.499	5.405
8	.706	1.397	1.860	2.306	2.896	3.355	5.041
9	.703	1.383	1.833	2.262	2.821	3.250	4.781
10	.700	1.372	1.812	2.228	2.764	3.169	4.587
11	.697	1.363	1.796	2.201	2.718	3.106	4.437
12	.695	1.356	1.782	2.179	2.681	3.055	4.318
13	.694	1.350	1.771	2.160	2.650	3.012	4.221
14	.692	1.345	1.761	2.145	2.624	2.977	4.140
15	.691	1.341	1.753	2.131	2.602	2.947	4.073
16	.690	1.337	1.746	2.120	2.583	2.921	4.015
17	.689	1.333	1.740	2.110	2.567	2.898	3.965
18	.688	1.330	1.734	2.101	2.552	2.878	3.922
19	.688	1.328	1.729	2.093	2.539	2.861	3.883
20	.687	1.325	1.725	2.086	2.528	2.845	3.850
21	.686	1.323	1.721	2.080	2.518	2.831	3.819
22	.686	1.321	1.717	2.074	2.508	2.819	3.792
23	.685	1.319	1.714	2.069	2.500	2.807	3.767
24	.685	1.318	1.711	2.064	2.492	2.797	3.745
25	.684	1.316	1.708	2.060	2.485	2.787	3.725
26	.684	1.315	1.706	2.056	2.479	2.779	3.707
27	.684	1.314	1.703	2.052	2.473	2.771	3.690
28	.683	1.313	1.701	2.048	2.467	2.763	3.674
29	.683	1.311	1.699	2.045	2.462	2.756	3.659
30	.683	1.310	1.697	2.042	2.457	2.750	3.646
40	.681	1.303	1.684	2.021	2.423	2.704	3.551
60	.679	1.296	1.671	2.000	2.390	2.660	3.460
120	.677	1.289	1.658	1.980	2.358	2.617	3.373
∞	.674	1.282	1.645	1.960	2.326	2.576	3.291

[a] $P(t$ r.v. with k degrees of freedom $\leq$ tabled value$) = \gamma$. Source: Table abridged from Table III of Fisher, R. A. and Yates, F. (1938). *Statistical Tables for Biological, Agricultural, and Medical Research.* Oliver & Boyd, Edinburgh, by permission of the authors and publishers.

Table A9.4

Critical Values of Fisher's F-Distribution[a]

γ	l \ k	1	2	3	4	5	6	7	8	9	10	12	15	20	30	60	120	∞
.90	1	39.9	49.5	53.6	55.8	57.2	58.2	58.9	59.4	59.9	60.2	60.7	61.2	61.7	62.3	62.8	63.1	63.3
.95	1	161	200	216	225	230	234	237	239	241	242	244	246	248	250	252	253	254
.975	1	648	800	864	900	922	937	948	957	963	969	977	985	993	1000	1010	1010	1020
.99	1	4,050	5,000	5,400	5,620	5,760	5,860	5,930	5,980	6,020	6,060	6,110	6,160	6,210	6,260	6,310	6,340	6,370
.995	1	16,200	20,000	21,600	22,500	23,100	23,400	23,700	23,900	24,100	24,200	24,400	24,600	24,800	25,000	25,200	25,400	25,500
.90	2	8.53	9.00	9.16	9.24	9.29	9.33	9.35	9.37	9.38	9.39	9.41	9.42	9.44	9.46	9.47	9.48	9.49
.95	2	18.5	19.0	19.2	19.2	19.3	19.3	19.4	19.4	19.4	19.4	19.4	19.4	19.5	19.5	19.5	19.5	19.5
.975	2	38.5	39.0	39.2	39.2	39.3	39.3	39.4	39.4	39.4	39.4	39.4	39.4	39.4	39.5	39.5	39.5	39.5
.99	2	98.5	99.0	99.2	99.2	99.3	99.3	99.4	99.4	99.4	99.4	99.4	99.4	99.4	99.5	99.5	99.5	99.5
.995	2	199	199	199	199	199	199	199	199	199	199	199	199	199	199	199	199	199
.90	3	5.54	5.46	5.39	5.34	5.31	5.28	5.27	5.25	5.24	5.23	5.22	5.20	5.18	5.17	5.15	5.14	5.13
.95	3	10.1	9.55	9.28	9.12	9.01	8.94	8.89	8.85	8.81	8.79	8.74	8.70	8.66	8.62	8.57	8.55	8.53
.975	3	17.4	16.0	15.4	15.1	14.9	14.7	14.6	14.5	14.5	14.4	14.3	14.3	14.2	14.1	14.0	13.9	13.9
.99	3	34.1	30.8	29.5	28.7	28.2	27.9	27.7	27.5	27.3	27.2	27.1	26.9	26.7	26.5	26.3	26.2	26.1
.995	3	55.6	49.8	47.5	46.2	45.4	44.8	44.4	44.1	43.9	43.7	43.4	43.1	42.8	42.5	42.1	42.0	41.8
.90	4	4.54	4.32	4.19	4.11	4.05	4.01	3.98	3.95	3.93	3.92	3.90	3.87	3.84	3.82	3.79	3.78	3.76
.95	4	7.71	6.94	6.59	6.39	6.26	6.16	6.09	6.04	6.00	5.96	5.91	5.86	5.80	5.75	5.69	5.66	5.63
.975	4	12.2	10.6	9.98	9.60	9.36	9.20	9.07	8.98	8.90	8.84	8.75	8.66	8.56	8.46	8.36	8.31	8.26
.99	4	21.2	18.0	16.7	16.0	15.5	15.2	15.0	14.8	14.7	14.5	14.4	14.2	14.0	13.8	13.7	13.6	13.5
.995	4	31.3	26.3	24.3	23.2	22.5	22.0	21.6	21.4	21.1	21.0	20.7	20.4	20.2	19.9	19.6	19.5	19.3
.90	5	4.06	3.78	3.62	3.52	3.45	3.40	3.37	3.34	3.32	3.30	3.27	3.24	3.21	3.17	3.14	3.12	3.11
.95	5	6.61	5.79	5.41	5.19	5.05	4.95	4.88	4.82	4.77	4.74	4.68	4.62	4.56	4.50	4.43	4.40	4.37
.975	5	10.0	8.43	7.76	7.39	7.15	6.98	6.85	6.76	6.68	6.62	6.52	6.43	6.33	6.23	6.12	6.07	6.02
.99	5	16.3	13.3	12.1	11.4	11.0	10.7	10.5	10.3	10.2	10.1	9.89	9.72	9.55	9.38	9.20	9.11	9.02
.995	5	22.8	18.3	16.5	15.6	14.9	14.5	14.2	14.0	13.8	13.6	13.4	13.1	12.9	12.7	12.4	12.3	12.1
.90	6	3.78	3.46	3.29	3.18	3.11	3.05	3.01	2.98	2.96	2.94	2.90	2.87	2.84	2.80	2.76	2.74	2.72
.95	6	5.99	5.14	4.76	4.53	4.39	4.28	4.21	4.15	4.10	4.06	4.00	3.94	3.87	3.81	3.74	3.70	3.67
.975	6	8.81	7.26	6.60	6.23	5.99	5.82	5.70	5.60	5.52	5.46	5.37	5.27	5.17	5.07	4.96	4.90	4.85
.99	6	13.7	10.9	9.78	9.15	8.75	8.47	8.26	8.10	7.98	7.87	7.72	7.56	7.40	7.23	7.06	6.97	6.88
.995	6	18.6	14.5	12.9	12.0	11.5	11.1	10.8	10.6	10.4	10.2	10.0	9.81	9.59	9.36	9.12	9.00	8.88
.90	7	3.59	3.26	3.07	2.96	2.88	2.83	2.78	2.75	2.72	2.70	2.67	2.63	2.59	2.56	2.51	2.49	2.47
.95	7	5.59	4.74	4.35	4.12	3.97	3.87	3.79	3.73	3.68	3.64	3.57	3.51	3.44	3.38	3.30	3.27	3.23
.975	7	8.07	6.54	5.89	5.52	5.29	5.12	4.99	4.90	4.82	4.76	4.67	4.57	4.47	4.36	4.25	4.20	4.14
.99	7	12.2	9.55	8.45	7.85	7.46	7.19	6.99	6.84	6.72	6.62	6.47	6.31	6.16	5.99	5.82	5.74	5.65
.995	7	16.2	12.4	10.9	10.1	9.52	9.16	8.89	8.68	8.51	8.38	8.18	7.97	7.75	7.53	7.31	7.19	7.08
.90	8	3.46	3.11	2.92	2.81	2.73	2.67	2.62	2.59	2.56	2.54	2.50	2.46	2.42	2.38	2.34	2.31	2.29
.95	8	5.32	4.46	4.07	3.84	3.69	3.58	3.50	3.44	3.39	3.35	3.28	3.22	3.15	3.08	3.01	2.97	2.93
.975	8	7.57	6.06	5.42	5.05	4.82	4.65	4.53	4.43	4.36	4.30	4.20	4.10	4.00	3.89	3.78	3.73	3.67
.99	8	11.3	8.65	7.59	7.01	6.63	6.37	6.18	6.03	5.91	5.81	5.67	5.52	5.36	5.20	5.03	4.95	4.86
.995	8	14.7	11.0	9.60	8.81	8.30	7.95	7.69	7.50	7.34	7.21	7.01	6.81	6.61	6.40	6.18	6.06	5.95

.90		3.36	3.01	2.81	2.69	2.61	2.55	2.51	2.47	2.44	2.42	2.38	2.34	2.30	2.25	2.21	2.18	2.16
.95		5.12	4.26	3.86	3.63	3.48	3.37	3.29	3.23	3.18	3.14	3.07	3.01	2.94	2.86	2.79	2.75	2.71
.975	9	7.21	5.71	5.08	4.72	4.48	4.32	4.20	4.10	4.03	3.96	3.87	3.77	3.67	3.56	3.45	3.39	3.33
.99		10.6	8.02	6.99	6.42	6.06	5.80	5.61	5.47	5.35	5.26	5.11	4.96	4.81	4.65	4.48	4.40	4.31
.995		13.6	10.1	8.72	7.96	7.47	7.13	6.88	6.69	6.54	6.42	6.23	6.03	5.83	5.62	5.41	5.30	5.19
.90		3.29	2.92	2.73	2.61	2.52	2.46	2.41	2.38	2.35	2.32	2.28	2.24	2.20	2.15	2.11	2.08	2.06
.95		4.96	4.10	3.71	3.48	3.33	3.22	3.14	3.07	3.02	2.98	2.91	2.84	2.77	2.70	2.62	2.58	2.54
.975	10	6.94	5.46	4.83	4.47	4.24	4.07	3.95	3.85	3.78	3.72	3.62	3.52	3.42	3.31	3.20	3.14	3.08
.99		10.0	7.56	6.55	5.99	5.64	5.39	5.20	5.06	4.94	4.85	4.71	4.56	4.41	4.25	4.08	4.00	3.91
.995		12.8	9.43	8.08	7.34	6.87	6.54	6.30	6.12	5.97	5.85	5.66	5.47	5.27	5.07	4.86	4.75	4.64
.90		3.18	2.81	2.61	2.48	2.39	2.33	2.28	2.24	2.21	2.19	2.15	2.10	2.06	2.01	1.96	1.93	1.90
.95		4.75	3.89	3.49	3.26	3.11	3.00	2.91	2.85	2.80	2.75	2.69	2.62	2.54	2.47	2.38	2.34	2.30
.975	12	6.55	5.10	4.47	4.12	3.89	3.73	3.61	3.51	3.44	3.37	3.28	3.18	3.07	2.96	2.85	2.79	2.72
.99		9.33	6.93	5.95	5.41	5.06	4.82	4.64	4.50	4.39	4.30	4.16	4.01	3.86	3.70	3.54	3.45	3.36
.995		11.8	8.51	7.23	6.52	6.07	5.76	5.52	5.35	5.20	5.09	4.91	4.72	4.53	4.33	4.12	4.01	3.90
.90		3.07	2.70	2.49	2.36	2.27	2.21	2.16	2.12	2.09	2.06	2.02	1.97	1.92	1.87	1.82	1.79	1.76
.95		4.54	3.68	3.29	3.06	2.90	2.79	2.71	2.64	2.59	2.54	2.48	2.40	2.33	2.25	2.16	2.11	2.07
.975	15	6.20	4.77	4.15	3.80	3.58	3.41	3.29	3.20	3.12	3.06	2.96	2.86	2.76	2.64	2.52	2.46	2.40
.99		8.68	6.36	5.42	4.89	4.56	4.32	4.14	4.00	3.89	3.80	3.67	3.52	3.37	3.21	3.05	2.96	2.87
.995		10.8	7.70	6.48	5.80	5.37	5.07	4.85	4.67	4.54	4.42	4.25	4.07	3.88	3.69	3.48	3.37	3.26
.90		2.97	2.59	2.38	2.25	2.16	2.09	2.04	2.00	1.96	1.94	1.89	1.84	1.79	1.74	1.68	1.64	1.61
.95		4.35	3.49	3.10	2.87	2.71	2.60	2.51	2.45	2.39	2.35	2.28	2.20	2.12	2.04	1.95	1.90	1.84
.975	20	5.87	4.46	3.86	3.51	3.29	3.13	3.01	2.91	2.84	2.77	2.68	2.57	2.46	2.35	2.22	2.16	2.09
.99		8.10	5.85	4.94	4.43	4.10	3.87	3.70	3.56	3.46	3.37	3.23	3.09	2.94	2.78	2.61	2.52	2.42
.995		9.94	6.99	5.82	5.17	4.76	4.47	4.26	4.09	3.96	3.85	3.68	3.50	3.32	3.12	2.92	2.81	2.69
.90		2.88	2.49	2.28	2.14	2.05	1.98	1.93	1.88	1.85	1.82	1.77	1.72	1.67	1.61	1.54	1.50	1.46
.95		4.17	3.32	2.92	2.69	2.53	2.42	2.33	2.27	2.21	2.16	2.09	2.01	1.93	1.84	1.74	1.68	1.62
.975	30	5.57	4.18	3.59	3.25	3.03	2.87	2.75	2.65	2.57	2.51	2.41	2.31	2.20	2.07	1.94	1.87	1.79
.99		7.56	5.39	4.51	4.02	3.70	3.47	3.30	3.17	3.07	2.98	2.84	2.70	2.55	2.39	2.21	2.11	2.01
.995		9.18	6.35	5.24	4.62	4.23	3.95	3.74	3.58	3.45	3.34	3.18	3.01	2.82	2.63	2.42	2.30	2.18
.90		2.79	2.39	2.18	2.04	1.95	1.87	1.82	1.77	1.74	1.71	1.66	1.60	1.54	1.48	1.40	1.35	1.29
.95		4.00	3.15	2.76	2.53	2.37	2.25	2.17	2.10	2.04	1.99	1.92	1.84	1.75	1.65	1.53	1.47	1.39
.975	60	5.29	3.93	3.34	3.01	2.79	2.63	2.51	2.41	2.33	2.27	2.17	2.06	1.94	1.82	1.67	1.58	1.48
.99		7.08	4.98	4.13	3.65	3.34	3.12	2.95	2.82	2.72	2.63	2.50	2.35	2.20	2.03	1.84	1.73	1.60
.995		8.49	5.80	4.73	4.14	3.76	3.49	3.29	3.13	3.01	2.90	2.74	2.57	2.39	2.19	1.96	1.83	1.69
.90		2.75	2.35	2.13	1.99	1.90	1.82	1.77	1.72	1.68	1.65	1.60	1.54	1.48	1.41	1.32	1.26	1.19
.95		3.92	3.07	2.68	2.45	2.29	2.18	2.09	2.02	1.96	1.91	1.83	1.75	1.66	1.55	1.43	1.35	1.25
.975	120	5.15	3.80	3.23	2.89	2.67	2.52	2.39	2.30	2.22	2.16	2.05	1.94	1.82	1.69	1.53	1.43	1.31
.99		6.85	4.79	3.95	3.48	3.17	2.96	2.79	2.66	2.56	2.47	2.34	2.19	2.03	1.86	1.66	1.53	1.38
.995		8.18	5.54	4.50	3.92	3.55	3.28	3.09	2.93	2.81	2.71	2.54	2.37	2.19	1.98	1.75	1.61	1.43
.90		2.71	2.30	2.08	1.94	1.85	1.77	1.72	1.67	1.63	1.60	1.55	1.49	1.42	1.34	1.24	1.17	1.00
.95		3.84	3.00	2.60	2.37	2.21	2.10	2.01	1.94	1.88	1.83	1.75	1.67	1.57	1.46	1.32	1.22	1.00
.975	∞	5.02	3.69	3.12	2.79	2.57	2.41	2.29	2.19	2.11	2.05	1.94	1.83	1.71	1.57	1.39	1.27	1.00
.99		6.63	4.61	3.78	3.32	3.02	2.80	2.64	2.51	2.41	2.32	2.18	2.04	1.88	1.70	1.47	1.32	1.00
.995		7.88	5.30	4.28	3.72	3.35	3.09	2.90	2.74	2.62	2.52	2.36	2.19	2.00	1.79	1.53	1.36	1.00

[a] $P(F$ r.v. with k, l degrees of freedom $\leq$ tabled value$) = \gamma$. Source: Merrington, M., and Thompson, C. M. (1943). "Tables of percentage points of the inverted beta distribution." *Biometrika* 33.

Table A9.5 *Upper and Lower Confidence Limits For Multiple Coherence* R^2 [a]

	$(n-q)$ Values											
$\hat{R}$[b]	5		10		20		40		80		160	
0.0	0.00	0.00	0.00	0.00	0.00	0.00	0.00	0.00	0.00	0.00	0.00	0.00
0.1	0.42	0.00	0.38	0.00	0.32	0.00	0.26	0.00	0.21	0.00	0.18	0.03
0.2	0.57	0.00	0.52	0.00	0.45	0.00	0.39	0.01	0.34	0.06	0.30	0.10
0.3	0.67	0.00	0.61	0.00	0.55	0.00	0.49	0.07	0.44	0.14	0.40	0.18
0.4	0.75	0.00	0.69	0.00	0.64	0.05	0.58	0.15	0.54	0.23	0.50	0.28
0.5	0.80	0.00	0.75	0.00	0.71	0.14	0.66	0.25	0.62	0.33	(0.60)	0.39
0.6	0.86	0.00	0.82	0.07	0.78	0.25	0.74	0.37	0.71	0.45	(0.69)	0.50
0.7	0.89	0.00	0.87	0.21	0.84	0.39	0.81	0.50	0.79	0.57	(0.77)	(0.62)
0.8	0.93	0.09	0.92	0.39	0.89	0.56	0.88	0.64	(0.84)	0.70	(0.85)	(0.74)
0.9	0.96	0.38	0.96	0.64	0.95	0.75	0.94	0.80	(0.93)	(0.84)	(0.92)	(0.87)
1.00	1.00	1.00	1.00	1.00	1.00	1.00	1.00	1.00	1.00	1.00	1.00	1.00
$\hat{R}$[c]												
0.0	0.00	0.00	0.00	0.00	0.00	0.00	0.00	0.00	0.00	0.00	0.00	0.00
0.1	0.24	0.00	0.26	0.00	0.24	0.00	0.21	0.00	0.18	0.02	0.16	0.04
0.2	0.43	0.00	0.41	0.00	0.37	0.00	0.33	0.04	0.30	0.08	0.27	0.12
0.3	0.55	0.00	0.52	0.00	0.48	0.04	0.44	0.11	0.40	0.17	0.38	0.21
0.4	0.64	0.00	0.61	0.00	0.57	0.11	0.54	0.21	0.50	0.27	0.47	0.31
0.5	0.72	0.00	0.69	0.07	0.66	0.21	0.62	0.31	0.59	0.37	(0.58)	0.41
0.6	0.79	0.00	0.76	0.18	0.74	0.33	0.71	0.43	0.68	0.49	(0.67)	(0.52)
0.7	0.85	0.08	0.83	0.33	0.81	0.47	0.79	0.56	0.76	0.61	(0.75)	(0.64)
0.8	0.90	0.29	0.89	0.51	0.87	0.63	0.86	0.69	(0.85)	0.73	(0.84)	(0.76)
0.9	0.96	0.57	0.95	0.73	0.94	0.80	0.93	0.84	(0.93)	(0.86)	(0.92)	(0.88)
1.0	1.00	1.00	1.00	1.00	1.00	1.00	1.00	1.00	1.00	1.00	1.00	1.00

[a] Values in parentheses are interpolated, not actually computed, but are believed accurate to at worst ±0.02. Source: Groves, G. W., and Hannan, E. J. (1968). "Time series regression of sea level on weather," *Rev. Geophys. 6*, 129–174. Copyright by American Geophysical Union. [b] Values are for $q = 2$ and $1 - \alpha = 0.99$. [c] Values are for $q = 2$ and $1 - \alpha = 0.95$.

Table A9.5 (*cont.*)

$\hat{R}^d$	(n − q) Values											
	5		10		20		40		80		160	
0.0	0.00	0.00	0.00	0.00	0.00	0.00	0.00	0.00	0.00	0.00	0.00	0.00
0.1	0.00	0.00	0.17	0.00	0.21	0.00	0.21	0.00	0.19	0.00	0.17	0.02
0.2	0.32	0.00	0.37	0.00	0.37	0.00	0.35	0.00	0.32	0.04	0.29	0.09
0.3	0.49	0.00	0.51	0.00	0.49	0.00	0.46	0.04	0.43	0.12	0.39	0.18
0.4	0.61	0.00	0.61	0.00	0.59	0.00	0.56	0.12	0.52	0.21	0.50	0.27
0.5	0.70	0.00	0.70	0.00	0.68	0.08	0.64	0.22	0.61	0.32	(0.58)	0.38
0.6	0.78	0.00	0.77	0.00	0.75	0.20	0.72	0.35	0.70	0.43	(0.67)	0.49
0.7	0.84	0.00	0.83	0.11	0.82	0.35	0.80	0.48	0.78	0.56	(0.76)	(0.60)
0.8	0.90	0.00	0.89	0.32	0.89	0.53	0.87	0.64	(0.85)	0.69	(0.84)	(0.72)
0.9	0.96	0.25	0.95	0.59	(0.95)	(0.72)	(0.94)	(0.81)	(0.92)	(0.85)	(0.91)	(0.86)
1.0	1.00	1.00	1.00	1.00	1.00	1.00	1.00	1.00	1.00	1.00	1.00	1.00
$\hat{R}^e$												
0.0	0.00	0.00	0.00	0.00	0.00	0.00	0.00	0.00	0.00	0.00	0.00	0.00
0.1	0.00	0.00	0.03	0.00	0.13	0.00	0.16	0.00	0.16	0.00	0.14	0.03
0.2	0.13	0.00	0.25	0.00	0.29	0.00	0.29	0.00	0.28	0.07	0.26	0.11
0.3	0.33	0.00	0.40	0.00	0.42	0.00	0.41	0.08	0.39	0.15	0.37	0.20
0.4	0.48	0.00	0.52	0.00	0.53	0.05	0.51	0.18	0.49	0.25	0.46	0.30
0.5	0.60	0.00	0.62	0.00	0.62	0.16	0.61	0.29	0.58	0.36	(0.56)	0.41
0.6	0.70	0.00	0.71	0.07	0.71	0.29	0.69	0.40	0.67	0.47	(0.66)	(0.52)
0.7	0.79	0.00	0.79	0.24	0.79	0.43	0.77	0.54	0.76	0.60	(0.74)	(0.64)
0.8	0.86	0.11	0.86	0.44	0.86	0.60	0.85	0.68	(0.84)	0.72	(0.83)	(0.76)
0.9	0.93	0.47	0.93	0.69	0.93	0.79	0.93	0.83	(0.92)	(0.86)	(0.92)	(0.88)
1.00	1.00	1.00	1.00	1.00	1.00	1.00	1.00	1.00	1.00	1.00	1.00	1.00

[d] Values are for $q = 4$ and $1 - \alpha = 0.99$.
[e] Values are for $q = 4$ and $1 - \alpha = 0.95$.

Table A9.5 (*cont.*)

$\hat{R}^f$	(n − q) Values: 5		10		20		40		80		160	
0.0	0.00	0.00	0.00	0.00	0.00	0.00	0.00	0.00	0.00	0.00	0.00	0.00
0.1	0.00	0.00	0.00	0.00	0.00	0.00	0.11	0.00	0.14	0.00	0.14	0.00
0.2	0.00	0.00	0.07	0.00	0.21	0.00	0.27	0.00	0.28	0.01	0.27	0.07
0.3	0.08	0.00	0.29	0.00	0.37	0.00	0.39	0.00	0.39	0.08	0.38	0.16
0.4	0.31	0.00	0.44	0.00	0.50	0.00	0.50	0.06	0.50	0.18	0.48	0.25
0.5	0.48	0.00	0.57	0.00	0.60	0.00	0.61	0.17	0.59	0.29	(0.58)	0.36
0.6	0.62	0.00	0.68	0.00	0.69	0.10	0.69	0.29	0.68	0.41	(0.67)	0.48
0.7	0.74	0.00	0.77	0.00	0.78	0.26	0.78	0.44	(0.77)	0.54	(0.76)	(0.60)
0.8	0.83	0.00	0.86	0.16	0.86	0.46	0.86	0.61	(0.85)	0.68	(0.84)	(0.72)
0.9	0.92	0.00	0.93	0.50	(0.94)	(0.68)	(0.94)	(0.78)	(0.93)	(0.83)	(0.93)	(0.86)
1.00	1.00	1.00	1.00	1.00	1.00	1.00	1.00	1.00	1.00	1.00	1.00	1.00
$\hat{R}^g$												
0.0	0.00	0.00	0.00	0.00	0.00	0.00	0.00	0.00	0.00	0.00	0.00	0.00
0.1	0.00	0.00	0.00	0.00	0.00	0.00	0.05	0.00	0.11	0.00	0.12	0.00
0.2	0.00	0.00	0.00	0.00	0.13	0.00	0.21	0.00	0.24	0.03	0.25	0.09
0.3	0.00	0.00	0.15	0.00	0.29	0.00	0.34	0.01	0.36	0.12	0.36	0.18
0.4	0.13	0.00	0.33	0.00	0.43	0.00	0.46	0.11	0.46	0.22	0.46	0.29
0.5	0.34	0.00	0.47	0.00	0.54	0.04	0.56	0.22	0.56	0.33	(0.55)	0.39
0.6	0.50	0.00	0.61	0.00	0.64	0.18	0.66	0.36	0.65	0.45	(0.65)	(0.50)
0.7	0.65	0.00	0.71	0.06	0.74	0.35	0.75	0.50	0.75	0.58	(0.75)	(0.62)
0.8	0.78	0.00	0.82	0.31	0.83	0.54	0.83	0.65	(0.83)	0.71)	(0.83)	(0.74)
0.9	0.89	0.27	0.91	0.61	0.92	0.75	0.92	0.82	(0.92)	(0.86)	(0.92)	(0.87)
1.00	1.00	1.00	1.00	1.00	1.00	1.00	1.00	1.00	1.00	1.00	1.00	1.00

[f] Values are for $q = 8$ and $1 - \alpha = 0.99$.
[g] Values are for $q = 8$ and $1 - \alpha = 0.95$.

Table A9.5 (*cont.*)

$\hat{R}^h$	(n − q) Values											
	5		10		20		40		80		160	
0.0	0.00	0.00	0.00	0.00	0.00	0.00	0.00	0.00	0.00	0.00	0.00	0.00
0.1	0.00	0.00	0.00	0.00	0.00	0.00	0.00	0.00	0.04	0.00	0.10	0.00
0.2	0.00	0.00	0.00	0.00	0.00	0.00	0.11	0.00	0.20	0.00	0.23	0.04
0.3	0.00	0.00	0.00	0.00	0.12	0.00	0.27	0.00	0.32	0.02	0.35	0.12
0.4	0.00	0.00	0.08	0.00	0.30	0.00	0.40	0.00	0.44	0.12	0.45	0.22
0.5	0.02	0.00	0.30	0.00	0.45	0.00	0.52	0.06	0.55	0.24	(0.55)	0.34
0.6	0.29	0.00	0.48	0.00	0.58	0.00	0.63	0.20	0.64	0.36	(0.64)	0.46
0.7	0.50	0.00	0.63	0.00	0.70	0.09	0.73	0.36	0.74	0.50	(0.74)	(0.59)
0.8	0.69	0.00	0.77	0.00	0.81	0.33	0.82	0.55	(0.83)	0.65	(0.83)	(0.72)
0.9	0.86	0.00	0.89	0.31	(0.91)	(0.59)	(0.91)	(0.74)	(0.91)	(0.80)	(0.91)	(0.85)
1.0	1.00	1.00	1.00	1.00	1.00	1.00	1.00	1.00	1.00	1.00	1.00	1.00
$\hat{R}^i$												
0.0	0.00	0.00	0.00	0.00	0.00	0.00	0.00	0.00	0.00	0.00	0.00	0.00
0.1	0.00	0.00	0.00	0.00	0.00	0.00	0.00	0.00	0.00	0.00	0.07	0.00
0.2	0.00	0.00	0.00	0.00	0.00	0.00	0.04	0.00	0.16	0.00	0.20	0.05
0.3	0.00	0.00	0.00	0.00	0.03	0.00	0.21	0.00	0.29	0.05	0.32	0.15
0.4	0.00	0.00	0.00	0.00	0.21	0.00	0.35	0.00	0.40	0.16	0.43	0.25
0.5	0.00	0.00	0.18	0.00	0.38	0.00	0.47	0.11	0.51	0.28	(0.53)	0.36
0.6	0.11	0.00	0.38	0.00	0.52	0.00	0.59	0.26	0.62	0.40	(0.64)	0.48
0.7	0.38	0.00	0.56	0.00	0.65	0.19	0.70	0.43	0.72	0.54	(0.73)	(0.60)
0.8	0.61	0.00	0.71	0.04	0.78	0.43	(0.81)	(0.60)	(0.83)	(0.69)	(0.84)	(0.72)
0.9	0.82	0.00	0.86	0.46	0.89	0.69	(0.91)	(0.80)	(0.92)	(0.85)	(0.93)	(0.86)
1.0	1.00	1.00	1.00	1.00	1.00	1.00	1.00	1.00	1.00	1.00	1.00	1.00

[h] Values are for $q = 16$ and $1 - \alpha = 0.99$.
[i] Values are for $q = 16$ and $1 - \alpha = 0.95$.

Table A9.5 (*cont.*)

$\hat{R}^j$	(n − q) Values											
	5		10		20		40		80		160	
0.0	0.00	0.00	0.00	0.00	0.00	0.00	0.00	0.00	0.00	0.00	0.00	0.00
0.1	0.00	0.00	0.00	0.00	0.00	0.00	0.00	0.00	0.00	0.00	0.00	0.00
0.2	0.00	0.00	0.00	0.00	0.00	0.00	0.00	0.00	0.03	0.00	0.14	0.00
0.3	0.00	0.00	0.00	0.00	0.00	0.00	0.00	0.00	0.19	0.00	0.28	0.06
0.4	0.00	0.00	0.00	0.00	0.00	0.00	0.19	0.00	0.33	0.00	0.39	0.16
0.5	0.00	0.00	0.00	0.00	0.14	0.00	0.36	0.00	0.46	0.12	(0.51)	0.29
0.6	0.00	0.00	0.07	0.00	0.36	0.00	0.50	0.00	0.58	0.27	(0.62)	0.41
0.7	0.04	0.00	0.36	0.00	0.54	0.00	0.64	0.21	0.69	0.43	(0.72)	(0.55)
0.8	0.41	0.00	0.59	0.00	0.71	0.07	0.77	0.43	(0.80)	(0.61)	(0.83)	(0.69)
0.9	0.72	0.00	0.81	0.00	(0.87)	(0.48)	(0.90)	(0.68)	(0.91)	(0.80)	(0.93)	(0.84)
1.0	1.00	1.00	1.00	1.00	1.00	1.00	1.00	1.00	1.00	1.00	1.00	1.00
$\hat{R}^k$												
0.0	0.00	0.00	0.00	0.00	0.00	0.00	0.00	0.00	0.00	0.00	0.00	0.00
0.1	0.00	0.00	0.00	0.00	0.00	0.00	0.00	0.00	0.00	0.00	0.00	0.00
0.2	0.00	0.00	0.00	0.00	0.00	0.00	0.00	0.00	0.00	0.00	0.12	0.00
0.3	0.00	0.00	0.00	0.00	0.00	0.00	0.00	0.00	0.15	0.00	0.25	0.08
0.4	0.00	0.00	0.00	0.00	0.00	0.00	0.13	0.00	0.29	0.04	0.37	0.19
0.5	0.00	0.00	0.00	0.00	0.04	0.00	0.30	0.00	0.43	0.17	0.48	0.32
0.6	0.00	0.00	0.00	0.00	0.27	0.00	0.46	0.07	0.55	0.32	(0.60)	0.44
0.7	0.00	0.00	0.24	0.00	0.47	0.00	0.61	0.28	0.67	0.47	(0.70)	(0.58)
0.8	0.26	0.00	0.51	0.00	0.66	0.19	(0.76)	(0.43)	(0.80)	(0.64)	(0.81)	(0.72)
0.9	0.65	0.00	0.77	0.16	0.84	0.57	(0.89)	(0.70)	(0.91)	(0.82)	(0.91)	(0.86)
1.0	1.00	1.00	1.00	1.00	1.00	1.00	1.00	1.00	1.00	1.00	1.00	1.00

[j] Values are for $q = 32$ and $1 - \alpha = 0.99$.
[k] Values are for $q = 32$ and $1 - \alpha = 0.95$.

Table A9.6

Critical Values for the Test of Hypothesis H: $\rho = 0$ vs A; $\rho > 0$[a]

n	$\alpha = 0.05$	$\alpha = 0.10$
2	0.975	0.948
3	0.881	0.827
5	0.726	0.662
10	0.533	0.475
15	0.440	0.390
20	0.383.	0.338
30	0.314	0.276
50	0.244	0.215
70	0.206	0.181
100	0.173	0.152
200	0.122	0.107

[a] Table lists the values of v such that if $\rho = 0$, $P(\hat{\rho} > v) = \alpha$, where $\hat{\rho}$ is the estimate of coherence based on $2n$ degrees of freedom. Source: Amos, D. E. and Koopmans, L. H. (1963). "Tables of the distribution of the coefficient of coherence for stationary bivariate Gaussian processes." Sandia Corp. Monograph SCR-483. Sandia Corp., Albuquerque, New Mexico.

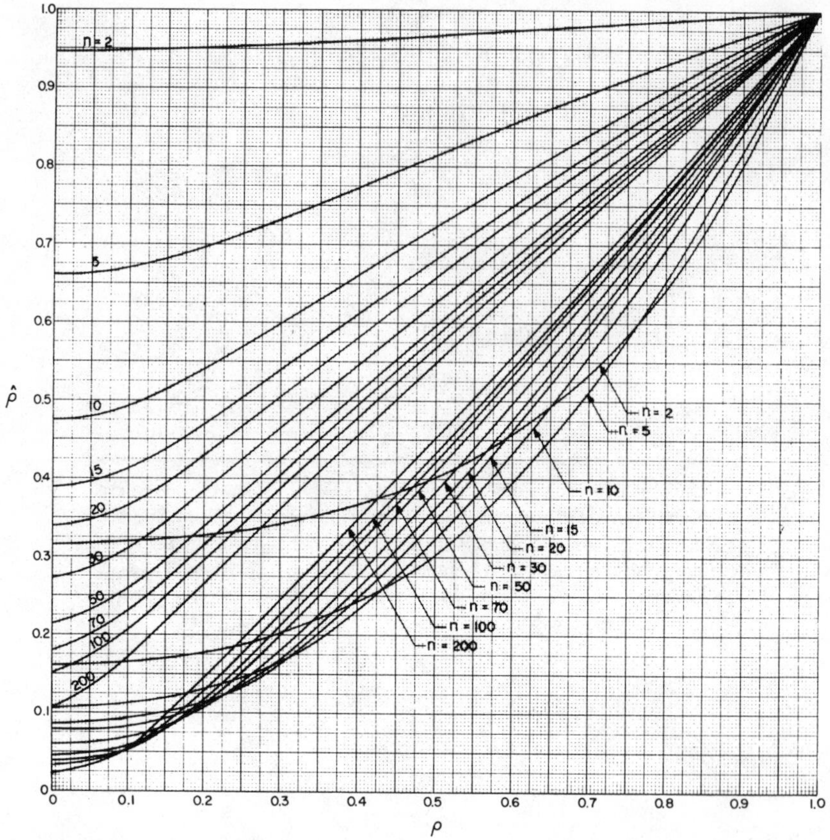

Fig. A9.1 *Graphs of upper and lower confidence limits for* 80% *confidence intervals for coherence. Degrees of freedom* = 2*n*. Source: Amos and Koopmans (1963).

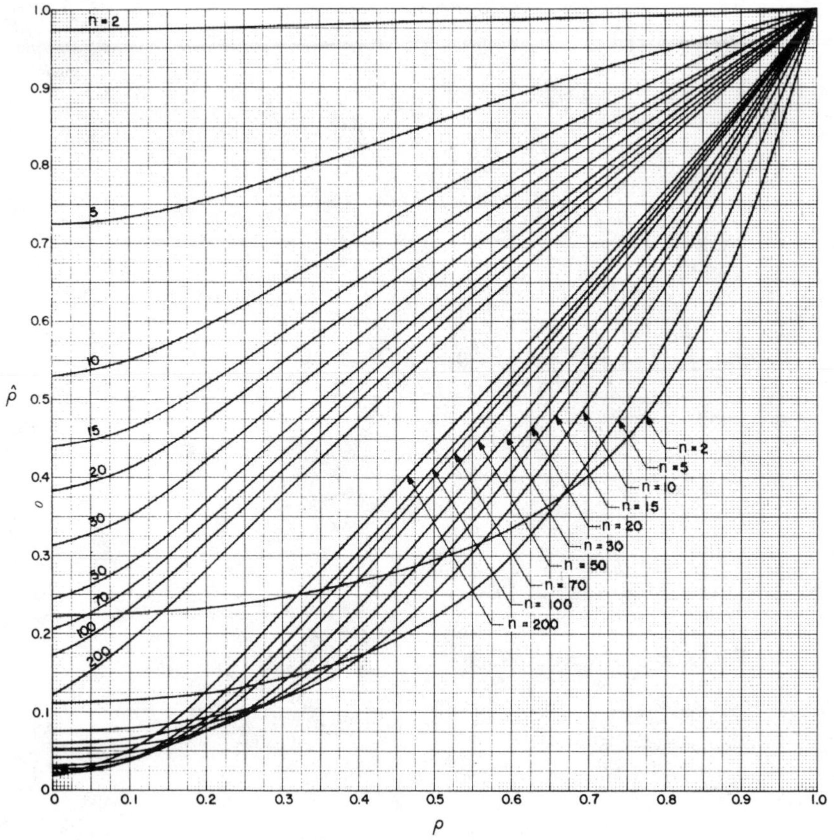

Fig. A9.2 *Graphs of upper and lower confidence limits for 90% confidence intervals for coherence. Degrees of freedom* = 2*n*. Source: Amos and Koopmans (1963).

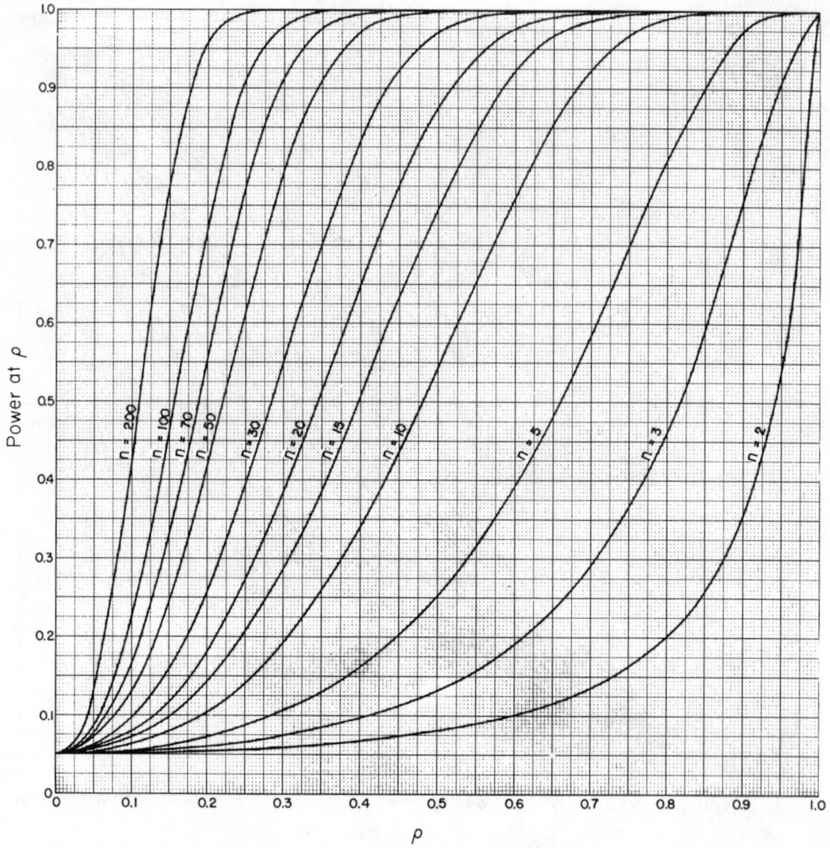

Fig. A9.3 *Graphs of power functions for the test of zero coherence. Significance level* $\alpha = 0.05$. *Degrees of freedom* $= 2n$. Source: Amos and Koopmans (1963).

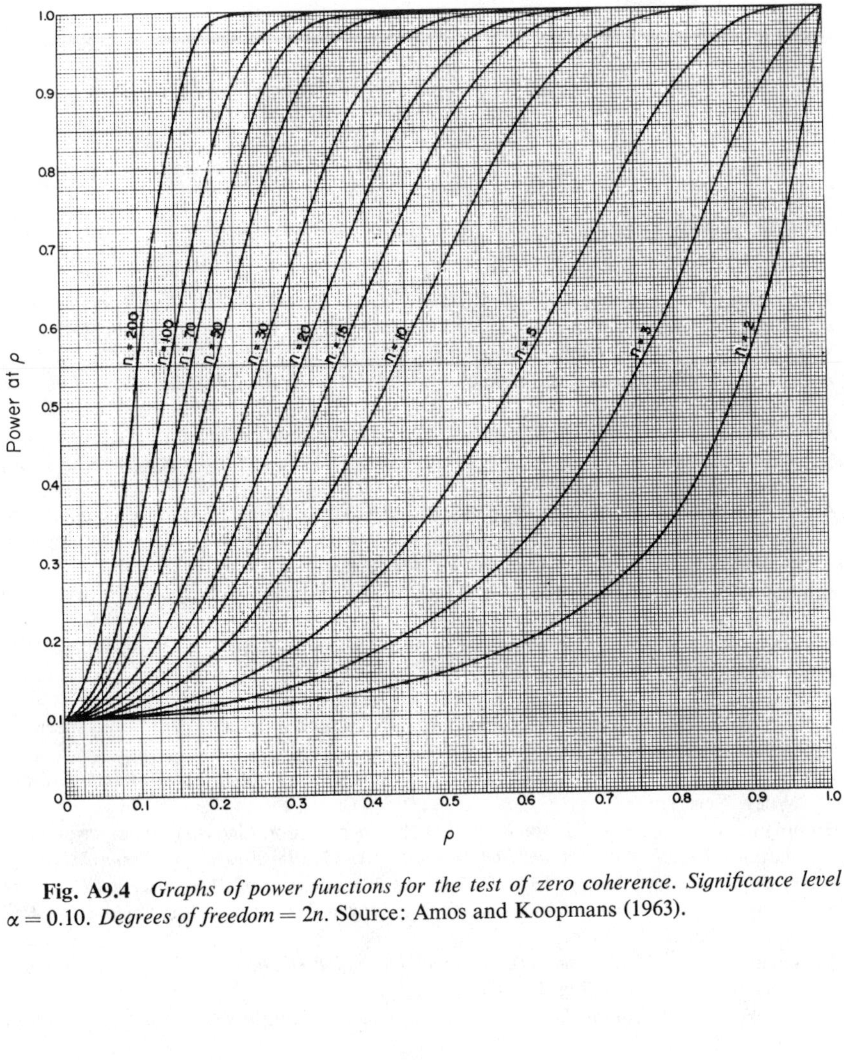

Fig. A9.4 *Graphs of power functions for the test of zero coherence. Significance level* $\alpha = 0.10$. *Degrees of freedom* $= 2n$. Source: Amos and Koopmans (1963).

References

Abramowitz, M., and Stegun, I. A. (eds.) (1964). *Handbook of Mathematical Functions with Formulas, Graphs and Mathematical Tables*, U.S. Nat. Bur. Stand. Appl. Math. Ser. 55. U.S. Dept. of Commerce, Washington, D.C.

Akaike, H. (1970). "Statistical predictor identification." *Ann. Inst. Statist. Math.* **22**, 203–217.

Alexander, M. J., and Vok, C. A. (1963). "Tables of the cumulative distribution of sample multiple coherence." Res. Memo. 972-351. Rocketdyne Div., North Amer. Aviation, Inc. Los Angeles, California.

Amos, D. E., and Koopmans, L. H. (1963). "Tables of the distribution of the coefficient of coherence for stationary bivariate Gaussian processes." Sandia Corp. Monograph SCR-483. Sandia Corp., Albuquerque, New Mexico.

Anderson, R. Y., and Koopmans, L. H., (1963). "Harmonic analysis of varve time series." *J. Geophys. Res.* **68**, 877–893.

Anderson, T. W. (1958). *An Introduction to Multivariate Statistical Analysis*. Wiley, New York.

Anderson, T. W. (1971). *The Statistical Analysis of Time Series*. Wiley, New York.

Bartlett, M. S. (1948). "Smoothing periodograms from time-series with continuous spectra." *Nature (London)* **161**, 686–687.

Bendat, J. S., and Piersol, A. G., (1966). *Measurement and Analysis of Random Data*. Wiley, New York.

Benningus, V. A. (1969). "Estimation of coherence of non-Gaussian time series populations." *IEEE Trans. Audio Electroacoust.* **AU-17**, 198–201.

Blackman, R. B., and Tukey, J. W. (1959). *The Measurement of Power Spectra from the Viewpoint of Communications Engineering*. Dover, New York. Reprinted from *Bell System Tech. J.* **37** (1958)

Blackman, R. B. (1965). *Linear Data Smoothing and Prediction in Theory and Practice*. Addison-Wesley, Reading, Massachusetts.

Box, G. E. P., and Jenkins, G. M. (1970). *Time Series Analysis Forecasting and Control*. Holden-Day, San Francisco, California.

Breiman, L. (1969). *Probability and Stochastic Processes with a View toward Applications*, Houghton, Boston, Massachusetts.

Brillinger, D. R. (1965). "A property of low pass filters." *SIAM Rev.* **7**, 65–67.

Brillinger, D. R. (1969). "A search for a relationship between monthly sunspot numbers and certain climatic series." *Bull. Inst. Int. Statist.* **43**, 293–306.

Brillinger, D. R. (1970). "The frequency analysis of relations between stationary spatial series." *Proc. Biennial Sem. Canad. Math. Congr., 12th* (R. Pyke, ed.). Canad. Math. Congr., Montreal.

Brown, R. G. (1962). *Smoothing, Forecasting and Prediction of Discrete Time Series*. Prentice-Hall, Englewood Cliffs, New Jersey.

Bucy, R. S., and Follin, J. W., Jr. (1962). "Adaptive finite time filtering." *IRE Trans.*

Automatic Control **AC-7**, 10–19.

Cogburn, R. F., and Davis, H. T. (1972). "Periodic splines and spectral estimation." Tech. Rep. 253. Dept. of Math. and Statist., Univ. of New Mexico, Albuquerque, New Mexico.

Cooley, J. W. and Tukey, J. W. (1965). "An algorithm for the machine calculation of complex Fourier series." *Math. Comput.* **19**, 297–301.

Cooley, J. W., Lewis, P. A. W., and Welch, P. D. (1967). "The fast Fourier transform algorithm and its applications." Res. Monograph RC 1743. I.B.M. Watson Res. Center, Yorktown Heights, New York.

Cramér, H. (1942). "On harmonic analysis in certain functional spaces." *Ark. Mat. Astron. Fys.* **283**, No. 12, 17 pp.

Cramér, H. (1951a). *Mathematical Methods of Statistics.* Princeton Univ. Press, Princeton, New Jersey.

Cramér, H. (1951b). A contribution to the theory of stochastic processes." *Proc. Symp. Math. Statist. and Probability, 2nd, Berkeley, 1950*, pp. 329–339. Univ. of Calif. Press, Berkeley, California.

Daniell, P. J. (1946). Discussion following "On the theoretical specification and sampling properties of autocorrelated time series," by M. S. Bartlett. *J. Roy. Statist. Soc. Suppl.* **8**, 27–41.

Davis, H. T. (1972). "Some applications of spline functions to time series analysis." *Proc. Ann. Symp. Interface Comput. Sci. and Statist., 5th, 1972.* Oklahoma State Univ. Press, Stillwater.

Davis, H. T., and Koopmans, L. H. (1970). "Adaptive prediction of stationary time series." Tech. Rep. 208. Dept. Math. and Statist., Univ. New Mexico, Albuquerque, New Mexico. To appear in Sankhya.

Dixon, W. J. (ed.) (1969). *BMD Biomedical Computer Programs*, X-Ser. Suppl. Univ. of California Press, Berkeley.

Dixon, W. J. (ed.) (1970). *BMD Biomedical Computer Programs, 2nd ed.* (third printing revised). Univ. of California Press, Berkeley.

Doob, J. L. (1953). *Stochastic Processes.* Wiley, New York.

Edge, B. L., and Liu, P. C. (1970). "Comparing power spectra computed by Blackman-Tukey and fast Fourier transform." *Water Resour. Res.* **6**, 1601–1610.

Enochson, L. D., and Goodman, N. R. (1965). "Gaussian approximation to the distribution of sample coherence." AFFDL TR 65-57. Res. and Technol. Div., AFSC, Wright-Patterson AFB, Ohio.

Enochson, L. D., and Otnes, R. K. (1968). *Programming and Analysis for Digital Time Series Data.* The Shock and Vibration Information Center, U.S. Dept. of Defense, Washington, D.C.

Feller, W. (1968). *An Introduction to Probability Theory and Its Applications*, 3rd ed., Vol. 1. Wiley, New York.

Fisher, R. A. (1929). "Tests of significance in harmonic analysis." *Proc. Roy. Soc. London Ser. A* **125**, 54–59.

Fishman, G. S. (1969). *Spectral Methods in Econometrics.* Harvard Univ. Press, Cambridge, Massachusetts.

Frisch, R. (1933). "Propagation problems and impulse problems in dynamic economics." *Economic Essays in Honor of Gustav Cassal*, pp. 171–205. Allen & Unwin, London.

Gardner, Jr., L. A. (1962). "Adaptive predictors." *Proc. Conf. Information Theory, Statist. Decision Functions, Stochastic Processes, 3rd, Prague*, pp. 123–192.

Gentleman, W. M., and Sande, G. (1966). "Fast Fourier transform for fun and profit." *Proc. Fall Joint Comput. Conf., AFIPS, New York, 1965*, **29**, p. 563. Spartan Books, Washington, D.C.

Gersh, W. (1970). "Spectral analysis of EEG's by autoregressive decomposition of time series." *Math. Biosci.* **7**, 205–222.

Gersh, W. and Goddard, G. V. (1970). "Epileptic focus location: spectral analysis method." *Science* **169**, 701–702.

Goldberg, R. R. (1961). *Fourier Transforms.* Cambridge Univ. Press, London and New York.

Goodman, N. R. (1957). "On the joint estimation of the spectra, cospectrum and quadrature spectrum of a two-dimensional stationary gaussian process." Sci. Paper No. 10. Engrng. Statist. Lab., New York Univ., New York.

Goodman, N. R. (1963). "Statistical analysis based upon a certain multivariate complex Gaussian distribution (an introduction)." *Ann. Math. Statist.* **34**, 152–177.

Goodman, N. R., and Katz, S. (1958). "Calculating open loop transfer functions from closed loop measurements." *J. Assoc. Comput. Mach.* **3**, 289–297.

Granger, C. W. J., and Hatanaka, M. (1964). *Spectral Analysis of Economic Time Series.* Princeton Univ. Press, Princeton, New Jersey.

Graybill, F. (1969). *Introduction to Matrices with Applications in Statistics.* Wadsworth Belmont, California.

Grenander, U., and Rosenblatt, M. (1957). *Statistical Analysis of Stationary Time Series.* Wiley, New York.

Groves, G. W., and Hannan, E. J. (1968). "Time series regression of sea level on weather." *Rev. Geophys.* **6**, 129–174.

Halmos, P. R. (1948). *Finite Dimensional Vector Spaces.* Princeton Univ. Press, Princeton, New Jersey.

Halmos, P. R. (1951). *Introduction to Hilbert Space.* Chelsea, Bronx, New York.

Hamon, B. V., and Hannan, E. J. (1963). "Estimating relations between time series." *J. Geophys. Res.* **68**, 6033–6041.

Hannan, E. J. (1970). *Multiple Time Series.* Wiley, New York.

Hannan, E. J., and Thomson, P. J. (1971). "The estimation of coherence and group delay." *Biometrika* **58**, 469–482.

Jenkins, F. A., and White, H. A. (1950). *Fundamentals of Optics.* McGraw-Hill, New York.

Jones, R. H. (1962). "Spectral analysis with regularly missed observations." *Ann. Math. Statist.* **33**, 455–461.

Jones, R. H. (1965). "A reappraisal of the periodogram in spectral analysis." *Technometrics* **7**, 531–542.

Jones, R. H. (1969). "Phase free estimation of coherence." *Ann. Math. Statist.* **40**, 540–548.

Jones, R. H. (1971). "Spectrum estimation with missing observations." *Ann. Inst. Statist. Math.* **23** 387–398.

Kalman, R. E. (1960). "A new approach to linear filtering and prediction problems." *J. Basic Engrng.* **82**, 34–45.

Kalman, R. E. (1963). "New methods of Wiener filtering theory." *Proc. Symp. Engrng. Appl. Random Functions Theory and Probability, 1st, Purdue Univ., Lafayette, Indiana, 1962* (J. L. Bogdanoff and F. Kozin, eds.). Wiley, New York.

Kalman, R. E., and Bucy, R. S. (1961). "New results in linear filtering and prediction theory." *J. Basic Engrng.* **83**, 95–108.

Khatri, C. G. (1964). "Distribution of the 'generalized' multiple correlation matrix in the dual case." *Ann. Math. Statist.* **35**, 1801–1806.

Khintchine, A. (1934). "Korrelationstheorie der stationare stochastischen Processe." *Math. Ann.* **109**, 604–615.

Kolmogorov, A. N. (1933). "Grundbegriffe der Wahrscheinlichkeitrechnung." *Ergebnisse der Mathematik.* Published in English in 1950 as *Foundations of the Theory of Probability.* Chelsea, Bronx, New York.

Kolmogorov, A. N. (1941a). "Stationary sequences in Hilbert space." (Russian) *Bull. Math. Univ. Moscow* **2**, No. 6, 40 pp.

Kolmogorov, A. N. (1941b), "Interpolation und Extrapolation von stationaren zufalligen Folgen." (Russian, German summary) *Bull. Acad. Sci. U.R.S.S. Ser. Math.* **5**, 3–14.

Koopmans, L. H. (1961). "An evaluation of a signal summing technique for improving the signal-to-noise ratios for seismic events." *J. Geophys. Res.* **66**, 3879–3898.

Koopmans, L. H. (1964a). "On the coefficient of coherence for weakly stationary stochastic processes." *Ann. Math. Statist.* **35**, 532–549.

Koopmans, L. H. (1964b). "On the multivariate analysis of weakly stationary stochastic processes." *Ann. Math. Statist.* **35**, 1765–1780.

Koopmans, L. H., and Qualls, C. (1971). "Fixed length confidence intervals for parameters of the normal distribution based on two-stage sampling procedures." *Rocky Mountain J. Math.* **1**, 587–602.

Koopmans, L. H., and Qualls, C. (1972). "An exponential probability bound for the energy of a type of Gaussian process." *Ann. Math. Statist.* **43**, 1953–1960.

Koopmans, L. H., Qualls, C., and Yao, J. T. P. (1973). "An upper bound on the failure probability for linear structures." *J. Appl. Mech. Ser. E.* **40**, 181–185.

Lee, Y. W. (1964). "Contributions of Norbert Wiener to linear theory and nonlinear theory in engineering." In *Selected Papers of Norbert Wiener*. MIT Press, Cambridge, Massachusetts.

Monin, A. S., and Vulis, I. L. (1971). "On the spectra of long-period oscillations of geophysical parameters." *Tellus* **23**, 337–345.

Munk, W. H., Snodgrass, F. E., and Tucker, M. J. (1959). "Spectra of low-frequency ocean waves." *Bull. Scripps Inst. Oceanogr.* **7**, 283–362.

Pagano, M. (1973). "When is an autoregressive scheme stationary?" *Commun. Statist.* **2**, 533–544.

Parzen, E. (1961). "Mathematical considerations in the estimation of spectra." *Technometrics* **3**, 167–190. Reprinted in Parzen (1967).

Parzen, E. (1967). *Time Series Analysis Papers*. Holden-Day, San Francisco, California.

Parzen, E. (1972). "Some recent advances in time series analysis." In *Statistical Models and Turbulence* (M. Rosenblatt and C. Van Atta, eds.). Springer-Verlag, Berlin and New York.

Priestley, M. B. (1965). "Evolutionary spectra and nonstationary processes." *J. Roy. Statist. Soc. Ser. B* **27**, 204–237.

Reed, J. W. (1971). "Low-frequency periodicities in Panama rainfall runoff." *J. Appl. Meterol.* **10**, 666–673.

Riesz, F., and Nagy, B. Sz. (1955). *Functional Analysis*. Ungar, New York.

Robinson, E. A. (1967). *Multichannel Time Series Analysis with Digital Computer Programs*. Holden-Day, San Francisco, California.

Rozanov, Yu. A. (1967). *Stationary Random Processes*. Holden-Day, San Francisco California.

Schuster, A. (1898). "On the investigation of hidden periodicities with application to a supposed 26-day period of meteorological phenomena." *Terr. Magn.* **3**, 13–41.

Schuster, A. (1906). "On the periodicities of sunspots." *Philos. Trans. Roy. Soc. London Ser. A* **206**, 69–100.

Shannon, C. (1949). "Communication in the presence of noise." *Proc. IRE* **37**, 10–21.

Shapiro, A. (1962). "Estimation of coherence between signal and signal plus echoes." Bell Telephone Lab., Memo. Bell Telephone Lab., Inc., Whippany, New Jersey.

Slutsky, E. (1927). "The summation of random causes as the source of cyclic processes." (Russian) *Problems of Economic Conditions* **3**; Engl. transl. *Econometrica* **5**, 105–146 (1937).

Szegö, G. (1939). *Orthogonal Polynomials*. A.M.S. Colloq. Publ.

Tick, L. J. (1955). "Statistical time series analysis of blast furnace variables." Publ. of Res. Div., College of Engrng. New York Univ., New York.

Tick, L. J. (1967). "Estimation of coherency." *Advanced Sem. Spectral Anal. of Time Ser., 1966* (B. Harris, ed.), pp. 133–152. Wiley, New York.

Tinbergen, J. (1937). *An Economic Approach to Business Cycle Problems*. Hermann, Paris.

Tintner, G. (1940). *The Variate Difference Method*, Principia Press, Bloomington, Indiana.

Titchmarsh, E. C. (1939). *The Theory of Functions*, 2nd ed. Oxford Univ. Press, London and New York.

Tucker, H. G. (1962). *An Introduction to Probability and Mathematical Statistics*. Academic Press, New York.

Tukey, J. W. (1948-approx.) "Measuring Noise Color." Unpublished.

Tukey, J. W. (1949). "The sampling theory of power spectrum estimates." *Symp. Appl. Autocorrelation Anal. Phys. Problems, Woods Hole, Massachusetts, 1949*. NAVEXOS-P-735, Office of Naval Research, Washington, D.C.

Tukey, J. W. (1959). "An introduction to the measurement of spectra." In *Probability and Statistics, The Harold Cramér Volume* (U. Grenander, ed.). Wiley, New York.

Tukey, J. W. (1961). "Discussion emphasizing the connection between analysis of variance and spectral analysis." *Technometrics* **3**, 191–219.

Walter, D. O. (1969). "The method of complex demodulation." *Electroencephal. Clin. Neurophysiol. Suppl.* **27**, 53–57.

Walter, D. O., Rhodes, J. M., Brown, D., and Adey, W. R. (1966). "Comprehensive spectral analysis of human EEG generators in posterior cerebral regions." *Electroencephal. Clin. Neurophysiol.* **20**, 224–237.

Whittle, P. (1963). *Prediction and Regulation*. Van Nostrand-Reinhold, Princeton, New Jersey.

Wiener, N. (1930). "Generalized harmonic analysis." *Acta Math.* **55**, 117–258. Reprinted in *Selected Papers of Norbert Wiener* MIT Press, Cambridge, Massachusetts, 1964.

Wiener, N., and Masani, P. (1957). "The prediction theory of multivariate stochastic processes, I." *Acta Math.* **98**, 111–150.

Wiener, N., and Masani, P. (1958). "The prediction theory of multivariate stochastic processes, II." *Acta Math.* **99**, 93–137.

Wirshing, P. W., and Yao, J. T. P. (1971). "Monte Carlo study of seismic structural safety." *J. Struct. Div. ASCE* **97**, 1497–1519.

Wold, H. (1938). *A Study in the Analysis of Stationary Time Series*. Almqvist & Wiksell, Stockholm. 2nd ed. with Appendix by P. Whittle, 1954.

Wold, H. (1959). "Ends and means in econometric model building." In *Probability and Statistics, the Harold Cramér Volume* (U. Grenander, ed.). Wiley, New York.

Yaglom, A. M. (1955). "The correlation theory of processes whose nth differences constitute a stationary process." *Mat. Sb.* **37**, 141.

Yaglom, A. M. (1962). *An Introduction to the Theory of Stationary Random Functions*, translated from the Russian by R. A. Silverman. Prentice Hall, Englewood Cliffs, New Jersey.

Yule, G. U. (1921). "On the time-correlation problem, with especial reference to the variate-difference correlation method." *J. Roy. Statist. Soc.* **84**, 497–526.

Yule, G. U. (1927). "On a method of investigating periodicites in disturbed series with special reference to Wolfer's sunspot numbers." *Philos. Trans. Roy. Soc. London, Ser. A.* **226**, 267–298.

Index

J

K

L

M

N

Y

Z